A Note from the Author

Congratulations on your decision to take the AP Biology exam! Whether or not you're completing a year-long AP Biology course, this book can help you prepare for the exam. In it you'll find information about the exam as well as Kaplan's test-taking strategies, a targeted review that highlights important concepts on the exam, and practice tests. Take the diagnostic test to see which subject you should review most, and use the two full-length exams to get comfortable with the testing experience. The review chapters include summaries of the 12 AP Biology labs, so that even if you haven't completed them in class, you won't be surprised on Test Day. Don't miss the strategies for answering the free-response questions: you'll learn how to cover the key points AP graders will want to see.

By studying college-level biology in high school, you've placed yourself a step ahead of other students. You've developed your critical-thinking and time-management skills, as well as your understanding of the practice of biological research. Now it's time for you to show off what you've learned on the exam.

Best of luck,

Linda Brooke Stabler

Mark Metz

Paul Gier

RELATED TITLES

AP Calculus AB & BC

AP Chemistry

AP English Language & Composition

AP English Literature & Composition

AP Macroeconomics/Microeconomics

AP Physics B & C

AP Psychology

AP Statistics

AP U.S. Government and Politics

AP U.S. History

AP World History

SAT Subject Test: Biology E/M

SAT Subject Test: Chemistry

SAT Subject Test: Literature

SAT Subject Test: Mathematics Level 1

SAT Subject Test: Mathematics Level 2

SAT Subject Test: Physics

SAT Subject Test: Spanish

SAT Subject Test: U.S. History

SAT Subject Test: World History

Test Prep and Admissions

AP® Biology

2006 Edition

Linda Brooke Stabler

Mark Metz

Paul Gier

Simon & Schuster

NEW YORK · LONDON · SYDNEY · TORONTO

AP® is a registered trademark of the College Entrance Examination Board, which neither sponsors nor endorses this book.

Kaplan Publishing
Published by SIMON & SCHUSTER
Rockefeller Center
1230 Avenue of the Americas
New York, NY 10020

Copyright © 2006 by Anaxos, Inc.

All rights reserved. No part of this book may be reproduced or transmitted in any form or by any means, electronic or mechanical, including photocopying, recording, or by any information storage and retrieval system, without the written permission of the Publisher. Except where permitted by law.

Kaplan® is a registered trademark of Kaplan, Inc.

SIMON & SCHUSTER and colophon are registered trademarks
of Simon & Schuster, Inc.

Contributing Editor: Jon Zeitlin
Editorial Director: Jennifer Farthing
Project Editor: Anne Kemper
Production Manager: Michael Shevlin
Content Manager: Patrick Kennedy
Interior Page Layout: Renée Mitchell
Cover Design: Mark Weaver

Manufactured in the United States of America.
Published simultaneously in Canada.

10 9 8 7 6 5 4 3 2

January 2006

ISBN-13: 978-0-7432-6560-7
ISBN-10: 0-7432-6560-2

For information regarding special discounts for bulk purchases, please contact Simon & Schuster Special Sales at 1-800-456-6798 or business@simonandschuster.com.

Table of Contents

About the Authors ix
Kaplan Panel of AP Experts xi

PART ONE: The Basics
Chapter 1: Inside the AP Biology Exam 3
    Introduction 3
    Overview of the Test Structure 4
    How the Exam Is Scored 6
    Registration and Fees 7
    Additional Resources 7

Chapter 2: Strategies for Success: It's Not Always How Much You Know 11
    General Test-Taking Strategies 12
    How to Approach the Multiple-Choice Questions 13
    How to Approach the Ten-Minute Reading Period 21
    How to Approach the Free-Response Questions 22
    Stress Management 24
    Countdown to the Test 31

PART TWO: Diagnostic Test
    AP Biology Diagnostic Test 37
    Answers and Explanations 43
    Diagnostic Test Correlation Chart 45

PART THREE: AP Biology Review
Chapter 3: Chemistry of Life 49
    Starting Small: Atoms, Molecules, and Reactions Central to Biology 49
    The Importance of Water in Biology 52
    Structure and Function of Biologically Significant Functional
        Groups and Macromolecules 54
    Enzyme Catalysis Laboratory 56
    Applying the Concepts: Questions and Explanations 61

Chapter 4: Living Cells 77
- Life's Little Compartments: Types of Cells and How They Work 77
- Diffusion and Osmosis Laboratory 83
- Applying the Concepts: Questions and Explanations 86

Chapter 5: Cellular Energetics 103
- It's All About the Fuel: How Cells Use and Make Energy 103
- Cell Respiration Laboratory 107
- Plant Pigments and Photosynthesis Laboratory 108
- Applying the Concepts: Questions and Explanations 111

Chapter 6: Heredity 125
- The Roots of the Family Tree: Understanding Inheritance 125
- Mitosis and Meiosis Laboratory 133
- Genetics of Organisms Laboratory 133
- Applying the Concepts: Questions and Explanations 137

Chapter 7: Molecular Genetics 153
- The Double Helix: DNA and the Mechanisms of Inheritance 153
- Molecular Biology Laboratory 158
- Applying the Concepts: Questions and Explanations 162

Chapter 8: Evolutionary Biology 175
- A Little Change Can Add Up: Origins of Life and Evolution 175
- Population Genetics and Evolution Laboratory 180
- Applying the Concepts: Questions and Explanations 184

Chapter 9: Diversity of Organisms 199
- Cataloging Biodiversity: The Linnean System 199
- Applying the Concepts: Questions and Explanations 206

Chapter 10: Structure and Function of Plants and Animals 219
- Amazing Biological Machines: Physiological Processes and Responses 219
- Transpiration Laboratory 231
- Physiology of the Circulatory System Laboratory 231
- Applying the Concepts: Questions and Explanations 234

Chapter 11: Ecology 249

    It's Not a Vacuum Out There: How Organisms Interact with Each Other and
      Their Environment 249

    Animal Behavior Laboratory 257

    Dissolved Oxygen and Aquatic Primary Productivity Laboratory 258

    Applying the Concepts: Questions and Explanations 261

PART FOUR: Practice Tests

    How to Take the Practice Tests 276

    How to Compute Your Score 277

Practice Test 1 279

    Answer Key 304

    Answers and Explanations 305

Practice Test 2 323

    Answer Key 347

    Answers and Explanations 348

Glossary 363

ABOUT THE AUTHORS

Anaxos Inc.
Founded in 1999 by Drew and Cynthia Johnson, Anaxos is a leading provider of educational content for print and electronic media.

Linda Brooke Stabler holds a Ph.D. in Biology from Arizona State University and teaches biology at Redlands Community College in Oklahoma.

Mark Metz works for the U.S. Department of Agriculture and holds a Ph.D. in Environmental Science from the University of Illinois. He taught biology at the university level for four years.

Paul Gier is Professor of Biology at Huntingdon College in Alabama.

TEACHERS AND STUDENTS—GO ONLINE!

FOR TEACHERS ONLY:
HOW TO USE THIS BOOK IN YOUR CLASSROOM

simonsaysteach.com

Visit our resource area for teachers. We've provided ideas about how this book can be used effectively in and outside of your AP classroom. Easy to implement, these tips are designed to complement your curriculum plans and help you create a classroom full of high-scoring students.

FOR STUDENTS AND TEACHERS

kaptest.com/publishing

The material in this book is up-to-date at the time of publication, but the College Board may have instituted changes in the test since that date. Be sure to carefully read the materials you receive when you register for the test. If there are any important late-breaking developments—or any changes or corrections to the Kaplan test preparation materials in this book—we will post that information online at kaptest.com/publishing.

FEEDBACK AND COMMENTS

kaplansurveys.com/books

If you have comments and suggestions about this book, we invite you to fill out our online survey form at kaplansurveys.com/books. Your feedback is extremely helpful as we continue to develop high-quality resources to meet your needs.

Kaplan Panel of AP Experts

Congratulations—you have chosen Kaplan to help you get a top score on your AP exam.

Kaplan understands your goals, and what you're up against—achieving college credit and conquering a tough test—while participating in everything else that high school has to offer.

You expect realistic practice, authoritative advice, and accurate, up-to-the-minute information on the test. And that's exactly what you'll find in this book, as well as every other in the AP series. To help you (and us!) reach these goals, we have sought out leaders in the AP community. Allow us to introduce our experts:

AP Biology Experts

LARRY CALABRESE has taught biology for 36 years, and AP Biology for 19 years at Palos Verdes Peninsula High School in CA. He also teaches Anatomy and Physiology at Los Angeles Harbor College in Wilmington, CA. He has been an AP Biology reader since 1986, and a table leader since 1993. For the past 10 years, he has taught College Board workshops.

CHERYL G. CALLAHAN has been teaching biology for over 20 years, first at the college level and now at the high school level at Savannah Country Day School, Savannah, GA. She has been an AP Biology reader and table leader since 1993. She also moderates the AP Biology Electronic Discussion Group.

AP Calculus Expert

William Fox has been teaching calculus at the university level for over 20 years. He has been at Francis Marion University in Florence, SC since 1998. He has been a reader and table leader for the AP Calculus AB and BC exams since 1992.

AP Chemistry Experts

Lenore Hoyt teaches chemistry at Idaho State University. She has done post-doctoral studies at Yale University, and holds a Ph.D. from University of Tennessee. She has been a reader for the AP Chemistry exam for 5 years.

Lisa Zuraw has been a professor in the Chemistry Department at The Citadel in Charleston, SC for 12 years. She has served as a faculty reader for the AP Chemistry exam as well as a member of the AP Chemistry Test Development Committee.

AP English Language & Composition Experts

Natalie Goldberg recently retired from St. Ignatius College Prep in Chicago, IL, where she helped develop a program to prepare juniors for the AP English Language and Composition exam and taught AP English Literature. She was a reader for the AP English Language and Composition exam for 6 years, and has been a consultant with the Midwest Region of the College Board since 1997.

Ronald Sudol has been a reader of AP English Language and Composition for 20 years and a workshop consultant for 12. He is Professor of Rhetoric at Oakland University in MI, where he is also Acting Dean of the College of Arts and Sciences.

AP English Literature & Composition Experts

Mitchell S. Billings has taught for 35 years and began the AP program at Catholic High School in Baton Rouge, LA 12 years ago. He has been an Endorsed College Board Consultant for 12 years. He has also been a reader for the AP exam for 10 years, and for the Alternate AP exam for 2 years.

William H. Pell has taught AP English for the past 20 years, and chairs the Language Arts Department at Spartanburg High School in Spartanburg, SC. He has been an adjunct instructor of English at the University of South Carolina Upstate, and has been active in many programs for the College Board, including the reading, workshops, institutes, conferences, and vertical teams.

AP Macroeconomics/Microeconomics Experts

Linda M. Manning is a visiting professor and researcher in residence at the University of Ottawa in Ontario, Canada. She has worked with the AP Economics Program and the Educational Testing Service as a faculty consultant for almost 15 years, and served on the test development committee from 1997–2000.

Bill McCormick has been teaching AP Economics at Spring Valley High School and Richland Northeast High School in Columbia, SC for the past 26 years. He has served as a reader of AP Macroeconomics and Microeconomics exams for the past 8 years.

Peggy Pride has taught AP Economics at St. Louis University High School in St. Louis, MO for 15 years. She served for 5 years on the AP Economics Test Development Committee, and was the primary author for the new AP Economics Teacher's Guide published by College Board in the spring of 2005.

AP Physics B & C Experts

Jeff Funkhouser has taught high school physics since 1988, and since 2001 has taught AP Physics B and C at Northwest High School in Haslet, TX. He has been a reader and table leader for the AP Physics B & C exams for the past 3 years.

Martin Kirby has taught AP Physics B and tutored students for Physics C at Hart High School in Newhall, CA for the last 18 years. He has been a reader for the AP Physics B & C exams and a workshop presenter for the College Board for the past 10 years.

AP Psychology Experts

Ruth Ault has taught psychology for the past 26 years at Davidson College in Davidson, NC. She was a reader for the AP Psychology exam from 1998–2001 and has been a table leader since 2001.

Nancy Homb has taught AP Psychology for the past 7 years at Cypress Falls High School in Houston, TX. She has been a reader for the AP Psychology exam since 2000, and began consulting for the College Board in 2005.

AP Statistics Experts

Lee Kucera has been teaching Statistics at Capistrano Valley High School in Mission Viejo, CA for 12 years, where she has been a teacher for 26 years. Lee has also been a reader for the AP Statistics exam since its inception in 1997.

Mary Mortlock has been teaching statistics since 1996, beginning at Thomas Jefferson High School for Science & Technology in Alexandria, VA, and continuing on to her current position at Cal Poly State University, CA, where she has taught since 2001. Mary has also been a reader for the AP Statistics exam since 2000.

AP U.S. Government & Politics Expert

Chuck Brownson teaches AP U.S. Government & Politics and AP Economics at Stephen F. Austin High School in Sugar Land, TX. He is currently a graduate student working on his Master's Degree in Political Science at the University of Houston. He has been teaching AP U.S. Government & Politics and AP Economics classes for 2 years.

AP U.S. History Experts

Steven Mercado has taught AP U.S. History and AP European History at Chaffey High School in Ontario, CA for the past 12 years. He has also served as a reader for the AP U.S. History exam, and as a reader and table leader for the AP European History exam.

Diane Vecchio has taught in the History Department at Furman University in Greenville, SC since 1996. She has been Chief Faculty Consultant for the AP U.S. History exam for the past four years, and also served as a member of the AP U.S. History Test Development Committee.

AP World History Expert

Jay Harmon has taught world history for the past 22 years at Catholic High School in Baton Rouge, LA. He has been a table leader for the AP World History exam since 2002, and also serves on the test development committee.

| PART ONE |

The Basics

Part One: The Basics | 3

Chapter 1: Inside the AP Biology Exam

- Introduction
- Overview of the Test Structure
- How the Exam Is Scored
- Registration and Fees
- Additional Resources

INTRODUCTION

There's a Good Way and a Bad Way to skip the Introduction to Biology class in college. Many students take the Bad Way, which consists of going to sleep ridiculously late every night with the Xbox controller still wedged in their sweaty hands, setting the alarm for 1:30 P.M., then waking up and asking a roommate, "What did I miss?" Not exactly the sort of behavior that will land you on the Dean's List.

Then there's the Good Way. Skip the whole Introduction to Biology experience entirely—hundreds of students crammed into an auditorium, the tiny dot that is the professor just visible down in front of an ocean of seats—by getting a good score on the Advanced Placement (AP) Biology exam. Depending on the college, a score of 4 or 5 on the AP Biology exam will allow you to leap over the freshman intro course and jump right into more advanced classes. These advanced classes are usually smaller in size, better focused, more intellectually stimulating, and simply put, just more interesting than a basic course. If you are just concerned about fulfilling your science requirement so you can get on with your study of pre-Columbian art or Elizabethan music or some such non-biological area, the AP exam can help you there, too. Ace the AP Biology exam and, depending on the requirements of the college you choose, you may never have to take a science class again.

Test Prep ≠ Studying

If you're holding this book, chances are you are already gearing up for the AP Biology exam and probably completing the AP Biology course. Your teacher has spent the year cramming your head full of the biology know-how you will need to have at your disposal. But there is more to the AP Biology exam than biology know-how. You have to be able to work around the challenges and pitfalls of the test—and there are many—if you want your score to reflect your abilities. You see, studying biology and preparing for the AP Biology exam are not the same thing. Rereading your textbook is helpful, but it's not enough.

That's where this book comes in. We'll show you how to marshal your knowledge of biology and put it to brilliant use on Test Day. We'll explain the ins and outs of the test structure and question format so you won't experience any nasty surprises. We'll even give you answering strategies designed specifically for the AP Biology exam.

Preparing effectively for the AP Biology exam means doing some extra work. You need to review your text *and* master the material in this book. Is the extra push worth it? If you have any doubts, keep in mind that you can always sleep until 1:30 P.M. with the Xbox controller in your hand on the weekend.

OVERVIEW OF THE TEST STRUCTURE

Advanced Placement exams have been around for half a century. While the format and content have changed over the years, the basic goal of the AP program remains the same: to give high school students a chance to earn college credit or advanced placement. To do this, a student needs to do two things:

- Find a college that accepts AP scores
- Do well enough on the exam

The first part is easy, since a majority of colleges accepts AP scores in some form or another. The second part requires a little more effort. If you have worked diligently all year in your coursework, you've laid the groundwork. The next step is familiarizing yourself with the test.

What's on the Test

The Educational Testing Service (ETS)—the company that creates the AP exams—releases a list of the topics covered on the exam. ETS even provides the percentage of the test questions drawn from each topic. Since this information is useful to anyone considering taking the test, check out the breakdown on the next page. The College Board is the organization that administers the Advanced Placement program. ETS is the company that generates the actual exams.

Topics Covered on the AP Biology Exam

I. Molecules and Cells (25%)

A. Chemistry of Life (7%)
1. Water
2. Organic molecules in organisms
3. Free energy changes
4. Enzymes

B. Cells (10%)
1. Prokaryotic and eukaryotic cells
2. Membranes
3. Subcellular organization
4. Cell cycle and its regulation

C. Cellular Energetics (8%)
1. Coupled reactions
2. Fermentation and cellular respiration
3. Photosynthesis

II. Heredity and Evolution (25%)

A. Heredity (8%)
1. Meiosis and gametogenesis
2. Eukaryotic chromosomes
3. Inheritance patterns

B. Molecular Genetics (9%)
1. RNA and DNA structure and function
2. Gene regulation
3. Mutation
4. Viral structure and replication
5. Nucleic acid technology and applications

C. Evolutionary Biology (8%)
1. Early evolution of life
2. Evidence for evolution
3. Mechanisms of evolution

III. Organisms and Populations (50%)

A. Diversity of Organisms (8%)
1. Evolutionary patterns
2. Survey of the diversity of life
3. Phylogenetic classification
4. Evolutionary relationships

B. Structure and Function of Plants and Animals (32%)
1. Reproduction, growth, and development
2. Structural, physiological, and behavioral adaptations
3. Response to the environment

C. Ecology (10%)
1. Population dynamics
2. Communities and ecosystems
3. Global issues

In addition to factual knowledge, there is a different way to look at the study of biology by understanding the way we synthesize information into broader concepts. Two main goals of the College Board are (1) to help students develop a conceptual framework for modern biology, and (2) to help students gain an appreciation of science as a process. To this end, the AP Biology course is designed to expose the student to eight overarching themes.

I. Science as a Process
II. Evolution
III. Energy Transfer
IV. Continuity and Change
V. Relationship of Structure to Function
VI. Regulation
VII. Interdependence in Nature
VIII. Science, Technology, and Society

For example, both the concept of cellular respiration and the concept of a food web in an ecosystem apply to the theme of Energy Transfer. This approach to scientific discovery is about thinking, not just memorization. It's about learning concepts and how they relate, not just facts. Because of this, **the College Board plans to increase the emphasis on themes and concepts and place less weight on specific facts in both the AP Biology course and exam.** The chapters in the review section of this book are designed to take advantage of this design by focusing on concepts and synthesizing information from different concepts to better understand, and learn, the AP Biology course and exam content.

Now that you know what's on the test, let's talk about the test itself. The AP Biology exam consists of three parts, or, more precisely, two parts and one intermission. In Section I, you have 80 minutes to answer 100 multiple-choice questions with five answer choices each. This section is worth 60 percent of your total score.

After this section is completed, you get a ten-minute "reading period." This doesn't mean you get to pull your favorite novel out of your backpack and finish that chapter you started earlier. Instead, you get 10 minutes to pore over Section II of the exam, which consists of four "free-response" questions that are worth 40 percent of your score. The term "free-response" means roughly the same thing as "large, multistep, and involved," since you will spend the 90 minutes of Section II answering these four questions. Although these free-response questions are long and often broken down into multiple parts, they usually don't cover an obscure topic. Instead, they take a fairly basic biology concept and ask you a *bunch* of questions about it. Sometimes diagrams are required, or experiments must be set up properly. It's a lot of biology work, but it's fundamental biology work.

HOW THE EXAM IS SCORED

When your 200 minutes of testing are up, your exam is sent away for grading. The multiple-choice part is handled by a machine, while qualified graders—a group that includes biology teachers and professors, both current and former—grade your responses to Section II. After an interminable wait, your composite score will arrive by mail. (For information on rush score reports and other grading options, visit collegeboard.com or ask your AP Coordinator.) Your results will be placed into one of the following categories, reported on a five-point scale:

5 Extremely well qualified (to receive college credit or advanced placement)
4 Well qualified
3 Qualified
2 Possibly qualified
1 No recommendation

Some colleges will give you credit for a score of 3 or higher, but it's much safer to get a 4 or a 5. If you have an idea of where you will be applying to college, check out the schools' websites or call the admissions offices to find out their particular rules regarding AP scores. For information on scoring the practice tests in this book, see "How to Compute Your Score" in Section IV, p. 277.

REGISTRATION AND FEES

You can register for the exam by contacting your guidance counselor or AP Coordinator. If your school doesn't administer the exam, contact AP Services for a listing of schools in your area that do. The fee for each AP exam is $82. For students with financial need, a $22 reduction is available. To learn about other sources of financial aid, contact your AP Coordinator.

For more information on all things AP, visit collegeboard.com or contact AP Services:

AP Services
P.O. Box 6671
Princeton, NJ 08541-6671
Phone: 1-609-771-7300 or 1-888-225-5427 (toll-free in the U.S. and Canada)
Email: apexams@info.collegeboard.org

ADDITIONAL RESOURCES

Books

Biology (Seventh Edition)
Neil A. Campbell and Jane B. Reece, 2004
Benjamin Cummings, San Francisco, CA
ISBN: 0-8053-7146-X

Introduction to Organic and Biochemistry (Fifth Edition)
Frederick A. Bettelheim, William H. Brown, and Jerry March, 2003
Brooks/Cole, Pacific Grove, CA
ISBN: 0-5344-0188-0

Biology (Fourth Edition)
Karen Arms and Pamela S. Camp, 1995
Saunders College Publishing, Philadelphia, PA
ISBN: 0-0301-5434-0

Plant Physiology (Fourth Edition)
Frank Salisbury and Cleon Ross, 1991
Wadsworth Publishing, Inc., Belmont, CA
ISBN: 0-5341-5162-0

The Cartoon Guide to Genetics, Revised Edition
Larry Gonick and Mark Wheelis 1991
HarperResource (HarperCollins), New York, NY
ISBN: 0-0627-3099-1

Insect Molecular Genetics: An Introduction to Principles and Applications (Second Edition)
Marjorie A. Hoy, 2002
Academic Press/Elsevier, San Diego, CA
ISBN: 0-1235-7031-X

Evolutionary Biology (Third Edition)
Douglas J. Futuyma, 1998
Sinauer Associates, Inc., Sunderland, MA
ISBN: 0-8789-3189-9

Human Physiology (Eighth Edition)
Stuart Ira Fox, 2003
McGraw-Hill Science/Engineering/Math, New York, NY
ISBN: 0-0724-4082-1

Silent Spring (Reprint Edition)
Rachel Carson, 1994 (1962)
Mariner Books (Houghton-Mifflin), New York, NY
ISBN: 0-3956-8329-7

Websites

"The pH Scale"
Dr. Paul Decelles
staff.jccc.net/pdecell/chemistry/phscale.html

"Enzymes"
Mark Rothery's Biology Web Site
mrothery.co.uk/enzymes/enzymes.htm

"The Cell Cycle"
James A. Sullivan, cellsalive.com
cellsalive.com/cell_cycle.htm

"Mitosis"
James A. Sullivan, cellsalive.com
cellsalive.com/mitosis.htm

Handwritten at top: www.phschool.com/science/biology—Phce/labbench/index.html

"AS Guru—Biology"
BBC Television
bbc.co.uk/education/asguru/biology/01cellbiology/index.shtml

"Cellular Respiration"
Regina Bailey, Biology on About.com
biology.about.com/library/weekly/aa090601a.htm

"Cellular Processes: Cellular Respiration"
Cellupedia provided by ThinkQuest by Oracle
library.thinkquest.org/C004535/cellular_respiration.html

"Photosynthesis—Autotrophic Metabolism"
Dr. Charles Mallery, University of Miami
fig.cox.miami.edu/~cmallery/150/phts/phts.htm

"Photosynthesis"
Western Kentucky University, Biology 120 course tutorial
bioweb.wku.edu/courses/Biol120/Web/Photosynth1x.asp

"Plant Physiology Online"
Sinauer Associates, Inc., Sundarland, MA
plantphys.net/index.php

"Genetics" Section
Steve Lubey, Lubey's BioHELP
borg.com/~lubehawk/biotopcs.htm

"Mendelian Genetics"
The Biology Project, The University of Arizona
biology.arizona.edu/mendelian_genetics/mendelian_genetics.html

"A Monk's Flourishing Garden: The Basics of Molecular Biology Explained"
BioTeach
bioteach.ubc.ca/MolecularBiology/AMonksFlourishingGarden

"Gel Electrophoresis Simulator"
Physlets Web Pages. By John Cowan, revised and modified by Wolfgang Christian.
webphysics.davidson.edu/applets/biogel/biogel.html

"Origin of Life"
New York Center for Studies on the Origins of Life
origins.rpi.edu

"The Origin of Species by Charles Darwin"
Online Literature Library
literature.org/authors/darwin-charles/the-origin-of-species

"Understanding Evolution"
The University of California Museum of Paleontology, Berkeley
evolution.berkeley.edu

"Tree of Life Web Project"
David R. Maddison and K.S. Schulz (ed.), 2004
tolweb.org

"Linné on line"
Uppsala University, Sweden
linnaeus.uu.se/online/index-en.html

"The Biosphere: Life on Earth"
The University of California Museum of Paleontology, Berkeley
ucmp.berkeley.edu/alllife/threedomains.html

"The Online Biology Book" (Chapters 20–36)
Dr. Michael J. Farabee, Estrella Mountain Community College, Avondale, Arizona
emc.maricopa.edu/faculty/farabee/BIOBK/BioBookTOC.html

"Welcome to the Dynamics of Development!"
Dr. Jeff Hardin, University of Wisconsin, Department of Zoology
worms.zoology.wisc.edu/embryology_main.html

"Checks on Population Growth"
Kimball's Biology Pages, John W. Kimball
users.rcn.com/jkimball.ma.ultranet/BiologyPages/P/Populations2.html

"Introduction to Biomes"
Susan L. Woodward, Department of Geography, Radford University
radford.edu/~swoodwar/CLASSES/GEOG235/biomes/intro.html

Part One: The Basics | 11

Chapter 2: **Strategies for Success: It's Not Always How Much You Know**

- General Test-Taking Strategies
- How to Approach the Multiple-Choice Questions
- How to Approach the Ten-Minute Reading Period
- How to Approach the Free-Response Questions
- Stress Management
- Countdown to the Test

Even non-biologists know that the world constantly changes. Fifteen years ago there weren't that many cell phones. There weren't that many standardized tests, either. Nowadays, you can't go a semester of school without taking some letter-jumble exam like the PSAT, SAT, ACT, BLAM, ZORK, or FWOOSH (some of those tests are fake, some aren't). And right after leaving school, many students call their friends on their cell phones to talk about the test they both just took.

Rampant cell phone usage is a problem for another book; this one will concentrate on standardized testing. Since everyone reading this has taken a standardized test of one kind or another, you are all probably familiar with some of the general strategies that help students increase their scores on a standardized exam. Let's review some of these.

GENERAL TEST-TAKING STRATEGIES

1. **Pacing.** Since many tests are timed, proper pacing allows someone to attempt every question in the time allotted. Poor pacing causes students to spend too much time on some questions to the point where they run out of time before completing all the questions.

2. **Process of Elimination.** On every multiple-choice test you ever take, the answer is given to you. The only difficulty resides in the fact that the correct answer is hidden among incorrect choices. Even so, the multiple-choice format means you don't have to pluck the answer out of the air. Instead, if you can eliminate answer choices you know are incorrect, and only one choice remains, then that must be the correct answer.

3. **Knowing When to Guess.** Some tests, including the AP Biology exam, deduct one-quarter of a point for each wrong answer, while questions left unanswered receive zero points. That means that whenever you can eliminate one or two answer choices, you should guess on a question. Over time, if you practice educated guessing, you'll see your score rise.

4. **Patterns and Trends.** The key word here is the *standardized* in "standardized testing." Being standardized means that tests don't change greatly from year to year. Sure, each question won't be the same, and different topics will be covered from one administration to the next, but there will also be a lot of overlap from one year to the next. That's the nature of *standardized* testing: if the test changed wildly each time it came out, it would be useless as a tool for comparison. Because of this, certain patterns can be uncovered regarding any standardized test. Learning about these trends and patterns can help students taking the test for the first time.

5. **The Right Approach.** Having the right mindset plays a large part in how well people do on a test. Those students who are nervous about the exam and hesitant to make guesses often fare much worse than students with an aggressive, confident attitude. Students who start with the first question and plod on from there don't score as well as students who deal with the easy questions before tackling the harder ones. People who take a test cold have more problems than those who take the time to learn about the test beforehand. In the end, factors like these make the difference between people who are good test takers and those who struggle even when they know the material.

These points are all valid for every standardized test, but they are quite broad in scope. The rest of this chapter will discuss how these general ideas can be modified to apply specifically to the AP Biology exam. These test-specific strategies and the factual information covered in your course as well as this book's review section are the one-two punch that will help you succeed on the exam.

HOW TO APPROACH THE MULTIPLE-CHOICE QUESTIONS

Since all biologists use "the scientific method" in some form or another, let's apply it to the test. Our hypothesis is that you can achieve a higher score by attacking the 100 multiple-choice questions in a specific order. Other students will just turn the page once the test begins and start with question 1. Let these drones be the control group.

All 100 questions are multiple choice, but there are three distinct question types.

1. **The Stand-Alones.** These are the first questions on each AP Biology test, and typically make up a little over half of the exam. Each Stand-Alone question covers a specific topic, and then the next Stand-Alone hits a different topic. Usually there are 5 to 60 words in the question stem, and these words provide you with the information you need to answer the question. The next question is a typical Stand-Alone.

    44. $AB + energy \rightarrow A + B$

        Which of the following best characterizes the reaction represented above?

        (A) Anabolism
        (B) Endergonic reaction
        (C) Exergonic reaction
        (D) Hydrolysis
        (E) Oxidation-reduction

    You get some information to start with, and then you're expected to answer the question. The number of the question, 44, makes no difference since there's no order of difficulty on the AP Biology exam. Tough questions are scattered between easy and medium questions.

2. **Cluster Questions.** Cluster Questions appear right after the Stand-Alones. With these problems, you get some initial piece of information (often visual), and this information has the letters (A) through (E) within it. Four to five problems without answer choices follow this information, and each question describes one of the choices (A–E) in the initial information. You pick the right letter choice for each problem. Here's an example.

    (A) Porifera
    (B) Echinodermata
    (C) Annelida
    (D) Arthropoda
    (E) Mollusca

56. Contains all the sponges
57. Contains sea cucumbers and sea urchins
58. Contains segmented worms
59. Contains sea stars

This shows a typical group of Cluster Questions. The initial information lists five phyla as answer choices, and then the four problems ask you to pick the right phyla.

3. **Data Questions**. Data Questions appear at the end of the AP Biology exam. Just as the name suggests, a group of two to four questions is preceded by data in one form or another. The data might be a simple sentence or two, but usually it is something more complex, like:

- A description of an experiment (50–200 words), often with an accompanying illustration
- A graph or series of graphs
- A large table
- A diagram

The next question is a sample Data Question.

Questions 98–100 refer to the following graphs. These graphs present the frequency of size classes for tail length of *Felis domesticus* as measured at four different sites around the world.

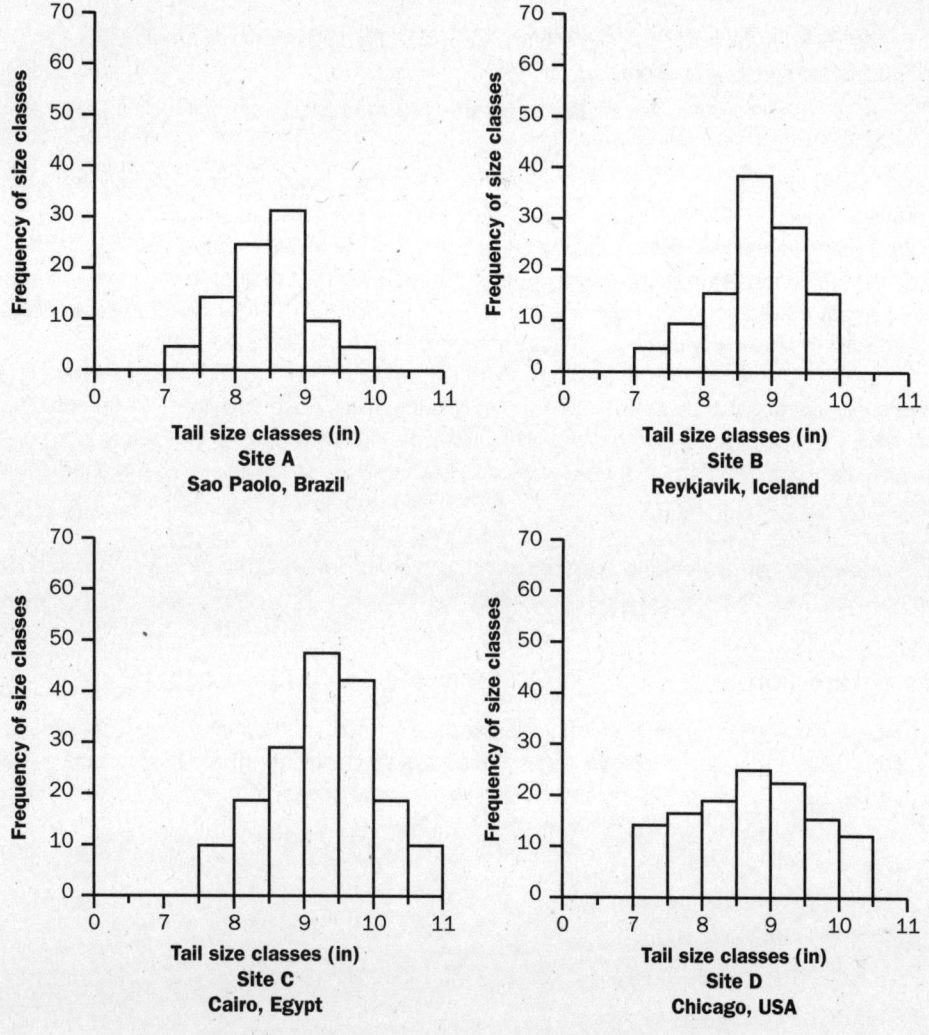

98. According to the data, the most common size class of tails among *Felis domesticus* is

   (A)   7–8 inches
   (B)   8–8.5 inches
   (C)   9–9.5 inches
   (D)   9–10 inches
   (E)   10–11 inches

You might think there is an advantage in the fact that the questions all concern the same set of data. However, Data Questions require a student to analyze the presented information carefully, and while that is occurring, time is ticking away and no questions are being answered.

Those are the three question types. Combine this knowledge with the fact that you must answer 100 questions in 80 minutes, and you'll see why this game plan makes sense:
- Always start with the Cluster Questions, since they take the least amount of time.
- Then tackle the Stand-Alones.
- Finally, use the time remaining to attempt some (but necessarily all) of the Data Questions.

You don't need to get every multiple-choice question right to score well on the AP Biology exam, so don't try for perfection. On an AP exam nothing is gained by a perfect score. All you need is a 4 or 5, and that means you need to get a large portion, but not all, of the multiple-choice questions right. If you don't have enough time to get to every question, make sure that the questions you skip are the longest, most involved ones (the Data Questions). If you understand the topic, you can answer five Cluster Questions in roughly a minute. That's a great use of your limited resource: time.

So while the control-group students around you are blithely starting with question 1 and proceeding from there, you are going to head to the Cluster Questions and answer them first. Once you're done with those easy pickings, it's off to the Stand-Alones, and finally the Data Questions. This approach will make the best use of your time and effort on the AP Biology exam.

There's more to it than just tackling questions in the right order, however. The more you know about each question type, the better equipped you will be to handle it.

Cluster Questions

You have two main options on how you should approach a group of short, sweet Cluster Questions. If you look at the cluster and know the topic well, start with the answer choices (A–E) and then find the question that matches them. For instance, if you are confident you know what each of these five phyla is . . .

(A) Porifera
(B) Echinodermata
(C) Annelida
(D) Arthropoda
(E) Mollusca

. . . then start with (A), Porifera, and seek out a question that matches it. You'll find question 56. One question down, three to go. The advantage to this approach is that it puts you in control of the test. The more you break away from that control group "answer every question the way everybody else does" mindset, the better your score will be.

Sometimes the Answers-Then-Question approach isn't the best. This is especially true if you don't know the topic or answer choices as well. Let's say you know what (A) and (C) are, but you're vague about the other three phyla. This means you get the answers to 56 (Porifera) and 58 (Annelida) correct, but you don't know the answers to 57 and 59. Should you leave them blank? No! For both questions you can eliminate the phyla you know are incorrect [(A) and (C)], so you should take a guess and hope for the best.

Anytime you can eliminate at least one answer choice from a problem, you should take a guess. You won't get every guess right, but over the course of the test this form of educated guessing will improve your score. Let's say you picked (B) for 57 and (D) for 59. It turns out the answer for both 57 and 59 is (B), so you got one right and one wrong. Overall, that boosts your score.

The same letter can be the correct answer for more than one problem in the same cluster. It doesn't happen very often, but it does occur, so don't worry if you find yourself picking the same letter twice.

The Stand-Alones

It's easier to talk about what isn't in the Stand-Alone Questions than what is there.

- There's no order of difficulty; that is, questions don't start out easy and gradually become tougher.
- There are no two questions connected to each other in any way.
- There's no pattern as to what biology concepts appear when.

The Stand-Alones look like a bunch of disconnected biology questions, one following another, and that's just what they are. A genetics question may follow a taxonomy question, which may follow a question about the Krebs cycle.

There's no overall pattern, so don't bother looking for one. But just because the section is randomly ordered, that doesn't mean you have to approach it on the same random terms. Instead, draw up two lists right now using the topics covered in chapter 1. Label one list "Concepts I Enjoy and Know About in Biology" and label the other list "Concepts That Are Not My Strong Points."

When you get ready to tackle the Stand-Alone section, keep these two lists in mind. On your first pass through the section, answer all the questions that deal with concepts you like and know a lot about. None of the questions will have a lot of text in front of it, so you should be able to figure out very quickly whether or not you have the factual chops needed to answer it. If you do, answer that problem and move on. If the question is on a subject that's not one of your strong points, skip it and come back later.

The overarching goal is to answer correctly the greatest possible number of questions in the time available. To do this, focus on your strengths during the first pass through the Stand-Alones. Some questions might be very difficult, even in a subject you're familiar with. Take a minute or so on a tough question, and if you can't come up with an answer, make a mark by the question number in your test booklet and move on. The first pass is about picking up easy points.

Once you've swept through and snagged all the easy questions, take a second pass and try the tougher ones. These tougher questions might cover subjects you're not strong in, or they might just be very difficult questions on subjects you are familiar with. Odds are high that you won't know the answer to some of these questions, but don't leave them blank. You should always take a stab at eliminating some answer choices, and then make an educated guess.

Admittedly, the AP Biology test is a test of specific knowledge, so picking the right answer from the incorrect answer choices is harder to do than it is other standardized tests. Still, it can be done. The following two key ideas are easy to remember and will come in handy on the tougher multiple-choice questions.

Comprehensive, Not Sneaky

Some tests are sneakier than others. They have convoluted writing, questions designed to trip you up mentally, and a host of other little tricks. Students taking a sneaky test often have the proper facts, but get the question wrong because of a trap in the question itself.

The AP Biology test is NOT a sneaky test. Its objective is to see how much biology knowledge you have stored in that skull of yours. To do this, it asks a wide range of questions from an even wider range of biology topics. The exam tries to cover as many different facts in biology as it can, which is why the problems jump around from DNA to food webs. The test works hard to be as comprehensive as it can be, so that students who only know one or two biology topics will soon find themselves struggling.

Understanding these facts about how the test is designed can help you answer questions on it. The AP Biology exam is comprehensive, not sneaky; it makes questions hard by asking about hard subjects, not by crafting hard questions. And you've probably taken an AP Biology course, so trust your instincts when guessing. If you think you know the right answer, chances are you dimly remember the topic being discussed in your AP course. The test is about knowledge, not traps, so trusting your instincts will help more often than not.

You don't have much time to ponder every tough question, so trusting your instincts can help keep you from getting bogged down and wasting time on a problem. You might not get every educated guess correct, but again, the point isn't about getting a perfect score. It's about getting a good score, and surviving hard questions by going with your gut feelings is a good way to achieve this.

On other problems, though, you might have no inkling of what the correct answer should be. In that case, turn to the second key idea.

Think "Good Science!"

The AP Biology test rewards good biologists. The test wants to foster future biologists by covering fundamental topics and sound laboratory procedure. What the test doesn't want is bad science. It doesn't want answers that are factually incorrect, too extreme to be true, or irrelevant to the topic at hand.

Yet these "bad science" answers invariably appear, because hey! It's a multiple-choice test and you have to have four incorrect answer choices around the one right answer. So if you don't know how to answer a question, look at the answer choices and think "Good Science." This may lead you to find some poor answer choices that can be eliminated.

44. AB + energy → A + B

    Which of the following best characterizes the reaction represented above?

    (A) Anabolism
    (B) Endergonic reaction
    (C) Exergonic reaction
    (D) Hydrolysis
    (E) Oxidation-reduction

Here's our trusty Stand-Alone question from before. Even if you don't know what the problem is asking, look at choice (D), *hydrolysis*. The prefix *hydro-* means "water." You don't even have to be a biologist to know that. Even if you don't know exactly what hydrolysis is, you should know that it has something to do with water.

Nothing in the question stem has anything to do with water, does it? The AP Biology exam is comprehensive, not sneaky, so how could choice (D) be the correct answer? It just can't. You could even make a similar case for choice (E), since the term *oxidation* implies oxygen is involved. That makes basic Good Science sense. Yet the question stem has nothing about oxygen, so (E) is probably also wrong. Thinking about Good Science, you can cross out (D) and (E). You have a one-in-three shot, so take a guess.

You would be surprised how many times the correct answer on a multiple-choice question is a simple, blandly worded fact like, "Cells come in variety of sizes and shapes." No breaking news there, but it is Good Science: a carefully worded statement that is factually accurate.

Thinking about Good Science in terms of the AP Biology exam can help you in two ways.
1. It helps you cross out extreme answer choices or choices that are untrue or out of place, and
2. It can occasionally point you toward the correct answer, since the correct answer will be a factual piece of information sensibly worded.

Neither the "Good Science" nor the "Comprehensive, Not Sneaky" strategy is 100-percent effective every time, but they do help more often than not. On a tough Stand-Alone Question, these techniques can make the difference between an unanswered question and a good guess.

Data Questions

Data Questions require the most time and effort, but they aren't worth more than other multiple-choice question types. This is why you should save them for last. When you finish the Stand-Alone Questions, ideally you should have at least one minute for every Data Question remaining. You might have more time, which would be good; if you have less, you'll have to make some choices. Either way, scan the section and look for the topics you are most comfortable with. If you see a Punnett square and you love heredity, by all means start with those problems. If you have to deal with a diagram, try to eliminate the most obvious answers first, then try to eliminate each of the harder answers one by one.

At least one—and most likely several—of the Data Questions will deal with experiments. If you don't like experiments, you probably won't enjoy this section (you probably also won't enjoy biology either, for that matter.) Make sure you understand all the basic points of an experiment: testing a hypothesis, setting up an experiment properly to isolate a particular variable, and so on, so that you will be able to breeze through this section when you come to it. The key to getting through the Data Questions section of the exam is to be able to quickly analyze and draw a conclusion from data presented.

Questions With Graphs

There aren't many graphs on the AP Biology test, but when they appear, they are usually in the Data Questions section. Most graph questions usually require a bit of biology knowledge to determine what the right answer is, but some graph questions only test whether or not you can read a graph properly. If you can make sense of the vertical and horizontal axes, then you can determine what the correct answer is. Granted, very, very few graph questions are this easy, but even so, it's nice to have a slam-dunk question or two. Therefore, if you see a graph, look at the problem and see if you can answer the question just by knowing how to read a graph.

That's all that can be said for the multiple-choice section of the AP Biology test. Be sure to practice these strategies on the practice test in this book so that you'll actually use them on the real test. Once you implement these techniques, your mindset, approach, and score should benefit.

Of course, the multiple-choice questions only account for 60 percent of your total score. To get the other 40 percent, you have to tackle the free-response questions. Before jumping to the free-response question strategies, take about a few minutes to read about the ten-minute reading period.

HOW TO APPROACH THE TEN-MINUTE READING PERIOD

There is a ten-minute reading period sandwiched between the multiple-choice section and the free-response (essay) portion of the AP Biology exam. Notice that this is a "reading" period, not a "nap" period. Ten minutes isn't much time, but it does give you an opportunity to read the essay questions. You won't be able to write in the test booklet, but you will be able to carefully read the questions and plan out how you are going to answer them.

This gives you 2 minutes, 30 seconds per question to plan your response. Take at least 30 seconds to read and reread each question. Make sure you understand what is being asked. Next you should jot down any thoughts you have about the answer on a piece of scratch paper. Write down key words you want to mention. Undoubtedly, many of you have brainstormed ideas when writing an essay for your English class. This is exactly what you want to do here as well: brainstorm ideas about the best way to answer each free-response question.

With the remaining time, make a quick outline of how you would answer each question. You don't need to write complete sentences; just jot down notes that you can understand. If drawings or diagrams are requested, make a brief, crude version of what they will look like.

The following is an example of a free-response question and some notes that could be taken. The actual answer should not be in this outline form, but in a more coherent essay, with more detailed descriptions of each concept where appropriate.

> Transcription is the process of generating RNA from a DNA blueprint. It occurs in both prokaryotes and eukaryotes, but the mechanism and results are different. **Describe** the differences between prokaryotic and eukaryotic transcription for the following cases.
>
> (a) Where does transcription occur and how does this affect translation?
> (b) What is the structure of a messenger RNA?
> (c) What enzymes and other molecules and sequences are necessary for transcription?

Things to Cover:

Prok — cytoplasm, translation → Christmas-tree structure

Euk — nucleus (rRNA in nucleolus), no translation until exits

Prok — many genes on one mRNA (operon/polycistronic), no introns, cap, or tail

Euk — single gene, introns, cap, poly A tail

Prok — same RNA pol for all RNA (sigma factor), –10 and –35 region of promoter, maybe rho for termination

Euk — RNA pol I, II, and III (rRNA, mRNA, tRNA), transcription factors, enhancers, TATA box or something like it, spliceosome

By the time you've written that, your 2.5 minutes on that question should be up. Move on to the next question. When the time comes to start writing your answers, you'll have a good set of notes on which to base your answer to this question.

HOW TO APPROACH THE FREE-RESPONSE QUESTIONS

For Part II, you have 90 minutes to jot down answers to only four questions. That's a little over 20 minutes per question, which gives you an idea of how much work each question may take. You might finish faster, but it wouldn't be good to finish each free-response question in less than five minutes, since each question counts as 10 percent of your final score. Take the time to make your answers as precise and detailed as possible.

The College Board states that the four questions will come from the following categories:
- 1 question covering Molecules and Cells
- 1 question covering Heredity and Evolution
- 2 questions covering Organisms and Populations

This is only slightly helpful since the categories are so broad. Free-response questions can come in any shape or size, but there are some things you can know about them beforehand.

Important Distinctions

Each free-response question will, of course, be about a distinct topic. However, this is not the only way in which these questions differ from one another. Each question will also need a certain kind of answer, depending on the type of question it is. Part of answering each question correctly is understanding what general type of answer is required. There are five important signal words that indicate the rough shape of the answer you should provide.

- Describe
- Discuss
- Explain
- Compare
- Contrast

Each of these words indicates that a specific sort of response is required; none of them mean the same thing. Questions that ask you to *describe*, *discuss*, or *explain* are testing your comprehension of a topic. A description is a detailed verbal picture of something; a description question is generally asking for "just the facts." This is not the place for opinions or speculation. Instead, you want to create a precise picture of something's features and qualities. A description question might, for example, ask you to describe the results you would expect from an experiment. A good answer here will provide a rich, detailed account of the results you anticipate.

A question that asks you to discuss a topic is asking you for something broader than a mere description. A discussion is more like a conversation among ideas, and—depending on the topic—this may be an appropriate place to talk about tension between competing theories and views. For example, a discussion question might ask you to discuss which of several theories offers the best explanation for a set of results. A good answer here would go into detail about why one theory does a better job of explaining the results, and it would talk about why the other theories cannot cope with the results as thoroughly.

A question that asks you to explain something is asking you to take something complicated or unclear and present it in simpler terms. For example, an explanation question might ask you to explain why an experiment is likely to produce a certain set of results, or how one might measure a certain sort of experimental result. A simple description of an experimental setup would not be an adequate answer to the latter question. Instead, you would need to describe that setup *and* talk about why it would be an effective method of measuring the result.

Questions that ask you to *compare* or *contrast* are asking you to analyze a topic in relation to something else. A question about comparison needs an answer that is focused on similarities between the two things. On the other hand, a question that focuses on contrast needs an answer emphasizing differences and distinctions.

Three Points to Remember about the Free-Response Questions

1. **Most Questions Are Stuffed with Smaller Questions**. You usually won't get one broad question like, "Are penguins really happy?" Instead, you'll get an initial setup followed by questions labeled (a), (b), (c), and so on. Expect to spend about one page writing about each lettered question.

2. **Writing Smart Things Earns You Points**! For each sub-question on a free-response question, points are given for saying the right thing. The more points you score, the better off you are on that question. Going into the details about how points are scored would make your head spin, but in general, the AP Biology people have a rubric, which acts as a blueprint for what a good answer should look like. Every subsection of a question has two to five key ideas attached to it. If you write about one of those ideas, you earn yourself a point. There's a limit to how many points you can earn on a single sub-question, and there are other strange regulations, but it boils down to: Writing smart things about each question will earn you points toward that question.

So don't be terse or in a hurry. You have 20 minutes to answer each free-response question. Use the time to be as precise as you can be for each sub-question. Part of being precise is presenting your answer in complete sentences. Do not simply make lists or outlines. Sometimes doing well on one sub-question will earn you enough points to cover up for another sub-question you're not as strong on. When all the points are tallied for that free-response question, you come out strong on total points, even though you didn't ace every single sub-question.

3. **Mimic the Data Questions**. Data Questions, which appear at the end of the multiple-choice section, often describe an experiment and provide a graph or table to present the information in visual form. On at least one free-response question, you will be asked about an experiment in some form or another. To score points on this question you must describe the experiment well and perhaps present the information in visual form.

So look over the sample Data Questions you see in this book and on the actual test, because you can use knowledge of this format when tackling the free-response questions. In a way, this is just another aspect of the "Good Science" idea. The AP Biology test wants to show you what good science looks like on the Data Questions. You can then use that information when crafting your free-response answers.

Beyond these points, there's a bit of a risk in the free-response section since there are only four questions. If you get a question on a subject you're weak in, things might look grim. Still, take heart. Quite often, you'll earn some points on every question since there will be some sub-questions or segments that you are familiar with. Remember, the goal is not perfection. If you can ace two of the questions and slug your way to partial credit on the other two, you will put yourself in position to get a good score on the entire test. That's the Big Picture, so don't lose sight of it just because you don't know the answer to one sub-question.

Be sure to use all the strategies discussed in this chapter when taking the practice exams. Trying out the strategies there will get you comfortable with them, and you should be able to put them to good use on the real exam.

STRESS MANAGEMENT

You can beat anxiety the same way you can beat the AP Biology exam—by knowing what to expect beforehand and developing strategies to deal with it.

Sources of Stress

In the space provided, write down your sources of test-related stress. The idea is to pin down any sources of anxiety so you can deal with them one by one. We have provided common examples—feel free to use them and any others you think of.

- I always freeze up on tests.
- I'm nervous about the molecular biology (or the evolutionary biology, ecology, etc.).
- I need a good/great score to get into my first-choice college.
- My older brother/sister/best friend/girlfriend/boyfriend did really well. I must match that score or do better.
- My parents, who are paying for school, will be quite disappointed if I don't do well.
- I'm afraid of losing my focus and concentration.
- I'm afraid I'm not spending enough time preparing.
- I study like crazy but nothing seems to stick in my mind.
- I always run out of time and get panicky.
- The simple act of thinking, for me, is like wading through refrigerated honey.

My Sources of Stress

_____
_____
_____
_____
_____
_____
_____
_____

Read through the list. Cross out things or add things. Now rewrite the list in order of most disturbing to least disturbing.

My Sources of Stress, in Order

_____
_____
_____
_____
_____
_____
_____
_____

Chances are, the top of the list is a fairly accurate description of exactly how you react to test anxiety, both physically and mentally. The later items usually describe your fears (disappointing Mom and Dad, looking bad, etc.). Taking care of the major items from the top of the list should go a long way toward relieving overall test anxiety. That's what we'll do next.

Strengths and Weaknesses

Take 60 seconds to list the areas of biology that you are good at. They can be general ("evolution") or specific ("protein synthesis"). Put down as many as you can think of, and if possible, time yourself. Write for the entire time; don't stop writing until you've reached the one-minute stopping point. Go.

Strong Test Subjects

_____
_____
_____
_____
_____
_____
_____

Now take one minute to list areas of the test you're not so good at, just plain bad at, have failed at, or keep failing at. Again, keep it to one minute, and continue writing until you reach the cutoff. Go.

Weak Test Subjects

_____
_____
_____
_____
_____
_____
_____

Taking stock of your assets and liabilities lets you know the areas you don't have to worry about, and the ones that will demand extra attention and effort. It helps a lot to find out where you need to spend extra effort. We mostly fear what we don't know and are probably afraid to face. You can't help feeling more confident when you know you're actively strengthening your chances of earning a higher overall score.

Now, go back to the "good" list, and expand on it for two minutes. Take the general items on that first list and make them more specific; take the specific items and expand them into more general conclusions. Naturally, if anything new comes to mind, jot it down. Focus all of your attention and effort on your strengths. Don't underestimate yourself or your abilities. Give yourself full credit. At the same time, don't list strengths you don't really have; you'll only be fooling yourself.

Expanding from general to specific might go as follows. If you listed "ecology" as a broad topic you feel strong in, you would then narrow your focus to include areas of this subject about which you are particularly knowledgeable. Your areas of strength might include population analysis, energy flow in communities, etc. Whatever you know well goes on your "good" list. OK. Check your starting time. Go.

Strong Test Subjects: An Expanded List

_____
_____
_____
_____
_____
_____
_____

After you've stopped, check your time. Did you find yourself going beyond the two minutes allotted? Did you write down more things than you thought you knew? Is it possible you know more than you've given yourself credit for? Could that mean you've found a number of areas in which you feel strong?

You just took an active step toward helping yourself. Enjoy your increased feelings of confidence, and use them when you take the AP Biology exam.

Visualize

This next little group of activities is a follow-up to the "good at" and "bad at" lists. Sit in a comfortable chair in a quiet setting. If you wear glasses, take them off. Close your eyes and breathe in a deep, satisfying breath of air. Really fill your lungs until your rib cage is fully expanded and you can't take in any more. Then, exhale the air completely. Imagine you're blowing out a candle with your last little puff of air. Do this two or three more times, filling your lungs to their maximum and emptying them totally. Keep your eyes closed, comfortably but not tightly. Let your body sink deeper into the chair as you become even more comfortable.

With your eyes shut you can notice something very interesting. You're no longer dealing with the worrisome stuff going on in the world outside of you. Now you can concentrate on what happens inside you. The more you recognize your own physical reactions to stress and anxiety, the more you can do about them. You may not realize it, but you've begun to regain a sense of being in control.

Let images begin to form on TV screens on the back of your eyelids. Allow the images to come easily and naturally; don't force them. Visualize a relaxing situation. It might be in a special place you've visited before or one you've read about. It can be a fictional location that you create in your imagination, but a real-life memory of a place or situation you know is usually better. Make it as detailed as possible and notice as much as you can.

Stay focused on the images as you sink farther into your chair. Breathe easily and naturally. You might have the sensations of any stress or tension draining from your muscles and flowing downward, out your feet and away from you.

Take a moment to check how you're feeling. Notice how comfortable you've become. Imagine how much easier it would be if you could take the test feeling this relaxed and in this state of ease. You've coupled the images of your special place with sensations of comfort and relaxation. You've also found a way to become relaxed simply by visualizing your own safe, special place.

Close your eyes and start remembering a real-life situation in which you did well on a test. If you can't come up with one, remember a situation in which you did something that you were really proud of—a genuine accomplishment. Make the memory as detailed as possible. Think about the sights, the sounds, the smells, even the tastes associated with this remembered experience. Remember how confident you felt as you accomplished your goal. Now start thinking about the AP Biology exam. Keep your thoughts and feelings in line with that prior, successful experience. Don't make comparisons between them. Just imagine taking the upcoming test with the same feelings of confidence and relaxed control.

This exercise is a great way to bring the test down to earth. You should practice this exercise often, especially when you feel burned out on test preparation. The more you practice it, the more effective the exercise will be for you.

Exercise

Whether it's jogging, walking, biking, mild aerobics, push-ups, or a pickup basketball game, physical exercise is a very effective way to stimulate both your mind and body and to improve your ability to think and concentrate. Lots of students get out of the habit of regular exercise when they're prepping for the exam. Also, sedentary people get less oxygen to the blood, and hence to the brain, than active people. You can watch TV fine with a little less oxygen; you just can't think as well.

Any big test is a bit like a race. Finishing the race strong is just as important as being quick early on. If you can't sustain your energy level in the last sections of the exam, you could blow it. Along with a good diet and adequate sleep, exercise is an important part of keeping yourself in fighting shape and thinking clearly for the long haul.

There's another thing that happens when students don't make exercise an integral part of their test preparation. Like any organism in nature, you operate best if all your "energy systems" are in balance. Studying uses a lot of energy, but it's all mental. When you take a study break, do something active. Take a five- to ten-minute exercise break for every 50 or 60 minutes that you study. The physical exertion helps keep your mind and body in sync. This way, when you finish studying for the night and go to bed, you won't lie there unable to sleep because your head is wasted while your body wants to run a marathon.

One warning about exercise: It's not a good idea to exercise vigorously right before you go to bed. This could easily cause sleep-onset problems. For the same reason, it's also not a good idea to study right up to bedtime. Make time for a "buffer period" before you go to bed. Take 30 to 60 minutes to take a long hot shower, to meditate, or to watch a TV show.

Stay Drug Free

Using drugs to prepare for or take a big test is not a good idea. Don't take uppers to stay alert. Amphetamines make it hard to retain information. Mild stimulants, such as coffee, cola, or over-the-counter caffeine pills can help you study longer since they keep you awake, but they can also lead to agitation, restlessness, and insomnia. Some people can drink a pot of coffee sludge and sleep like a baby. Others have one cup and start to vibrate. It all depends on your tolerance for caffeine. Remember, a little anxiety is a good thing. The adrenaline that gets pumped into your bloodstream helps you stay alert and think more clearly.

You can also rely on your brain's own endorphins. Endorphins have no side effects and they're free. It just takes some exercise to release them. Running, bicycling, swimming, aerobics, and power walking all cause endorphins to occupy the happy spots in your brain's neural synapses. In addition, exercise develops your mental stamina and increases the oxygen transfer to your brain.

To reduce stress you should eat fruits and vegetables (raw is best, or just lightly steamed or nuked), low-fat protein such as fish, skinless poultry, beans, and legumes (like lentils), or whole grains such as brown rice, whole wheat bread, and pastas (no bleached flour). Don't eat sweet, high-fat snacks. Simple carbohydrates like sugar make stress worse, and fatty foods lower your immunity. Don't eat salty foods either. They can deplete potassium, which you need for nerve functions. You can go back to your Combos-and-Dew diet after the AP Biology exam.

Isometrics

Here's another natural route to relaxation and invigoration. You can do it whenever you get stressed out, including during the test. Close your eyes. Starting with your eyes and—without holding your breath—gradually tighten every muscle in your body (but not to the point of pain) in the following sequence:

- Close your eyes tightly.
- Squeeze your nose and mouth together so that your whole face is scrunched up. (If it makes you self-conscious to do this in the test room, skip the face-scrunching part.)
- Pull your chin into your chest, and pull your shoulders together.
- Tighten your arms to your body, then clench your fists.
- Pull in your stomach.
- Squeeze your thighs and buttocks together, and tighten your calves.
- Stretch your feet, then curl your toes (watch out for cramping in this part).

At this point, every muscle should be tightened. Now, relax your body, one part at a time, in reverse order, starting with your toes. Let the tension drop out of each muscle. The entire process might take five minutes from start to finish (maybe a couple of minutes during the test). This clenching and unclenching exercise will feel silly at first, especially the buttocks part, but if you get good at it, you will feel very relaxed.

COUNTDOWN TO THE TEST

Study Schedule

The schedule presented here is the ideal. Compress the schedule to fit your needs. Do keep in mind, though, that research in cognitive psychology has shown that the best way to acquire a great deal of information about a topic is to prepare over a long period of time. Since you may have several months to prepare for this exam, it makes sense for you to use that time to your advantage. This book, along with your text, should be invaluable in helping you prepare for this test.

If you have two semesters to prepare, use the following schedule:

September:

Take the diagnostic test in this book and isolate areas in which you need help. The diagnostic will serve to familiarize you with the type of material you will be asked about on the AP exam.

Begin reading your biology textbook along with the class outline.

October–February:

Continue reading this book and use the summaries at the end of each chapter to help guide you to the most salient information for the exam.

March and April:

Take the two practice tests and get an idea of your score. Also, identify the areas in which you need to brush up. Then go back and review those topics in both this book and your biology textbook.

May:

Do a final review and take the exam.

If you only have one semester to prepare, you'll need a more compact schedule:

January:

Take the diagnostic test in this book.

February–April:

Begin reading this book and identify areas of strengths and weaknesses.

Late April:

Take the two practice tests and use your performance results to guide you in your preparation.

May:

Do a final review and take the exam.

Three Days Before the Test

It's almost over. Eat a Power Bar, drink some soda—do whatever it takes to keep going. Here are Kaplan's strategies for the three days leading up to the test.

Take a full-length practice test under timed conditions. Use the techniques and strategies you've learned in this book. Approach the test strategically, actively, and confidently.

WARNING: DO NOT take a full-length practice test if you have fewer than 48 hours left before the test. Doing so will probably exhaust you and hurt your score on the actual test. You wouldn't run a marathon the day before the real thing.

Two Days Before the Test

Go over the results of your practice test. Don't worry too much about your score, or about whether you got a specific question right or wrong. The practice test doesn't count. But do examine your performance on specific questions with an eye to how you might get through each one faster and better on the test to come.

The Night Before the Test

DO NOT STUDY. Get together an "AP Biology Exam Kit" containing the following items:

- A watch
- A few No. 2 pencils (pencils with slightly dull points fill the ovals better)
- Erasers
- Photo ID card
- Your admission ticket from ETS

Know exactly where you're going, exactly how you're getting there, and exactly how long it takes to get there. It's probably a good idea to visit your test center sometime before the day of the test, so that you know what to expect—what the rooms are like, how the desks are set up, and so on.

Relax the night before the test. Do the relaxation and visualization techniques. Read a good book, take a long hot shower, watch something on the WB. Get a good night's sleep. Go to bed early and leave yourself extra time in the morning.

The Morning of the Test

First, wake up. After that:

- Eat breakfast. Make it something substantial, but not anything too heavy or greasy.
- Don't drink a lot of coffee if you're not used to it. Bathroom breaks cut into your time, and too much caffeine is a bad idea.
- Dress in layers so that you can adjust to the temperature of the test room.
- Read something. Warm up your brain with a newspaper or a magazine. You shouldn't let the exam be the first thing you read that day.
- Be sure to get there early. Allow yourself extra time for traffic, mass transit delays, and/or detours.

During the Test

Don't be shaken. If you find your confidence slipping, remind yourself how well you've prepared. You know the structure of the test; you know the instructions; you've had practice with—and have learned strategies for—every question type.

If something goes really wrong, don't panic. If the test booklet is defective—two pages are stuck together or the ink has run—raise your hand and tell the proctor you need a new book. If you accidentally misgrid your answer page or put the answers in the wrong section, raise your hand and tell the proctor. He or she might be able to arrange for you to regrid your test after it's over, when it won't cost you any time.

After the Test

You might walk out of the AP Biology exam thinking that you blew it. This is a normal reaction. Lots of people—even the highest scorers—feel that way. You tend to remember the questions that stumped you, not the ones that you knew. We're positive that you will have performed well and scored your best on the exam because you followed the Kaplan strategies outlined in this section. Be confident in your preparation, and celebrate the fact that the AP Biology exam is soon to be a distant memory.

Now, continue your exam prep by taking the diagnostic test that follows this chapter. This short test will give you an idea of the format of the actual exam, and it will demonstrate the scope of topics covered. After the diagnostic test you'll find answers with detailed explanations. Be sure to read these explanations carefully, even when you got the question right, as you can pick up bits of knowledge from them. Use your score to learn which topics you need to review more carefully. Of course, all the strategies in the world can't save you if you don't know anything about biology. The chapters following the diagnostic test will help you review the primary concepts and facts that you can expect to encounter on the AP Biology exam.

| PART TWO |

Diagnostic Test

Part Two: Diagnostic Test | 37

AP Biology Diagnostic Test

Enough said about strategies; onto the review portion of our program. The following chapters contain a wealth of information and review questions about all the main topics covered on the exam. Ideally, you will have the time to go through every chapter and try every review question while working at a steady pace. You'll finish with enough time to take a practice test each week, and then follow that up by taking the real AP Biology exam.

This is the ideal scenario, but one thing often prevents students from following it. That one thing is the real world. The fact is, many students have schedules that are already chock-full of activities. Finding large chunks of time to devote to studying a test—one that isn't even part of your regular schoolwork—isn't just difficult. It's next to impossible.

If this is the case with you, take a moment for the following 20-question diagnostic test. The questions in this diagnostic test are designed to cover most of the topics you will encounter on the AP Biology exam. After you take it, you can use the results to give yourself a broad idea of what subjects you are strong in and what topics you need to review more. You can use this information to tailor your approach to the following review chapters. Hopefully you'll still have time to read all the chapters, but if pressed, you can start with the chapters and subjects you know you need to work on.

Time yourself, and take the entire test without interruption—you can always call your friend back *after* you finish. Also, no TV! You won't get to watch your favorite show while taking the real AP Biology exam, so you may as well get used to it now.

Be sure to read the explanations for all questions, even those you answered correctly. (This is something you should do on the two practice exams as well.) Even if you got the question right, reading another person's answer can give you insights that will prove helpful on the real exam.

Good luck on the diagnostic test!

Diagnostic Test Answer Grid

To compute your score for the diagnostic test, calculate the number of questions you got wrong, then deduct $\frac{1}{4}$ of that number from the number of right answers. So, if you got 5 questions wrong out of 20, subtract $\frac{1}{4}$ of that (1.25) from the number of questions you got right (15). The final score is 13.75. To set this equal to a score out of 100, set up a proportion:

$$\frac{13.75}{20} = \frac{n}{100}$$

$$20n = 1375$$

$$n = 68.75 = 69$$

The approximate score range is as follows:
5 = 90–100 (extremely well qualified)
4 = 80–89 (well qualified)
3 = 70–79 (qualified)
2 = 60–69 (possibly qualified)
1 = 0–59 (no recommendation)

A score of 69 is approximately a 2, so you can definitely do better. If your score is low, keep on studying to improve your chances of getting credit for the AP Biology exam.

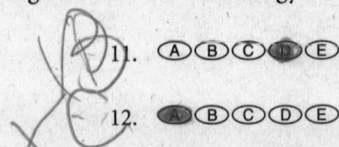

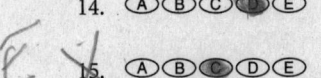

1. Ⓐ Ⓑ Ⓒ Ⓓ Ⓔ
2. Ⓐ Ⓑ Ⓒ Ⓓ Ⓔ
3. Ⓐ Ⓑ Ⓒ Ⓓ Ⓔ
4. Ⓐ Ⓑ Ⓒ Ⓓ Ⓔ
5. Ⓐ Ⓑ Ⓒ Ⓓ Ⓔ
6. Ⓐ Ⓑ Ⓒ Ⓓ Ⓔ
7. Ⓐ Ⓑ Ⓒ Ⓓ Ⓔ
8. Ⓐ Ⓑ Ⓒ Ⓓ Ⓔ
9. Ⓐ Ⓑ Ⓒ Ⓓ Ⓔ
10. Ⓐ Ⓑ Ⓒ Ⓓ Ⓔ
11. Ⓐ Ⓑ Ⓒ Ⓓ Ⓔ
12. Ⓐ Ⓑ Ⓒ Ⓓ Ⓔ
13. Ⓐ Ⓑ Ⓒ Ⓓ Ⓔ
14. Ⓐ Ⓑ Ⓒ Ⓓ Ⓔ
15. Ⓐ Ⓑ Ⓒ Ⓓ Ⓔ
16. Ⓐ Ⓑ Ⓒ Ⓓ Ⓔ
17. Ⓐ Ⓑ Ⓒ Ⓓ Ⓔ
18. Ⓐ Ⓑ Ⓒ Ⓓ Ⓔ
19. Ⓐ Ⓑ Ⓒ Ⓓ Ⓔ
20. Ⓐ Ⓑ Ⓒ Ⓓ Ⓔ

Diagnostic Test

Time: 16 Minutes
20 Questions

1. Amylase is an enzyme important in the digestion of starches. What kind of organic macromolecule is amylase?

    (A) A protein
    (B) A carbohydrate
    (C) A lipid
    (D) A nucleic acid
    (E) A phospholipid

2. Which two subcellular organelles contain unique DNA similar to that of bacteria and are thought to have evolved from prokaryotic symbionts of the first eukaryotic cells?

    (A) The nucleus and endoplasmic reticulum
    (B) The chloroplast and mitochondrion
    (C) The nucleolus and mitochondrion
    (D) The chloroplast and ribosomes
    (E) The ribosomes and endoplasmic reticulum

3. Which of the following structures is present in all cells?

    (A) Nucleus
    (B) Cell wall
    (C) Plasma membrane
    (D) Mitochondria
    (E) Golgi bodies

Questions 4–5 refer to the following equations and choices.

$$6CO_2 + 6H_2O + energy \rightarrow C_6H_{12}O_6 + O_2$$
$$C_6H_{12}O_6 + 6O_2 \rightarrow 6CO_2 + 6H_2O + ATP$$

   (A) CO_2
   (B) H_2O
   (C) $C_6H_{12}O_6$
   (D) O_2
   (E) ATP

4. Which of the molecules above is considered the energy currency of the cell?

5. Which of the molecules is reduced to form glucose?

6. Production of purple kernels in *Zea maize* (corn) is dominant over yellow kernels, and smooth kernels are dominant over wrinkled. The two traits are passed on independently of one another. If an ear of corn has 160 kernels, how many would you expect to be yellow and smooth?

    (A) 120
    (B) 90
    (C) 60
    (D) 30
    (E) 10

GO ON TO THE NEXT PAGE

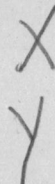

7. Sex-linked recessive disorders are usually carried on the X chromosome and most often affect males. This is because

    (A) mothers always pass the disorders on to their sons
    (B) it only takes one copy of the gene to affect males
    (C) it only takes one copy of the gene to affect females
    (D) it takes two copies of the gene to affect males
    (E) fathers always pass the disorders on to their sons

Questions 8–9 refer to the following.

Normal DNA strand:
5' TAC ACA GAA GGA GAG GGA ACA ATT 3'
Methionine Cystine Leucine Proline Leucine Proline Cystine Stop

Mutated DNA strand:
5' TAC GAC AGA AGG AGA GGG AAC AAT 3'
Methionine Leucine Serine Serine Serine Proline Leucine Phenylalanine

8. What type of mutation is shown?

    (A) Substitution
    (B) Deletion
    (C) Insertion
    (D) Translocation
    (E) Nondisjunction

9. What effect has the mutation had on the codon sequence?

    (A) Transposition
    (B) Trinucleotide repeat
    (C) Inversion
    (D) Frame shift
    (E) Duplication

10. While on the Galapagos Islands, Charles Darwin noted that several distinct kinds of finches had beak characteristics well suited to the various kinds of foods that they ate. This is an example of

    (A) analogous structures
    (B) Hardy-Weinberg equilibrium
    (C) genetic drift
    (D) stabilizing selection
    (E) adaptive radiation

11. Animals in the phylum Echinodermata such as sea stars and sand dollars are thought to be more closely related to the phylum Chordata (which includes humans and other vertebrates) than to other animal phyla because both groups

    (A) have radial symmetry
    (B) have deuterostome embryonic development
    (C) are segmented
    (D) have bony internal skeletons
    (E) lack a true coelem

12. *Naeglaria fowleri* causes a fatal form of meningitis. The infective form of this organism inhabits fresh water in warm climates, often in the sediment of lakes. It can infect humans when they swim in infested lakes, allowing entry through the nose. *N. fowleri* has a true membrane-bound nucleus and cellular organelles. It is a unicellular, heterotrophic organism that lacks a cell wall and moves via pseudopodia. What type of organism is it?

    (A) Bacteria
    (B) Virus
    (C) Protozoan
    (D) Fungus
    (E) Insect

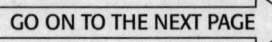

Questions 13–15 refer to the following list.

(A) Xylem
(B) Phloem
(C) Meristems
(D) Stomata
(E) Carpel

13. Vascular tissue in which water and mineral elements are transported

14. Pores formed by guard cells

15. Fruit-forming reproductive structure in angiosperms

16. Nerve impulses move along the axons of neurons via action potentials caused by

(A) movement of sodium across the cell membrane
(B) an electron transport chain
(C) release of calcium into the sarcomeres
(D) an ATP proton pump
(E) oxidation-reduction reactions

17. Which of the following is NOT part of the digestive tract?

(A) The esophagus
(B) The spleen
(C) The stomach
(D) The duodenum
(E) The colon

Questions 18–20 refer to the following paragraph and figures.

A scientist studying the ecology of cities found that in developed landscapes, plant roots were not colonized by mycorrhizal fungi to the same degree that they were in a nearby nature preserve. In addition, she found that rates of photosynthesis and root respiration were much higher in plants in the preserve than for plants in city landscapes. She conducted a controlled greenhouse experiment to see what effects mycorrhizal colonization had on plant photosynthesis and respiration. Her experimental design involved growing 10 plants in soil rich in mycorrhizal fungal elements and 10 in the same soil that had been sterilized to remove the fungi. She made periodic measurements of plant photosynthesis and root respiration and calculated the mean rates for each experimental treatment. Her results are shown below.

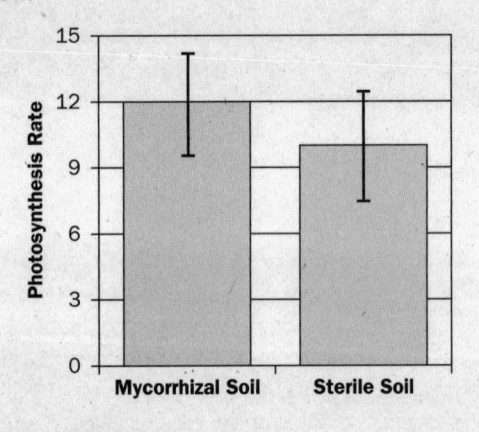

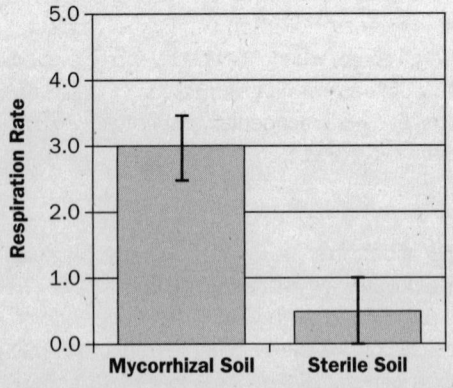

18. What conclusion can be drawn from the data?

    (A) Presence of mycorrhizae increased both photosynthesis and respiration rates significantly.
    (B) Presence of mycorrhizae increased photosynthesis but not respiration.
    (C) Presence of mycorrhizae increased respiration but not photosynthesis.
    (D) Presence of mycorrhizae had no effect on photosynthesis or respiration.
    (E) No conclusions can be drawn based on the data shown.

19. Why might mycorrhizae influence photosynthesis and/or respiration?

    (A) Mycorrhizae are important plant pathogens.
    (B) Mycorrhizae are important plant parasites.
    (C) Mycorrhizae are important plant predators.
    (D) Mycorrhizae are important plant symbionts.
    (E) Mycorrhizae would not be expected to influence plants.

20. Assuming that further experimentation showed conclusively that plants in cities had reduced rates of photosynthesis and respiration due to lack of colonization by mycorrhizae, what important biogeochemical cycle of an ecosystem would be most affected?

    (A) The nitrogen cycle
    (B) The water cycle
    (C) The hydrological cycle
    (D) The Krebs cycle
    (E) The carbon cycle

IF YOU FINISH BEFORE TIME IS CALLED, YOU MAY CHECK YOUR WORK. **STOP**

ANSWERS AND EXPLANATIONS

1. A

All known enzymes are proteins, including amylase.

2. B

Both chloroplasts and mitochondria have unique circular strands of DNA similar to that of bacteria, which are not found in the nucleus of the cell. It is thought that the earliest eukaryotic cells similar to unicellular protists formed symbioses with these bacteria. Endosymbiotic theory holds that energy-producing bacteria came to reside symbiotically within eukaryotic cells and ultimately formed the mitochondria, while photosynthetic bacteria inhabited the cells that later evolved into algae and plants by forming chloroplasts.

3. C

All cells are bound by a plasma membrane. Prokaryotic cells lack a true nucleus and organelles, ruling out (A), (D), and (E). Animal cells and many protist cells lack a cell wall, ruling out (B).

4. E

Adenosine triphosphate (ATP) is the form of energy used by cells to do work. It is formed from ADP when glucose is broken down via cellular respiration, the second equation shown.

5. A

The first equation shown is photosynthesis, the process by which plants harvest the sun's energy to fix carbon dioxide and produce glucose. An important step in the light-independent reactions of photosynthesis, the Calvin cycle, involves reduction of fixed carbon dioxide to the 3-carbon precursors of glucose. Water is oxidized during the light-dependent reactions of photosynthesis.

6. D

In a dihybrid cross such as the one described in the question, the expected phenotypic outcome is: 9/16 individuals will show both dominant traits, 3/16 will show one dominant and one recessive trait, 3/16 will show the *other* dominant and the *other* recessive trait, and 1/16 will show both recessive traits. The question asks about individuals showing one dominant and one recessive trait, so out of 160 offspring, 30 would be expected to fit the bill.

7. B

Sex-linked recessive disorders that are carried on the X chromosome are not always passed to offspring, since the mother has two copies of the X chromosome, ruling out (A). Boys always get a Y chromosome from their fathers, ruling out (E). Since males have one X and one Y chromosome, they can only have one copy of an X-linked gene, ruling out (D). For females to be affected, two copies of the gene must be present; since they are recessive disorders, one copy in a female would not be expressed, ruling out (C). In males, since there is only one X chromosome, one copy of the recessive gene is expressed because there is no dominant copy to mask its expression.

8. C

The mutated gene sequence has the single base guanine (G) inserted in the fourth position in the sequence; all of the previous and subsequent bases are the same as the normal strand. A substitution (A) involves replacement of one base by another, and deletion (B) involves loss of a single base. Translocation (D) involves movement of a segment of DNA, and nondisjunction (E) is an error in meiosis that affects chromosome number.

9. D

Insertions and deletions both cause frame shifts that can affect all of the codons that occur after the error, as in the example shown. Both are point mutations involving single nucleotides. Transposition (A) involves movement of a gene position on a chromosome. A trinucleotide repeat (B) involves sequences of three nucleotides repeated in a series on the same chromosome a number of times. Inversion (C) occurs when a DNA sequence is flipped, such that ATT would become TTA. Duplication (E) involves a section of DNA that is accidentally duplicated when a chromosome is copied.

10. E

The beaks of Darwin's finches represent homologous structures, not analogous ones, ruling out (A). Choices (B), (C), and (D) all describe how allele frequencies contribute to microevolution. Hardy-Weinberg equilibrium describes populations with stable allele frequencies. Genetic drift is a phenomenon associated with statistically unexpected changes in allele frequency, especially in small populations. Stabilizing selection occurs when heterozygous genotypes are favored. Darwin's finches are examples of adaptive radiation, an evolutionary process that results from adaptation to specific ecological niches.

11. B

Of the nine major animal phyla, only Echinodermata and Chordata have deuterostome embryonic development; the anus forms prior to the mouth in the gut cavity. Chordates have bilateral symmetry, ruling out (A), and echinoderms are not segmented, ruling out (C). Only vertebrates, a subphylum of chordates, have an internal bony skeleton (D). Both chordates and echinoderms have a true coelem (E), a fully enclosed internal body cavity.

12. C

The key to this question is in the last sentence before the question. Any organism that is a unicellular, heterotrophic eukaryote without a cell wall must be a protist. Bacteria (A) and viruses (B) lack a nucleus and cellular organelles, and fungi (D) have cell walls containing chitin. Insects (E) are multicellular animals.

13. A

Plants take up water and minerals from the soil and transport both upward through the plant via xylem. Phloem (B) is vascular tissue associated with transport of sugars produced in the leaves to all parts of the plant.

14. D

Meristems (C) are the sites at the tips of roots and shoots and in the vascular cambium where active cell division occurs. Stomata are pores in leaves that allow the uptake of carbon dioxide and loss of water via transpiration; they are formed by two guard cells that allow the pore to open and close.

15. E

The carpel houses the ovary of a flower and is one of the defining characteristics of flowering plants, called angiosperms. After fertilization, a seed is produced and the carpel develops into a fruit, which houses the seed and aids in dispersal.

16. A

The transmission of nerve impulses occurs when gated sodium channels are stimulated to open, depolarizing the membrane of the axon and setting off a chain reaction of depolarization/repolarization events. Electron transport chains (B) are associated with ATP production via the action proton pumps (D) in respiration and photosynthesis. Calcium functions in sarcomeres (C) to allow muscle contraction, and oxidation-reduction reactions (E) are associated with metabolic processes such as enzyme cascades.

17. B

The spleen is an organ associated with the lymphatic and immune systems. All of the other structures listed are part of the digestive tract, a continuous tube that extends from the mouth to the anus.

18. C

Although both figures show that rates of photosynthesis and respiration tended to be higher when mycorrhizae were present, the error bars indicate that there really was no difference in terms of photosynthesis, since the bars overlap the mean values for the two treatments. This rules out (A) and (B). The figures do show a difference between the two treatments in terms of respiration, ruling out (D) and (E).

19. D

Most land plants form mutualistic symbioses with mycorrhizae; it is thought that the two groups evolved together. Both organisms typically benefit from the association.

20. E

Both photosynthesis and respiration are important components of the carbon cycle. The nitrogen cycle (A) is predominantly moderated by bacteria. The water cycle (B) and hydrological cycle (C) are basically the same thing and would not necessarily be influenced by changes in photosynthesis and respiration. The Krebs cycle (D) is part of cellular respiration, but is not a biogeochemical cycle.

DIAGNOSTIC TEST CORRELATION CHART

Use the following table to determine which AP Biology topics you need to review most. After scoring your test, check to find out the areas of study covered by the questions you answered incorrectly.

Area of Study	Question Number
Organic molecules in organisms	1
Subcellular organization	2
Cells	3
Cellular energetics	4, 5, 18, 19, 20
Inheritance patterns	6, 7
Mutation	8, 9
Evolutionary patterns	10
Phylogenetic classification	11
Structure and function of plants and animals	12, 17
Reproduction, growth, and development	13, 14, 15
Chemistry of life	16

| PART THREE |

AP Biology Review

HOW TO USE THE REVIEW SECTION

Step 1: Review the concepts.

Each of the following review chapters begins by going over the main concepts that apply to the chapter's topic. The chapters will NOT include loads of factual material, but instead will help you tie all the facts together to understand the concepts and discuss briefly how they fit within the thematic emphases designed by the College Board.

Step 2: Review the labs.

The AP Biology course is designed with the expectation that students will conduct the 12 recommended laboratories, and there are questions on the AP Biology exam that apply directly to the experimental design and results obtained in these laboratories. To that end, the 12 laboratories are reviewed in the appropriate chapters. You will learn both the concepts associated with each laboratory experience and something about the scientific method itself. Keep in mind that the College Board expects that after completing the AP Biology course, you will be able to design an experiment yourself.

Step 3: Answer the sample questions.

Questions from both sections of the AP Biology exam, multiple choice (60 percent of the exam grade) and "free response" or essay (40 percent of the exam grade), are covered in each review chapter. There are three types of questions in the multiple-choice section: Stand-Alone Questions, Cluster Questions, and Data Questions (see the Strategies chapter for an explanation of question types). Essay questions must be answered in complete sentences. Answers that receive full credit by the essay evaluators are quite comprehensive. The College Board will expect you to be able to analyze and interpret data, synthesize information from your laboratory experience with what you learned in the course, and integrate different concepts.

Step 4: Review the answer explanations.

After completing all of the questions in the review chapters, you will be prepared to analyze the proper approach to each type of question and how to formulate the proper responses. You will be given the correct answer for each question and be informed of the thought processes you should go through to reach a correct answer. The sample essays in the review chapters are intended to help you learn how to analyze the questions and answer them thoroughly with a synthesis of knowledge. The quantity and type of essay questions in the review chapters do not represent the exact coverage of material expected on the AP Biology exam (as with the multiple-choice questions), but they are plausible examples of topics covered.

Read on, explore, discover, and learn!

Chapter 3: **Chemistry of Life**

- Starting Small: Atoms, Molecules, and Reactions Central to Biology
- The Importance of Water in Biology
- Structure and Function of Biologically Significant Functional Groups and Macromolecules
- Enzyme Catalysis Laboratory
- Applying the Concepts: Questions and Explanations

STARTING SMALL: ATOMS, MOLECULES, AND REACTIONS CENTRAL TO BIOLOGY

The world around us follows a hierarchy of organization. All life on earth is connected, from the smallest individual units of matter (atoms and molecules) to complex organisms. In this chapter we will discuss the basic building blocks that compose the living world. Although some organisms are more complex than others, different levels of organization do not correlate with levels of complexity. An individual cell with its multitude of chemical reactions is just as dynamic and complex as an entire community of species. This chapter will review the most crucial concepts of the building blocks of life before concluding with sample questions.

Elements Essential to Life and Organic Compounds

Ninety-nine percent of all living matter is made up of four elements. Only four! These elements are nitrogen (N), carbon (C), hydrogen (H), and oxygen (O). Phosphorus (P) and sulfur (S) account for almost all of the remaining 1 percent of living matter. The mnemonic device, N'CHOPS (for Nice chops!) or similar acronyms can be used to remember the main elements of life. All six elements are important in biochemistry, especially carbon (C).

A molecule containing carbon (C) is called an **organic** molecule or organic compound. Life wouldn't exist without organic compounds. Carbon earned its place as a staple in biology because it can bond to other atoms or to itself in four different, equally spaced directions, allowing complex molecules of almost unlimited size and shape to be formed. The structural properties of the carbon atom have allowed for the formation of molecular compounds—such as DNA and enzymes—that have unique chemical identities and functions. The message to take home is that there are only a few elements important to biology (N'CHOPS) and that carbon is the main element of life because of the variety of organic compounds it can form.

Chemical Bonds

The carbon atom shares its electrons fairly easily with other atoms, forming **covalent bonds**. In contrast, atoms like heavy metals—calcium (Ca), magnesium (Mg), potassium (K), and sodium (Na)—don't share electrons well. These metals form bonds that have charges between the atoms called **ionic bonds**. **Hydrogen bonds** are a third type of bond that occurs when a hydrogen atom in one molecule, such as water, is attracted to an electrostatic atom in another molecule, such as nitrogen, oxygen, or fluorine.

A compound that has mostly covalent bonds, with electrons evenly distributed around the molecules that are sharing electrons, is called **nonpolar**. A compound with ionic bonds has electrical charge and is said to be **polar**. Covalently bonded molecules in a compound can also be polar, if the electrons within the compound are unevenly shared.

Compounds that easily dissolve in and mix with water are called **hydrophilic**, or water loving. **Hydrophobic** compounds do not dissolve in water; they repel it. These compounds do not combine with water at all. An example of a hydrophobic compound would be oil, which sticks together and forms beads when dropped in water. A hydrophilic compound would be table salt, NaCl.

The key points to remember about chemical bonds are as follows:

- Polar compounds are composed of covalently bonded molecules in which electrons are unevenly shared, resulting in a negative charge on one part of a molecule in relation to another part of the same molecule.
- Nonpolar compounds share electrons equally, so that the molecules within the compound have no positive or negative poles.
- Nonpolar molecules do not dissolve in water, but polar and ionic molecules do dissolve in/mix with water.

Chemical Reactions

Several basic types of chemical reactions are presented throughout the AP Biology course, and the College Board expects you to know them for the exam. Chemical reactions fit into four categories.

1. Hydrolysis, Dehydration, and Ionic Reactions

Hydrolysis is the decomposition of something in the presence of water. It can involve creating ions in solution.

$$NaCl(s) \rightarrow Na^+(aq) + Cl^-(aq)$$

The (s) indicates that the substance is in a solid state and the (aq) indicates an aqueous state.

Hydrolysis can also involve donating a water molecule by splitting the molecule into H^+ and OH^- and attaching these to different molecules in a compound. Here is the hydrolysis of sucrose into glucose and fructose.

Sucrose + H₂O → Glucose + Fructose

Dehydration is the exact opposite reaction to hydrolysis, as it involves the release of water. Ionic reactions occur when one atom donates an electron to another atom, thereby becoming a positive ion, while the other atom attains a negative charge. The difference in charge between the two atoms holds them together.

$$NaCl(aq) + AgNO_3(aq) \rightarrow AgCl(s) + NaNO_3(aq)$$

2. Oxidation-Reduction Reactions

Oxidation-reduction reactions involve the gain (reduction) or loss (oxidation) of electrons, usually as a result of the action of an acid or a base on another substance. Since oxidation and reduction occur simultaneously during a reaction, the terms are coupled. Most importantly, oxidation-reduction reactions regularly take place between organic and inorganic compounds in living cells and provide the mechanism for energy transfer in biology. The loss or gain of electrons can be viewed by creating half reactions for the compounds involved in the full reaction. Creating half reactions is a concept from chemistry. Here is the half reaction converting ethanol to acetic acid:

$$CH_3CH_2OH + H_2O \rightarrow CH_3CO_2H + 4H^+ + 4e^-$$

The four electrons given up on the right side of the equation indicate that ethanol is being oxidized during the reaction. If electrons were being added on the left side of the equation we would know that the reaction was a reduction.

3. Anabolism and Catabolism Reactions

Anabolism and catabolism are different ways of describing oxidation-reduction reactions. Anabolism is the process of synthesizing simple substances into complex materials (like reduction). You can remember this by recalling that athletes take anabolic steroids to BUILD muscle. Catabolism is the opposite, breaking down complex material into simple parts (like oxidation). This is what happens during digestion when complex food particles are broken down into molecules like simple sugars, fatty acids, and amino acids.

4. Exergonic and Endergonic Reactions

Exergonic and endergonic reactions are used to explain the use or generation of energy. A reaction that releases energy is said to be exergonic. A reaction that requires energy is called endergonic.

THE IMPORTANCE OF WATER IN BIOLOGY

A water molecule is formed between two H atoms bonded covalently to a single O atom. The oxygen molecule tends to control all of the electrons by keeping them away from the hydrogen atoms, giving the oxygen a slightly negative charge, which is balanced by slightly positive charges on the hydrogen atoms. This is an example of a **polar covalent bond**. The water molecule looks like Mickey Mouse since it has a big head (the oxygen atom) and two big ears (the hydrogen atoms).

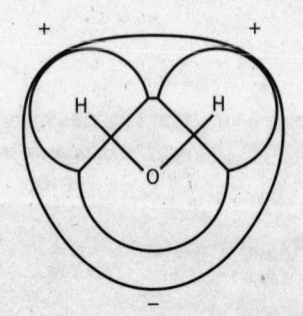

The head has a slightly negative charge and the ears have a slightly positive charge, which makes water a polar compound. The shape and charge of this molecule give it unique properties.

Water is the only substance on Earth that commonly exists in all three physical states (gas, liquid, and solid). The substance has a high specific heat so it serves as a temperature stabilizer for other compounds, vaporizes at a relatively high temperature, and expands instead of contracting when it freezes.

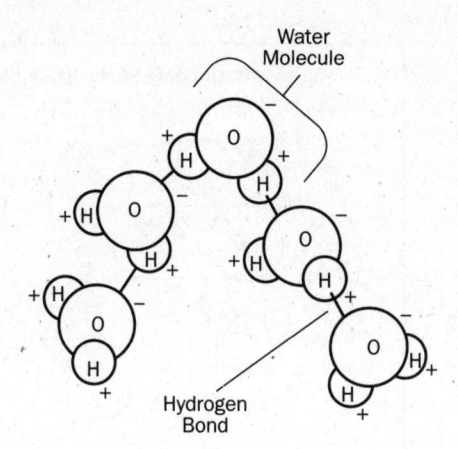

Water plays a key role in hydrolysis, condensation, and other chemical reactions that are essential to life. It is also fundamental to the biological activity of nucleotides, carbohydrates, and proteins. There are two specific characteristics of water that are particularly important to remember. First, the polar nature of water makes it "sticky." The positive charges on the hydrogen atoms cling to the negative charges on the oxygen atoms between molecules, i.e., water molecules attract one another, causing water to have high surface tension. This is what makes water bead on windshields and form round raindrops. Even matter with greater density can float on top of water if it doesn't break the surface tension.

This surface tension is also the force behind **capillary action**, by which water (and anything dissolved in it) will climb up a thin tube or move through the spaces of a porous material until it is overcome by gravity. It is as if each water molecule dragged the one behind it along, as well as any nutrients dissolved in the water. Capillary action plays an important role in moving nutrients and other metabolites through living things. Plants' roots, for example, take in water from the soil, full of minerals and dissolved nutrients; capillary action draws the water and its load through the plant.

Second, the polar nature of water makes it a good solvent. Capillary action would be a lot less useful if water lacked this property. Fortunately, water molecules are happy to surround positive and negative ions and readily dissolve other polar compounds. In addition, hydrogen bonds can form between the hydrogen and oxygen atoms of the water molecule and surrounding molecules.

| Solid NaCl | NaCl Being Dissolved in Water |

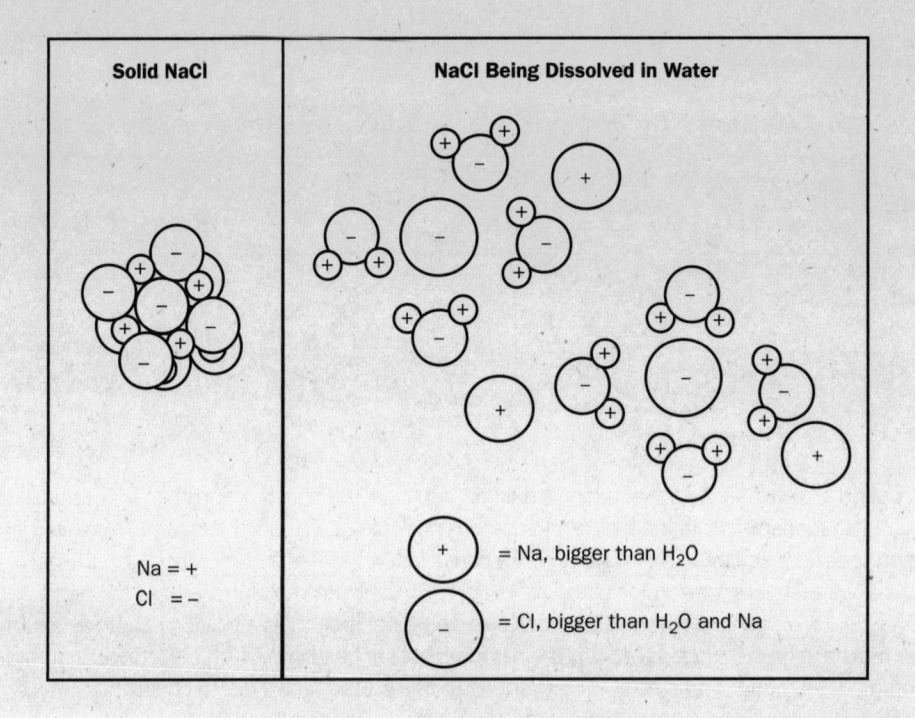

Na = +
Cl = −

(+) = Na, bigger than H_2O
(−) = Cl, bigger than H_2O and Na

Water's role as a solvent is another reason it is so important for living things. Chemicals that are dissolved by the water in blood can be carried around the body rapidly and easily; this is how sugar gets to muscles, for example. Each living cell is, in large part, a membrane surrounding chemicals dissolved in water. It is only by being dissolved in water that these chemicals can participate in many of the biological reactions that keep living things alive.

Remember that many organic compounds are nonpolar or have nonpolar regions and won't mix with water. This is an important aspect of segregating biological activity. All of life survives only because water exists on Earth. The chemistry that supports life occurs almost exclusively in water and relies on its presence.

STRUCTURE AND FUNCTION OF BIOLOGICALLY SIGNIFICANT FUNCTIONAL GROUPS AND MACROMOLECULES

Carbon forms bonds with various other elements to create characteristic functional groups. These functional groups are important for two reasons:

- The functional group(s) in a compound are the source of the compound's name.
- Functional groups are the sites of chemical reactivity for the compound.

Part Three: AP Biology Review
Chemistry of Life | 55

Below is an easy reference table for the biologically important functional groups. The functional groups listed are all moderately to very polar and important in biological reactions.

Functional Groups, Molecular Structures, and Their Properties

Functional Group	Structure	Name of Compound It Forms	Example	Some Important Properties
Amino	—C—N—H (with H on N)	Amines	CH_3NH_2 Methylamine	1. Acts as a base which affects electronegativity of amino acids as pH varies. 2. Formation of peptide bond between amino acids.
Carbonyl	—C—C(=O)—H	Aldehydes	CH_3CHO Acetaldehyde	1. Highly reactive carbonyl group. 2. Highly reactive carbon near carbonyl group. 3. Important intermediaries in several biological reactions.
Carbonyl	—C—C(=O)—C—	Ketones	$(CH_3)_2CO$ Acetone	
Carboxyl	—C—C(=O)—O—H	Carboxylic Acids	$CH_3(CH_2)_2CO_2H$ Butyric acid	1. Weak acids able to donate a H^+ ion to several biological reactions.
Hydroxyl	—C—O—H	Alcohols	CH_3CH_2OH Ethanol	1. Makes compounds soluble in water.
Phosphate	—C—O—P(=O)(OH)—OH	Organic Phosphates	$C_{10}H_8N_4O_2NH_2(OH)_2(PO_3H)_3H$ Adenosine triphosphate (ATP)	1. Storage and transfer of energy.
Sulfhydryl	—SH	Thiols	$CH_3H_7NO_2S$ Cysteine	1. Stabilizes protein structure.

ENZYME CATALYSIS LABORATORY

The enzyme catalysis lab is designed to allow you to explore structural proteins and environmental conditions that affect chemical activity in the natural world. In the lab, you will measure the chemical activity of the enzyme catalase as it breaks down hydrogen peroxide (H_2O_2) into water and oxygen gas. Specific experimental conditions—temperature, substrate and enzyme concentrations, pH, and so on—will be altered to help you understand the effects of environmental conditions on a specific chemical reaction. This experiment helps you to learn the experimental method through manipulation of experimental treatments compared to a control treatment.

Understanding the experiment itself can be encapsulated in just a few graphs. One or more of these graphs may be on the exam. The first graph is about free energy and enzyme activity.

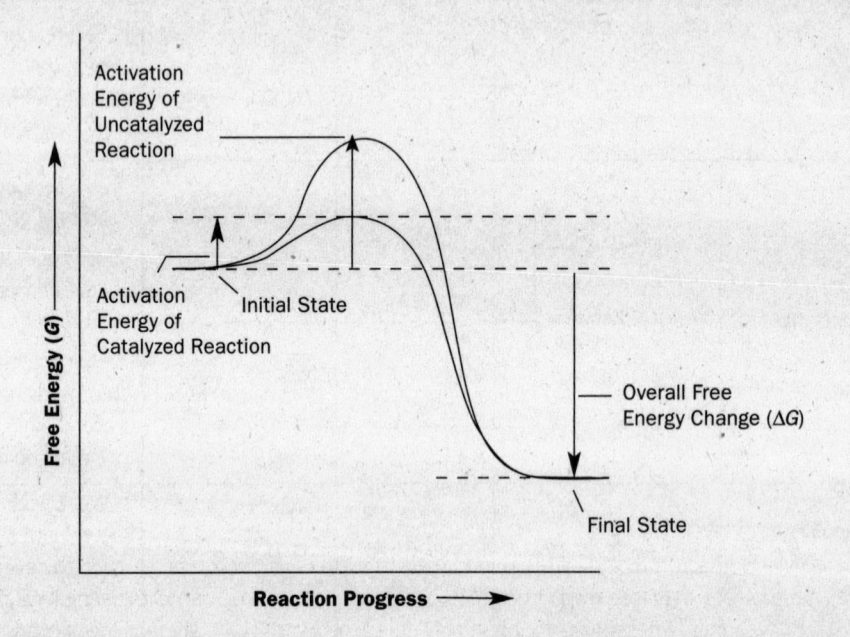

Before a reaction can take place it must reach a point called its **activation energy** by receiving enough energy from the environment, termed **free energy**, in the form of heat or kinetic energy. An enzyme catalyzes a reaction by lowering the activation energy needed (it requires less free energy) to allow the reaction to take place. Hydrogen peroxide breaks down into water and oxygen gas on its own, but at an incredibly slow rate. The enzyme catalase lowers the activation energy of the reaction and the reaction happens very quickly.

Many reactions that occur in biology can be sped up through the addition of enzymes that lower activation energy. Natural compounds readily metabolize and synthesize in the presence of certain enzymes, and biological systems control chemical activity largely through synchronizing the presence or absence of enzymes. Enzymes also control the release of energy so that cells don't combust during certain reactions from a sudden influx of energy.

Notice on the graph that the amount of free energy in the system changes over the progress of the reaction. Products of the reaction have an initial amount of energy, receive energy from the environment, then energy is released with the products. Remember that the energy of the reactants and the energy of the products is the same, whether the reaction is enzyme mediated or not. All the enzyme does is lower the activation energy for the molecule to do "what comes naturally" and makes the reaction happen. When the reaction is complete, the enzyme is free and helps to speed up the reaction for another H_2O_2 molecule. This laboratory demonstrates firsthand what occurs in all living systems on a regular basis and points out the importance of these specialized proteins.

The environment can also have a profound effect on the reaction. In this experiment, you measure the experimental effect of changing the reaction temperature and pH, as well as the effect of changing concentrations of the enzyme and substrate.

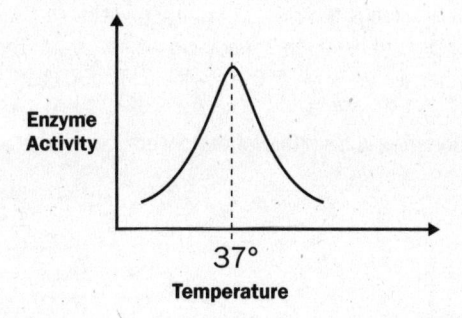

Catalytic activity is greatly affected by temperature and increases with increasing temperature. Since enzymes are proteins, they lose their structure at high temperatures, not only eliminating catalytic properties, but essentially destroying the protein. Different enzymes have different specific temperatures. The activity of animal catalase (catalase occurs in plants, too) peaks at about normal body temperature, or 35–40°C. Once the temperature increases beyond this temperature range the catalase proteins "die."

How does temperature change the reactivity? Remember that the reaction depends on free energy, so increasing temperature will also increase the amount of free energy in the form of both heat and kinetic energy (all the molecules will be moving faster). Since all of the molecules in the system will be moving faster at a higher temperature, there is an increased rate of collision between enzyme and substrate molecules. In essence, there are more chance meetings.

The pH of the environment affects the reactivity of enzymes in the same way as temperature in that there is a level of peak activity.

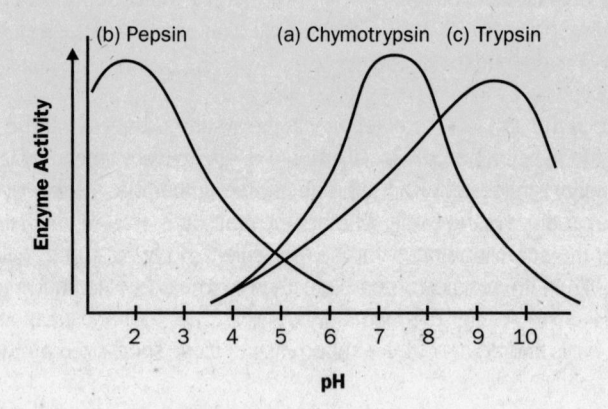

(b) Pepsin   (a) Chymotrypsin   (c) Trypsin

Enzyme Activity vs. pH graph showing curves from pH 2 to 10

In this graph you can see the peak activity for different common digestive enzymes. Pepsin, for example, has a peak activity at a very low pH because it catalyzes reactions in the stomach, which is very acidic. The peak activity for catalase occurs at a pH of 7 to 8, which is fairly neutral. The pH of a system affects protein activity by altering the structure of the protein. Proteins have a tertiary structure based on the electrochemical properties of the amino acids in the chain. The charges on these amino acids can change as the pH changes because of the available H^+ and OH^- ions.

The last things you need to understand from this laboratory are the effects of changing the amount of enzyme and substrate available.

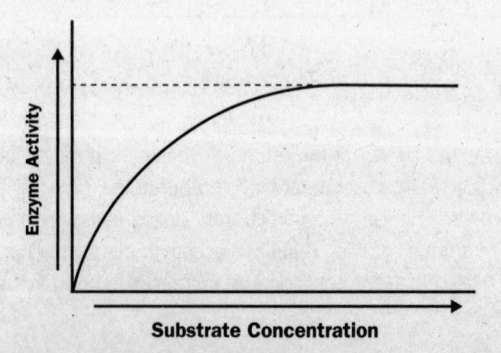

Enzyme Activity vs. Substrate Concentration graph

You may think that as you add more H_2O_2, you get more reaction. You would be right ... to a point. Eventually, no matter how much substrate you add, the reaction maxes out. This is because there are a limited number of enzyme molecules in the system. These enzymes can handle only one molecule of H_2O_2 at a time. Think of enzymes as cashiers at the grocery store. If a bunch of customers want to check out at the same time, they can only be helped one at a time by any single cashier. Once all of the cashiers are busy, a customer has to wait for the next available cashier. Thus, H_2O_2 has to wait for the next available enzyme to catalyze the reaction.

Changing the enzyme concentration affects the reaction in a similar way, except the reaction becomes almost instantaneous as the number of enzyme molecules becomes equal to the number of substrate molecules. An increase in enzyme concentration is like every customer in the grocery store having his own cashier. A customer would move through the line as quickly as the cashier could process her order. In the world of enzymes, orders are processed very quickly (catalase can convert almost 6 million molecules of H_2O_2 to H_2O and O_2 each minute!).

By completing this simple experiment with one enzyme catalyzing one reaction, you not only learn about catalase and H_2O_2, you learn about enzyme properties and the effect the environment has on chemical reactions as well. The type of reaction completed in the experiment was oxidation-reduction, where the active site of the enzyme catalase is a heme group like in hemoglobin, so it has a similar tertiary and quaternary structure. The reason the catalase enzyme exists in animal and plant tissue is to remove H_2O_2, an oxidizer that is dangerous to living cells and must be removed when it is created as a byproduct of metabolism. See how biology is so intricately connected with chemistry?

IF YOU LEARNED ONLY FIVE THINGS IN THIS CHAPTER . . . .

1. Organic compounds are molecules that contain carbon. All living matter is made up of carbon, hydrogen, oxygen, nitrogen, phosphorus, and sulphur (N'CHOPS).

2. Types of chemical bonds include covalent, in which atoms share electrons with other atoms, ionic, and hydrogen. Covalent bonds are polar; the other types are nonpolar. Polar compounds dissolve in water, and nonpolar compounds do not mix with water.

3. In hydrolysis reactions, a compound dissolves in water. Dehydration reactions release water. Oxidation-reduction reactions involve the gain (oxidation) or loss (reduction) of electrons. Anabolic reactions create larger molecules from smaller ones, and catabolic reactions break down larger molecules. Exergonic reactions release energy, while endergonic reactions require energy.

4. The water molecule's polar nature leads to surface tension (enabling capillary action) and makes it an effective solvent. It expands rather than contracting when it freezes. These properties make water essential to life on earth.

5. Enzymes are proteins that facilitate reactions by binding to the substrate and reducing the activation energy.

APPLYING THE CONCEPTS: QUESTIONS AND EXPLANATIONS

1. Which of the following functional groups characterizes the structure of an alcohol?

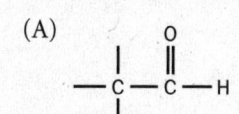

    (A) —C—C(=O)—H

    (B) —SH

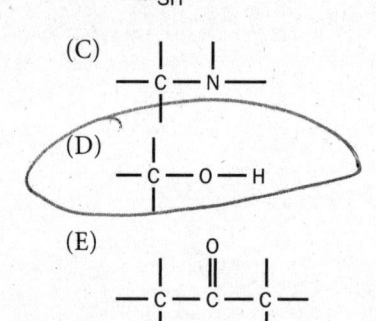

    (C) —C—N—

    (D) —C—O—H

    (E) —C—C(=O)—C—

2. Organic compounds, unlike inorganic compounds, contain

    (A) hydrogen
    (B) oxygen
    (C) nitrogen
    (D) sulfur
    (E) carbon

3. The bonding of two amino acid molecules to form a portion of a protein chain involves

    (A) the addition of a water molecule
    (B) the release of a water molecule
    (C) the addition of a nitrogen atom
    (D) the release of a carbon dioxide molecule
    (E) an increase in activation energy

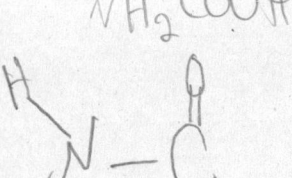

4. Starch and glycogen molecules are similar in that both are
   (A) water-soluble, simple sugars
   (B) polymers of glucose
   (C) intermediate products in the Krebs cycle
   (D) products of synthesis in plant cells
   (E) structural components of connective-tissue fibers

5. A change in pH from 5 to 3 indicates a change in concentration of H^+ ions by a factor of
   (A) 2
   (B) 20
   (C) 50
   (D) 100
   (E) 10,000

6. All of the following chemical compounds are organic EXCEPT which?
   (A) $Na_2S_2O_3$
   (B) CH_4
   (C) $(CH_2OH)_2$
   (D) $(NH_2)_2CO$
   (E) C_6H_5OH

7. All of the following are important characteristics of water in biology EXCEPT which?
   (A) Cohesion provides surface tension.
   (B) It is a nonpolar solvent.
   (C) It is a medium for complex chemical reactions in organisms.
   (D) It has a high specific heat that stabilizes ambient temperature.
   (E) It expands when frozen.

8. Which of the following statements about enzymes is NOT true?

    (A) Enzymes are almost always proteins.
    (B) Enzyme activity is affected by changes in pH.
    (C) Enzymes increase the rate of reaction.
    (D) Enzymes increase the activation energy.
    (E) Enzymes do not change the free energy of products.

9. In one type of enzyme regulation, the presence of the end product of a metabolic pathway inhibits an enzyme which functions in an early step in the pathway. This type of enzyme regulation is called

    (A) feedback inhibition
    (B) competitive inhibition
    (C) noncompetitive inhibition
    (D) irreversible inhibition
    (E) None of the above

10. Which of the following involves the binding of a molecule to the active site of the enzyme?

    (A) Noncompetitive inhibition
    (B) Allosteric inhibition
    (C) Competitive inhibition
    (D) A, B, and C
    (E) None of the above

11. If the free energy change of a reaction is greater than zero, then the reaction

    (A) is spontaneous
    (B) is nonspontaneous
    (C) is at equilibrium
    (D) is endothermic
    (E) is exothermic

12. Which is NOT a characteristic of proteins?

    (A) Can function as enzymes
    (B) Contain peptide bonds
    (C) Are important in cell signaling
    (D) May be used as an energy source
    (E) Contain nitrogenous bases

Questions 13–16 refer to the following graph. The top curve and the bottom curve represent alternate pathways for the same reaction. One pathway is enzyme catalyzed.

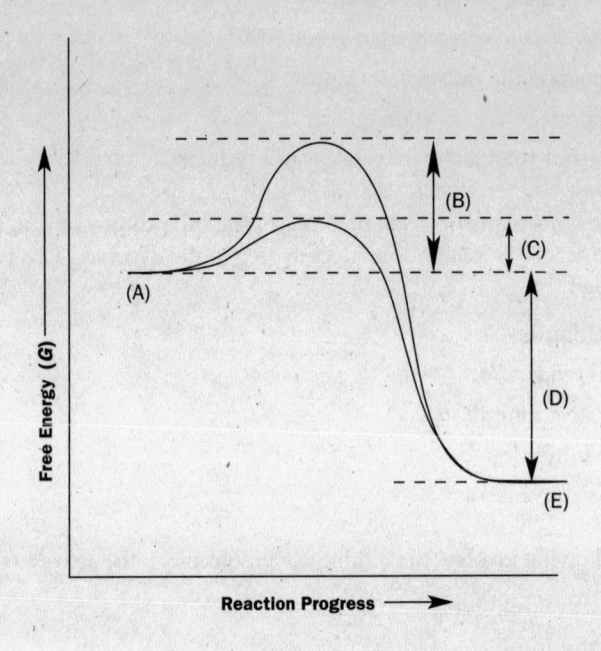

13. Represents the energy state of the reactants of the enzyme-catalyzed pathway
14. Represents the net energy change of the reaction
15. Represents the activation energy of the non-catalyzed reaction
16. Represents the energy state of the products of the enzyme-catalyzed pathway

Questions 17–20 refer to the following diagram showing the polypeptide structure of a protein.

17. Indicates the carboxyl group of an amino acid
18. Indicates a peptide bond between amino acids
19. Indicates an amine group of an amino acid
20. Indicates an α-carbon of an amino acid

Free-Response Question

Proteins are complex molecules important to the function of every living organism, and their chemical properties make them susceptible to conditions in the environment.

(a) **Discuss** the chemical composition of proteins and the chemical reactions involved in protein synthesis.

(b) **Discuss** protein structure and the structure and function of proteins in relation to environmental factors.

(c) **Discuss** the role of proteins in the intracellular and extracellular exchange of information.

ANSWERS AND EXPLANATIONS

1. D

There are two ways to approach this problem. First, if you remember that the functional group of an alcohol is a hydroxyl group (–OH) you know the obvious answer is (D) because it is the only choice of a functional group with a hydroxyl group. The second way to approach this problem is to think about how to construct molecular formulas, remember some examples of alcohols, and use process of elimination. Ethanol (CH_3CH_2OH) and isopropyl alcohol or 2-propanol ($CH_3CHOHCH_3$) are common alcohols and there is a good chance of your being familiar with them both.

So what do they have in common? These compounds are composed of only C, H, and O, so (B) and (C) can be eliminated because they have an S and an N atom respectively. Once you have eliminated these options you may remember that (B) is the functional group for sulfhydryl and that (C) is the functional group for amines. Choices (A) and (E) can be eliminated because the molecular formula of an alcohol always denotes the hydroxyl group (–OH) and the molecular formula for (A) would have to have a –CHO (as in acetic acid $C_2H_4O_2$) in order to qualify as an answer; the molecular formula of (E) would have to have a –CO (as in acetone $(CH_3)_2CO$) to be an alcohol.

2. E

Organic compounds contain carbon atoms. Each carbon atom is able to form four covalent bonds, accounting for the wide variety of organic molecules.

3. B

There's some memorization involved in this question. Amino acids form long polypeptide chains called proteins. These chains get their name from the peptide bond formed between adjacent amino acids.

Amino acids have four components: a central: α-carbon (A), amine (B) and carboxyl (C) functional groups, and a side chain (D) that dictates chemical and physical properties of the amino acid. The peptide bond forms between the amine group of one amino acid and the carboxyl group of another.

This reaction is called dehydration synthesis and it results in the production of a two-amino acid peptide chain and a water molecule.

4. B

Start this problem with the broad view that starch and glycogen are storage forms of sugar (carbohydrates) and are large molecules. Plants produce starch and animals produce glycogen. This rules out (D) and (E) because the correct answer can't be exclusive to either animals or plants. Only animals synthesize glycogen and plants don't have connective tissue (not to mention that connective tissue is proteinaceous). The Krebs cycle involves the metabolism of glucose by-products, which are smaller than glucose molecules so the answer can't be (C). Choice (A) isn't true. That leaves (B). These long chains of glucose are called polymers because they are repeating sequences of the same molecule.

5. D

Here is a test of your knowledge of the pH scale. The pH scale measures the amount of H^+ ions in a solution compared to pure water and ranges from 1 (high H^+ and very acidic) to 14 (very low H^+ and very basic). Pure water has a pH of 7. Your first response might be to pick (A) or (B) because they are obvious factors of 2, but these are not correct. The pH scale is logarithmic. Each step on the pH scale represents a tenfold change in H^+ concentration. In order to go from a benign pH of 7 for water to something very corrosive at a pH of 1 or 2, you need to add a lot of H^+ ions. Going two steps on the pH scale requires increasing the H^+ ion concentration by a factor of 10^2 or 100, which gives you (D).

6. A

The answer is easy to find as long as you note that the question asks for the choice that DOES NOT belong. Organic chemistry is about compounds containing the element carbon (represented in chemical formulas by a C). Organic compounds range from the simplest compounds like methane (CH_4) to the most complex like DNA. The only compound listed that doesn't include carbon is (A).

7. B

This is another "except" question so you have to find the choice that doesn't match. Recall that the water molecule has positive charges on the H atoms and a negative charge on the O atom. This makes it a polar molecule. Also, remember that polar and nonpolar compounds don't mix, and a nonpolar substance will not dissolve well in a polar solvent like water. Since water is polar and doesn't dissolve nonpolar compounds well, the answer is (B).

8. D

Enzymes increase the rate of a reaction by decreasing the activation energy. The other statements are true: enzymes are almost always proteins (A), enzyme activity is affected by changes in pH (B), enzymes increase the rate of reaction (C), and enzymes do not change the free energy of products (E).

9. A

This question and some of the following questions address aspects of enzyme and protein function and structure that were not covered in this chapter. You should be familiar with them from your course or outside reading. In feedback inhibition, the presence of the end product of a metabolic pathway inhibits an enzyme which functions in an early step in the pathway. In competitive inhibition, a molecule which resembles the normal substrate of an enzyme competes for the active site. In non-competitive inhibition, a molecule binds to a part of the enzyme other than the active site, causing a conformational change in the enzyme. This conformational change alters the effectiveness of the enzyme. In irreversible inhibition, a molecule covalently bonds to the active site of the enzyme, preventing substrate from accessing the active site.

10. C

Let's review the information from the explanation for question 9. In competitive inhibition (C), a molecule bonds to the active site of the enzyme, preventing substrate from accessing the active site. In noncompetitive inhibition (A), a molecule binds to a part of the enzyme other than the active site, causing a conformational change in the enzyme. This conformational change alters the effectiveness of the enzyme.

11. B

If the free energy change (ΔG) of a reaction is greater than zero, then the reaction is nonspontaneous. If the free energy change of a reaction is less than zero, then the reaction is spontaneous. If the free energy change of a reaction is equal to zero, then the reaction is at equilibrium. The terms exothermic and endothermic do not refer to the free energy change of a reaction, but to the enthalpy (change in heat) of a reaction. If the change in enthalpy (ΔH) is positive, then heat is absorbed and the reaction is endothermic. If the change in enthalpy is negative, then heat is released and the reaction is exothermic.

12. E

Proteins do not contain nitrogenous bases (nucleotides do contain nitrogenous bases). The other statements are true: proteins can function as enzymes, consist of amino acids joined by peptide bonds, are important in cell signaling (hormones, for example), and may be used as an energy source (amino acids are converted by enzymes into intermediates of glycolysis and the Krebs cycle).

13. A

Questions 13–16 are examples of Cluster Questions. You can actually answer three of these questions without knowing anything about enzyme activity and by using common sense. First notice that (A) and (E) indicate states of energy while (B), (C), and (D) indicate changes in energy. Question 13 can be answered by knowing that, in general:

   (a) graphs of data move from left to right (the diagram even shows the direction of the reaction with the arrow indicating progress of the reaction),
   (b) reactants are things you start with, and
   (c) the *y*-axis is the measure of change in energy.

Choice (A), all the way to the LEFT of the diagram, clearly indicates a stable state before energy change takes place, so it must be the answer.

14. D

This question can also be answered intuitively. The net change in anything is the amount you end with minus the amount you started with. This is clearly the difference between the energy state of the products and the energy state of the reactants and is indicated by (D).

15. B

This is the only question in the cluster that requires a little knowledge of enzymes and activation energy: enzymes lower the activation energy required to cause a reaction. There are two activation energies in the diagram and you need to choose the one that is non-catalyzed, the one that requires more energy. The answer must be (B) because it shows a greater need for energy.

16. E

This question is at the opposite end of the spectrum from question 13. If all the way to the left is the start of the reaction, where the reactants are, then all the way to the right is the end of the reaction, where the products are. Choice (E) indicates another state of energy and occurs after two changes in energy.

17. C

Questions 17–20 are Cluster Questions that require a little more memorization than the previous set, but they can be answered using logic. The carboxyl group of the amino acid is the carbon with the double-bonded O and an H atom that forms one side of the peptide bond between amino acids. This group is indicated by (C).

18. B

The question indicates that there must be at least two amino acids in the diagrammed structure. Try to separate structures in the diagram by drawing boxes around them. Notice that there is a repeating sequence of N-C-C along the center of the entire structure, an R attached to each center C, and an O atom attached to the other C. These subunits are amino acids.

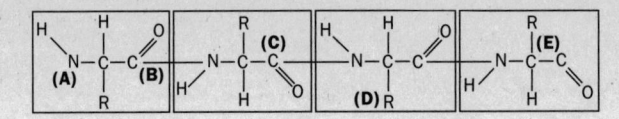

Notice that (B) falls on a line between two of the subunits. It is the correct answer.

19. A

The amine is the only functional group you need to remember that has an N atom. This group is indicated by (A).

20. E

The α-carbon of an amino acid is the center carbon from which all the other groups attach. This carbon is indicated by (E).

Free-Response Answer Explanation

Read and reread the question over again until you understand it. Parts (a) and (b) are about chemical and physical properties of proteins, which were covered in this chapter. Part (c) requires synthesis of information and a working knowledge of other themes and hierarchical levels of biology.

To answer the first two parts of the question, you need to organize information starting with tiny structures and ending with broader themes. Be sure to touch on all of the important points for each level of organization. Each level of organization has different chemical and physical properties that are affected by environmental conditions. These properties need to be discussed for part (b). Part (c) should include examples of "information exchange" and a general explanation of how information exchange is accomplished. Remember, your free-response answers must be written in complete sentences and formulated as a cohesive essay. You are allowed to use diagrams and figures, but they must be explained. Feel free to compose a quick outline, but keep in mind that you only have 90 minutes to complete four essays (22.5 minutes each). Readers are looking more for key bits of information in complete sentences than they are looking at the overall composition of the paper. While composition and coherence are important in the answer, the content of the answer is what is most important to the reader.

Here is a (good) potential answer to the free-response question:

(a) A chemical compound composed of a series of amino acids joined by peptide bonds is called a polypeptide or, when the polypeptide is long, a protein. Amino acids are composed of a backbone of a centralized α-carbon atom bonded to a carboxyl group and an amine group. The fourth bond on the α-carbon is to a side chain that gives each amino acid its unique identity and chemical and physical properties. For example, glycine has the simple side chain of only an H atom; alanine has a side chain of a methyl group (CH_3). The charges on the amine and carboxyl groups depend on the pH of the surrounding environment. Some side chains contain S atoms which allow disulfide bonds between some amino acids along the length of the protein.

(b) Protein structure is uniquely coded by regions of DNA through an mRNA intermediary in which triplet nucleic acids code for specific amino acids. Amino acids are added sequentially at the end of the polypeptide chain through a peptide bond between a carboxyl group on one amino acid and an amine group from the next amino acid. The addition of amino acids is mediated by a tRNA and is performed by a ribosome. The peptide bond is formed through dehydration synthesis, generating one molecule of H_2O.

The conformation of a protein is dictated by its primary structure, the sequence of amino acids that make the polypeptide. Bond angles and hydrogen bonding between atoms on amino acids contribute to the secondary structure that is the twisting and turning that occurs around the amino acid backbone. A tertiary structure develops from interactions between different side chains on the amino acids (e.g., disulfide bonds). A quaternary structure is also possible between different or similar polypeptides which form one functional macromolecule. Protein conformation is affected by environmental factors such as pH and temperature. The pH of the environment affects the electrical charges on the amine, carboxyl, and side chain groups of amino acids, which in turn affect hydrogen and tertiary bonding. Temperature can also cause a protein to unravel as free energy surpasses bond energy.

(c) The unique coding of the primary structure of proteins causes a great deal of specificity of attachment sites in the tertiary and quaternary structures. These highly specific proteins can exist on the external membrane of cells, allowing cellular recognition by the particular proteins that occur there. Transmembrane proteins can exchange information from the intracellular matrix to the extracellular matrix and vice versa via temporary changes in the protein conformation due to the presence or absence of a messenger at a receptor site (e.g., Acetyl CoA-mediated messenger system). Proteins in the cellular matrix can relay the same information from the cell nucleus to the cell membrane through specificity at a receptor site and changing bond affinities with other proteins.

Chapter 4: **Living Cells**

- Life's Little Compartments: Types of Cells and How They Work
- Diffusion and Osmosis Laboratory
- Applying the Concepts: Questions and Explanations

LIFE'S LITTLE COMPARTMENTS: TYPES OF CELLS AND HOW THEY WORK

After the last chapter, this one may be a little refreshing, since almost all you need to know about this topic can be summarized in a few tables and figures. This chapter is about **cells** and parts of cells, and it thematically builds upon the information presented in the last chapter. While reviewing the chapter, note that in cells, different types of reactions and products are produced in a compartmentalized world.

One way the living world stays compartmentalized is with **membranes**. Cells and cell **organelles** are surrounded by a membrane, a selectively permeable barrier that segregates cell contents from the outside world. In this chapter, you'll review some basics about the cell cycle. In the laboratory exercise, you'll learn how the membrane allows transport of certain materials between compartments. Cells and cell organelles come in many different sizes to form simple or complex organisms.

Membranes

All cells are surrounded by a **plasma membrane**. In eukaryotic cells, the **nucleus** and most of the organelles are surrounded by plasma membranes. Membranes are composed mostly of lipids, which is another word for fats. **Lipids** are full of nonpolar covalent bonds, so they are hydrophobic and do not dissolve in water. Since membranes are composed primarily of lipids, most of the material in membranes will not mix well with water. Most of the lipids in membranes have a phosphate group, or **phospholipid**, attached to one end. This charged end is polar and happy to be in water, which is why it is termed hydrophilic. The other end of the phospholipid is like a tail that is nonpolar, and it turns in toward the center of the membrane. Tails on phospholipids are hydrophobic, since the tails do not mix well with water. The attraction between the polar and nonpolar regions of these phospholipids creates the foundation for the bilayer of the membrane. The lipid ends group together like the insides of a sandwich surrounded by polar barriers.

Part Three: AP Biology Review
Chapter 4

Cytoplasm

Aqueous Environment

— Polar Phosphate Bead
— Nonpolar Lipid Chains
— Polar Phosphate Bead

Aqueous Environment

Extracellular Environment

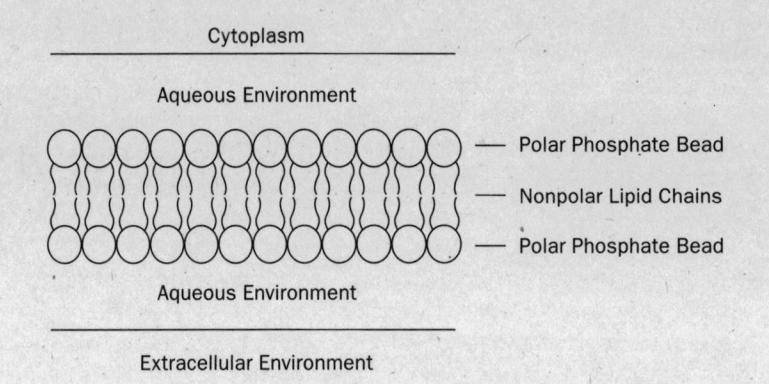

Phosphate and Lipid Chains

Embedded among all of these membrane lipids are **proteins**, **carbohydrates**, and **sterols** (cholesterol). Some proteins are embedded on the outer surface, some on the inner surface, and some span the entire width of the plasma membrane (these usually function as transport proteins). Some surface proteins have sugar groups attached to them, called glycoproteins. Within the cell membrane, each cell and organelle performs its own function. For example, some proteins that span the width of the cellular membrane act as channels for ion transport. Nerve cells have a higher density of these proteins than other cells. Each component of a cell membrane contributes to how the membrane functions. Proteins act as transport molecules, receptor sites, attachments to the cytoskeleton, and surface enzymes. Carbohydrates on the surface of the cell and glycoproteins contribute to cell recognition, particularly in the immune response. Cholesterol contributes fluidity to the membrane.

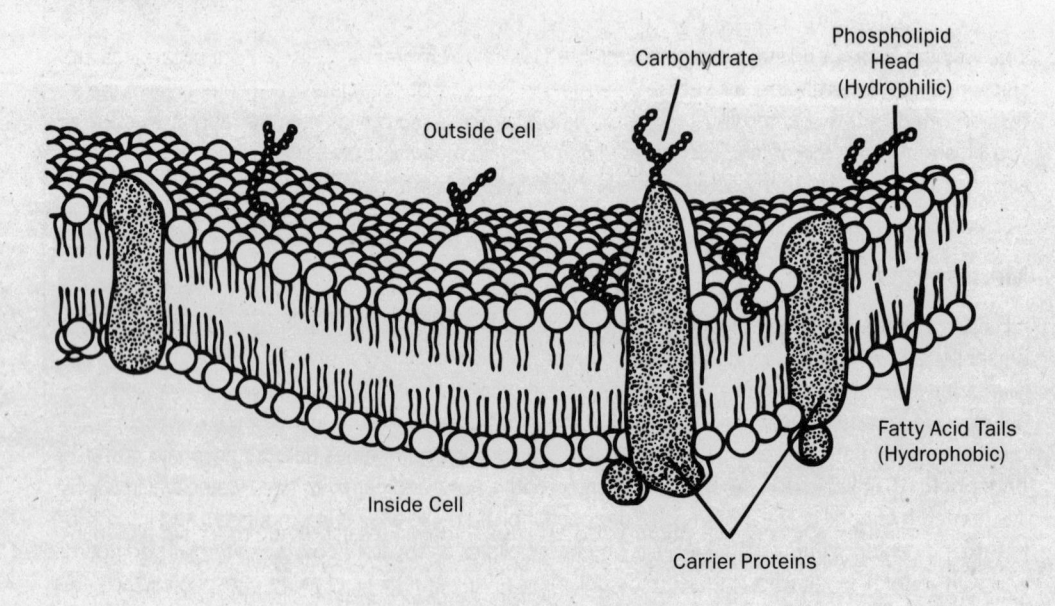

Lipid Bilayer Plasma Membrane

Membranes are **selectively permeable**, which means that they allow some things to pass through, but not others. Nonpolar molecules and small uncharged polar molecules (e.g., H_2O and CO_2) easily pass through the lipid matrix of the membrane. Charged ions and large uncharged polar molecules like glucose do not easily pass through the membrane. **Diffusion**, the movement of molecules from an area of higher concentration to an area of lower concentration, will be reviewed in the lab section of this chapter, but other kinds of transport are summarized in the table below. **Osmosis** is the diffusion of water across a semipermeable membrane.

Recall that molecules are in constant motion and tend to spread out if there is nothing in the way. These molecules move from areas of high concentration to areas of low concentration. This is called moving DOWN or with the **concentration gradient**. Often a cell expends energy to create or maintain a gradient between the cytoplasm and the extracellular environment, causing molecules to move AGAINST the concentration gradient.

Types of Transport Across the Cell Membrane

Type of Transport Gradient	Requires Energy?	Concentration Gradient
Passive (Diffusion)	No	Down
Osmosis (Diffusion of H_2O)	No	Down
Facilitated Diffusion (Ion transport via transmembrane carrier protein)	No	Down
Active Transport (ATP mediated via specific receptor proteins)	Yes	Against
Exocytosis (Vesicle fuses with cell membrane releasing molecules to extracellular environment)	Yes	N/A
Endocytosis — Phagocytosis ("cellular eating," membrane invaginates solid, making a vesicle)	Yes	N/A
Endocytosis — Pinocytosis ("cellular drinking," membrane invaginates liquid, making a vesicle)	Yes	N/A
Endocytosis — Receptor-mediated endocytosis (carrier protein binds to specific substances for invagination)	Yes	N/A

The Cell and Cellular Structures

There are two main types of cells—**prokaryotes** and **eukaryotes**. Prokaryotes are much simpler than eukaryotes; they are composed of a plasma membrane, cytoplasm, cell wall, DNA, ribosomes, and simple microtubules. Bacteria are an example of prokaryotic cells. Eukaryotes include both plant and animal cells. These cells are much more complex than prokaryotes, and contain numerous organelles. The next table describes in detail the properties of each type of cell.

Characteristics of Cells

	Prokaryotes	Eukaryotes		
		Plant Cells	Animal Cells	
Size	0.2–500µm, most 1–10 µm	Most 30–50µm	Most 10–20µm	
Structure				Properties
Cytoplasm	Yes	Yes	Yes	1. Intracellular matrix outside of nucleus
Nucleus	No	Yes	Yes	1. Contains DNA 2. Pores allow communication with cellular matrix
Plasma Membrane	Yes	Yes	Yes	1. Selective barrier around cell allowing the passage of some substances but excluding others 2. Phospholipid bilayer with proteins embedded
Cell Wall	Yes	Yes	No	1. Additional structural barrier around cell 2. Composed of cellulose and other polysaccharides and proteins
Chromosomes	One circular plasmid, only DNA	Multiple strands of DNA and protein	Multiple strands of DNA and protein	1. The cell's DNA
Ribosomes	Yes	Yes	Yes	1. Site of protein synthesis (translation)
Endoplasmic Reticulum (ER)	No	Yes	Yes	1. Site of attachment for ribosomes 2. Protein and membrane synthesis 3. Formation of vesicles for transport
Golgi Complex	No	Yes	Yes	1. Synthesis, accumulation, storage, and transport of products
Lysosomes	No	Some vacuoles function as lysosomes	Usually	1. Vesicle containing hydrolytic enzymes
Vacuoles or Vesicles	No	Not usually	Usually	1. Membrane-bound sacs in the cytoplasm
Mitochondria	No	Yes	Yes	1. Site of cellular respiration
Plastids	No	Yes	No	1. Group of plant organelles that includes chloroplasts 2. Site of photosynthesis 3. Carbohydrate storage
Microtubules (Cilia or Flagella)	Simple	On some sperm	Complex (9 + 2 arrangement)	1. Tubes of globular protein, tubulins 2. Provides structural framework for cell 3. Provides motility
Centrioles	No	No	Yes	1. Cell center for microtubule formation

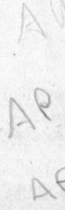

The Cell Cycle and Mitosis

The cell cycle is the life cycle of the cell, including reproduction. **Mitosis** is the process through which a cell replicates and divides. The following two figures provide details on these two processes. The first figure shows the cell cycle, with a table following it that describes each phase of the cycle.

S = DNA replicates
G2 = Gap 2 (cell gets ready to divide)
M = Mitosis (cell division)
G1 = Gap 1 (cell grows)

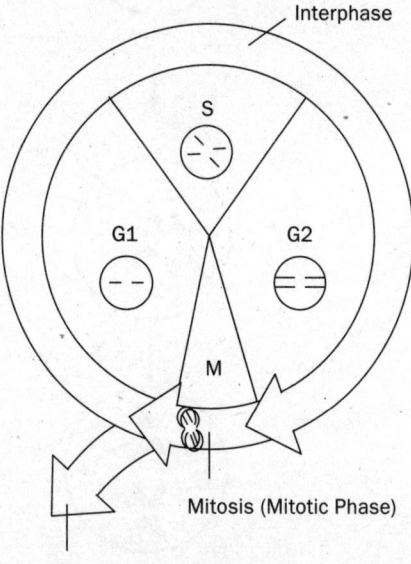

Mitosis (Mitotic Phase)

Cells that cease to divide

The Cell Cycle

Interphase (> 90% cell cycle)	G_1 (First Gap Phase)	Cell growth, protein and RNA synthesis
	S (Synthesis Phase)	DNA synthesis, duplication
	G_2 (Second Gap Phase)	Cell growth, protein synthesis
M phase (Mitotic Phase)		Cell division

The next figure provides details on each phase of mitosis, or nuclear division. The cell starts out in **interphase**, then proceeds to **prophase, metaphase, anaphase,** and **telophase. Cytokinesis** is the division of the cytoplasm in animal cells following division of the nucleus, when two distinct cells are produced.

82 | Part Three: AP Biology Review
Chapter 4

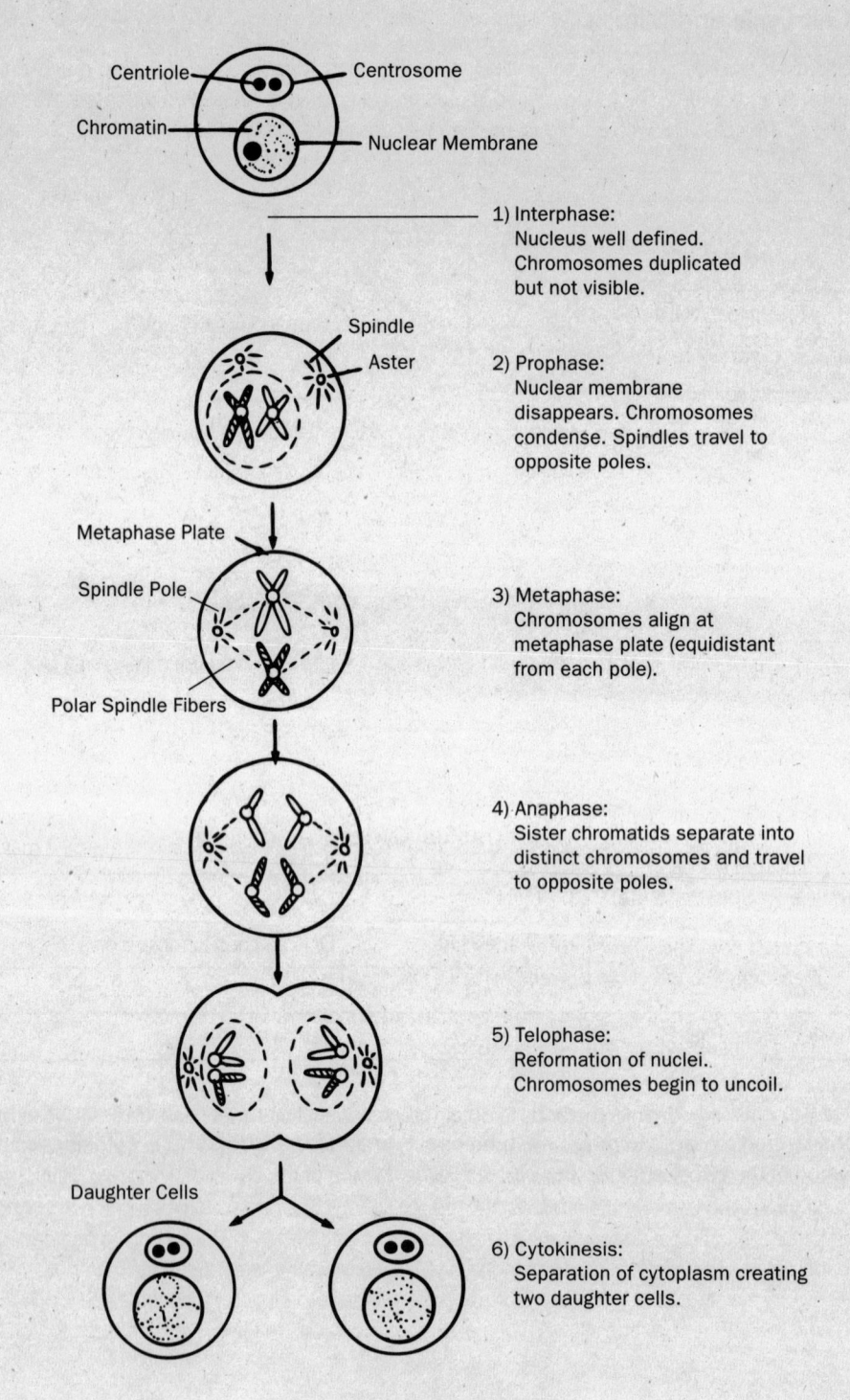

Labels on diagram:
- Centriole, Centrosome, Chromatin, Nuclear Membrane
- Spindle, Aster
- Metaphase Plate, Spindle Pole, Polar Spindle Fibers
- Daughter Cells

1) **Interphase:** Nucleus well defined. Chromosomes duplicated but not visible.

2) **Prophase:** Nuclear membrane disappears. Chromosomes condense. Spindles travel to opposite poles.

3) **Metaphase:** Chromosomes align at metaphase plate (equidistant from each pole).

4) **Anaphase:** Sister chromatids separate into distinct chromosomes and travel to opposite poles.

5) **Telophase:** Reformation of nuclei. Chromosomes begin to uncoil.

6) **Cytokinesis:** Separation of cytoplasm creating two daughter cells.

Mitosis and Cytokinesis

Cell division is primarily controlled through genetics. The DNA in a cell controls whether a cell divides at all and at what rate. Certain hormones or chemicals need to be present in order for a cell to divide. Cell division can be controlled in tissue cultures by inhibiting protein synthesis or affecting nutrient availability. Cell division is reduced by crowding through contact inhibition.

DIFFUSION AND OSMOSIS LABORATORY

The properties of diffusion and osmosis, discussed at the beginning of this chapter, will be explored in this lab. Recall that selectively permeable membranes let some molecules through, but not others. For this lab, it is important to know that:

- isotonic solutions are two solutions that have the same concentration of solute,
- a hypotonic solution has a lower solute concentration than another solution, and
- a hypertonic solution has a higher solute concentration than another solution (think *hypo-* "below" and *hyper-* "above").

Keep in mind that water potential is the measure of "force" a solution has for pulling or drawing water into it. The more negative the water potential, the stronger its pulling force.

Suppose you tied your dialysis bags and put them in solutions to complete the experiment, but didn't get the results you expected! Skewed results can be obtained by not tying the knot on your bag tight enough, not getting all the solution washed off the outside of your bag, or by some other slight oversight. Often one group gets usable results for one part of the lab and another group gets another part right, so everyone shares the "good" results. At times classes have really good luck and all the groups get "good" results. Either way, you need to know what should happen in order to score well on the exam. You can expect to see an osmosis question in one form or another on the exam. It may be in the multiple-choice section or it may be a free-response question.

Two key skills of the diffusion and osmosis lab are:
- Measuring the effects (e.g., weight change) of osmosis
- Determination of the osmotic concentration/water potential of an unknown tissue or solution using solutions of known concentrations

The first skill uses experimental methods to obtain results. Even if you know how osmosis works, it is important to know how to measure it. You can't really watch water molecules or sugar and starch molecules move since they're too small. In this lab, you learn how to measure things you don't see by observing things you can. To obtain results, the weight change of a dialysis bag or a piece of potato can be measured. An indicator color will appear if glucose or starch occurs in a solution. If a bag or piece of potato increased in weight, it gained water and had a more negative water potential than (was hypertonic to) the solution it was placed into, and vice versa. If a solution produced a color indicator, the semipermeable membrane allowed the passage of molecules that the dye is an indicator for.

The second key skill deals with using a standard to figure out the "identity" of an unknown and using observed data to interpolate expected data. The experiment begins with either a dialysis bag full of solution (the concentration of which is unknown) or with a piece of starchy vegetable such as potato. Questions to ask during the experiment include: what is the osmolarity/water potential of the unknown, and how do you measure it? The first parts of the lab teach you to observe the effects of osmosis by measuring the change in weight. Six identical samples of the unknown are weighed and placed into solutions with known concentrations, usually ranging from distilled water to a high concentration like 1.0 M. Percentage weight change can be plotted on a graph like this:

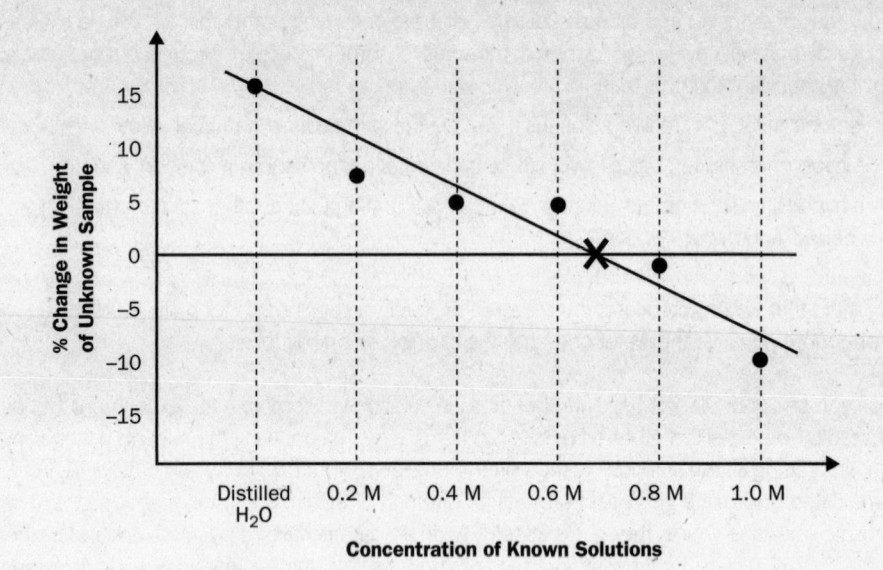

An imaginary line can be drawn that nearly bisects all data points. Note where the line crosses the *x*-axis. At this point there is no change in weight of the unknown sample, which represents the point at which the osmolarity/water potential of the unknown is the same as a known solution equal to that point on the *x*-axis (on the above graph this is about 0.675 M). By assuming change in weight is linear with respect to change in concentration, the expected concentration of a theoretical solution can be determined. The concentration of the unknown can be estimated by comparing the concentration of the solution that wasn't used (approximately 0.675 M on the graph). This type of experimental design and interpolation is very common in biological research. The College Board expects you to be able to design simple experiments like this to test simple hypotheses.

IF YOU LEARNED ONLY FOUR THINGS IN THIS CHAPTER . . . .

1. All cells are membrane-enclosed bodies of cytoplasm.

2. The cell membrane is selectively permeable, meaning it allows certain things through while keeping others out. Water diffuses across the membrane from areas of lesser to greater solute concentration (osmosis). While certain things can cross the membrane in the processes of diffusion or facilitated diffusion, which do not require energy, others require the expenditure of energy for active transport against the concentration gradient.

3. Prokaryotic cells (such as bacteria) lack a nucleus and contain ribosomes but no other organelles. Eukaryotic cells keep their DNA in a membrane-bound nucleus and have many types of organelles, including mitochondria, lysosomes, ribosomes, the Golgi complex, and others. Plant cells also have a cellulose cell wall and chloroplasts (the sites of photosynthesis).

4. Mitosis is the process in which a cell duplicates its genetic material in the production of two identical daughter cells. Its stages include interphase, prophase, metaphase, anaphase, and telophase. Cytokinesis occurs immediately following mitosis and refers to the splitting of the cell into two new cells.

APPLYING THE CONCEPTS: QUESTIONS AND EXPLANATIONS

1. Prokaryotic and eukaryotic cells share which of the following features?

    (A) A plasma membrane
    (B) Complex flagella (9 + 2)
    (C) Membrane-bound organelles
    (D) A membrane-bound nucleus
    (E) Linear chromosomes made of DNA and protein

2. Which of the following can only be observed by using an electron microscope?

    (A) A bacterium
    (B) A virus
    (C) The chromosomes of a eukaryotic cell at metaphase
    (D) A mitochondrion
    (E) A cell nucleus

3. Which of the following has a cell wall?

    (A) Animal cell
    (B) Plant cell
    (C) Some bacteria
    (D) B and C
    (E) A and B

4. Which of the following is a typical component of the plasma membrane of a eukaryotic cell?

    (A) DNA
    (B) mRNA
    (C) tRNA
    (D) Cholesterol
    (E) Actin

5. Which of the following examples of transport across a membrane requires energy?

    (A) The absorption of glucose into a muscle cell by facilitated diffusion
    (B) The flux of K^+ out of a neuron as a nerve impulse is propagated
    (C) The removal of waste from a central vacuole by exocytosis
    (D) The reabsorption of H_2O from across the wall of the large intestine
    (E) The diffusion of water across a cell membrane

6. Which of the following organelles translates the DNA code into proteins?

    (A) Chloroplast
    (B) Ribosome
    (C) Lysosome
    (D) Nucleolus
    (E) Vacuole

7. A mature plant cell can be distinguished from other eukaryotic cells because it has

    (A) energy-producing mitochondria
    (B) a rough endoplasmic reticulum
    (C) chloroplasts
    (D) a large central vacuole
    (E) a membrane-bound nucleus

8. ATP synthesis occurs in which of the following types of structures?

    I. Mitochondria
    II. Lysosomes
    III. Chloroplasts

    (A) I only
    (B) II only
    (C) III only
    (D) I and III only
    (E) I, II, and III

9. Which of the following cells would most likely have the greatest concentration of densely packed mitochondria?

    (A) A yeast cell in S phase of the cell cycle
    (B) A xylem cell in old wood of a tree
    (C) An oxygenated red blood cell
    (D) A smooth muscle cell in the diaphragm
    (E) A sensory nerve cell in the inner ear

10. Which of the following will most likely affect the rate of cell division?

    (A) Cell crowding
    (B) Cell size
    (C) Turgor pressure
    (D) Atmospheric pressure
    (E) Temperature

11. Which of the following could be identified by the presence of ribosomes, simple flagellae, and a cell wall, along with the absence of membrane-bound organelles?

    (A) An amoeba
    (B) A bacterium
    (C) A muscle cell
    (D) An algae
    (E) A virus

Questions 12–14 refer to the figure below in which a dialysis-tubing bag is filled with a mixture of 3 percent starch and 3 percent glucose and placed in a beaker of distilled water with KI indicator. After one hour, the solution inside the dialysis bag has turned a dark blue, while the solution in the beaker has remained clear.

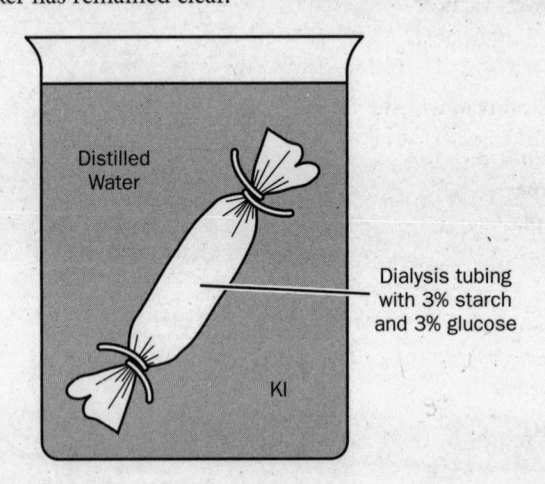

12. Which of the following is an accurate conclusion that can be made only from the observed results?

    (A) The dialysis-tubing bag weighs less.
    (B) Glucose has not diffused across the dialysis-tubing membrane.
    (C) The dialysis-tubing bag is selectively permeable.
    (D) A net movement of water into the beaker has occurred.
    (E) The water potential in the beaker is greater than in the bag.

13. The change in solution color to blue is a positive indicator by KI for the presence of starch. Based only on the results, which of the following conclusions can be made?

    (A) The dialysis-tubing bag is permeable to starch.
    (B) The pores in the dialysis tubing are larger than KI molecules.
    (C) The dialysis-tubing bag is not permeable to glucose.
    (D) A net movement of water into the dialysis-tubing bag has occurred.
    (E) The water potentials of the two solutions are equal.

14. Which of the following best describes the system after it reaches equilibrium?

    (A) Water will have a net movement into the dialysis-tubing bag.
    (B) The osmotic pressure inside the dialysis-tubing bag will be the same as the osmotic pressure in the surrounding solution.
    (C) KI will have a net movement into the dialysis-tubing bag.
    (D) Glucose will have a net movement out of the dialysis-tubing bag.
    (E) The solution inside of the dialysis-tubing bag will have the same starch concentration as the surrounding solution.

Questions 15–17 refer to the figure below of an animal cell undergoing mitosis.

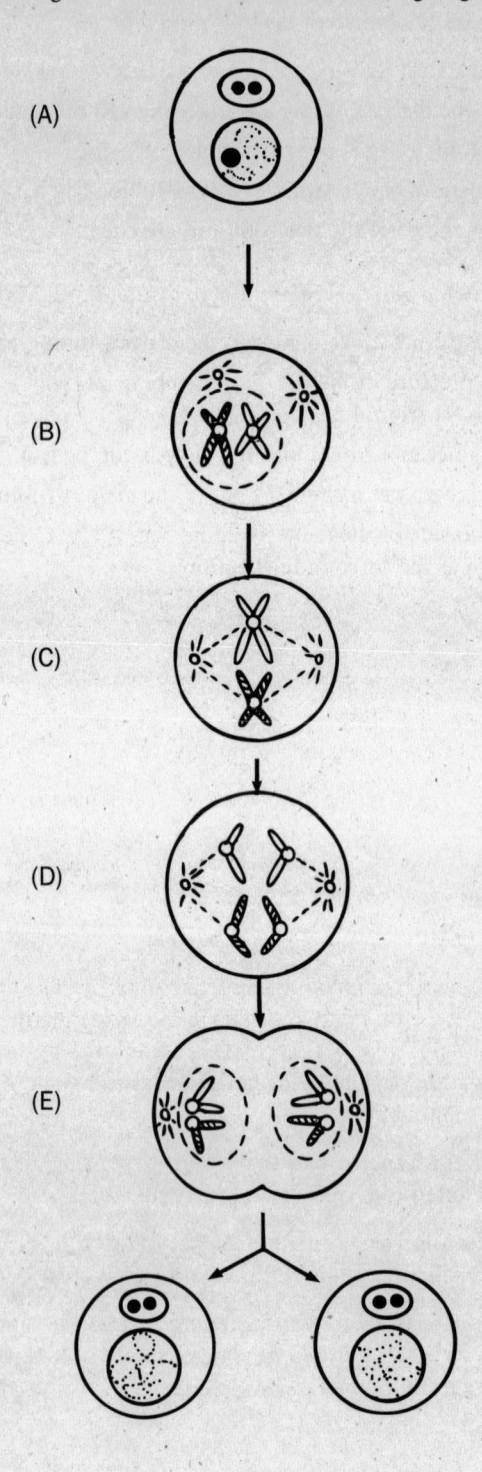

(A)

(B)

(C)

(D)

(E)

15. Anaphase D
16. Prophase B
17. DNA replication A

Questions 18–20 refer to the figure below of a plant cell.

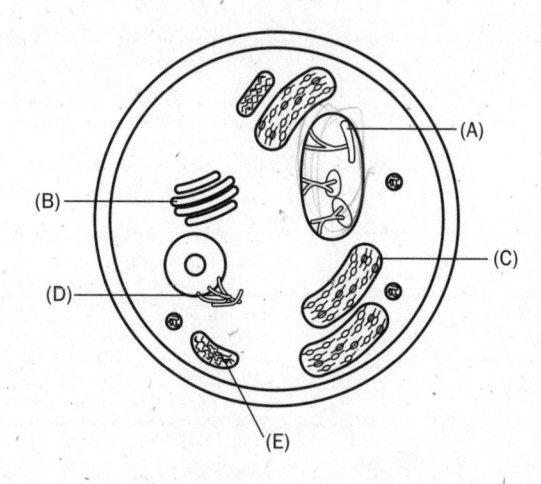

18. Chloroplast C
19. Mitochondrion
20. Central vacuole A

Free-Response Question

Fertilizers help plant growth by supplying important nutrients like nitrogen and phosphorous to plants. These nutrients are found in fertilizers in the form of ionic salts such as ammonium sulfate ($(NH_4)_2SO_4$) and phosphoric acid (H_3PO_4). Growers have to be careful to follow a careful regimen of irrigation after applying fertilizer to ensure that plants are not damaged by a high concentration of solutes in the soil.

(a) **Explain** why high concentrations of fertilizer in the soil might harm plants due to water movement into or out of root cells. Include a discussion of water potential in your answer.

(b) **Design** a simple diffusion experiment that would allow a farmer to know the approximate concentration of fertilizer to apply without damaging a crop.

(c) Some plants thrive in ecosystems with incredibly high solute concentrations, such as brine lakes or alkaline deserts. **Describe** some plant adaptations to severe solute concentrations that allow them to survive in their environment.

ANSWERS AND EXPLANATIONS

1. A

This question can be answered based on the information in the Characteristics of Cells table. The obvious answer is (A) because all cells have a plasma membrane. Since prokaryote cells are simpler than eukaryotes, they are the limiting factor for most answer choices to this question. Prokaryote cells have simple flagella composed of a single microtubule, so (B) can be ruled out. Prokaryotes don't have any membrane-bound organelles so that rules out (C) and (D). Prokaryotes have circular plasmids that contain their DNA, ruling out (E).

2. B

You will probably see a question like this on the exam, although the question might ask about the magnification possible from a light microscope instead. During your course you probably saw many images in your textbook of eukaryotic cells and organelles magnified through a transmission electron microscope, which include some of the answer choices above, so this might be a slightly confusing question. Light microscopes are less powerful than electron microscopes, but can still have very impressive magnifications, so the correct answer is probably going to be something really small. A mitochondrion (D) is still larger than 1μm and most bacteria (A) aren't much smaller than 1 μm. Viruses are less than 100 nm, and can only be viewed using an electron microscope.

3. D

Plant cells and some bacteria have a cell wall. Since both are correct, (D) has to be the answer. Animal cells have a cell membrane, but do not have the rigid structure of a cell wall.

4. D

Some of the answers contain nucleic acids and are misleading. DNA (A) is inside the nucleus of a eukaryotic cell and both mRNA (B) and tRNA (C) are inside the nucleus or in the cytoplasm. The correct answer, cholesterol (D), is a component of the plasma membrane that contributes fluidity. Actin (E) is a protein that exists in microfilaments.

5. C

This question at first seems like a synthesis of information from an extremely diverse collection of biological topics, but the answer can be narrowed down by process of elimination using only information from this chapter. Remember that nonpolar molecules and small, uncharged polar molecules easily pass through the plasma membrane. Choice (E) does not require any energy, since movement is not facilitated. Choice (D) refers to the movement of a small, uncharged polar molecule, so it can be ruled out as an answer. Choice (A) mentions diffusion, which is always energy-free, even if movement is facilitated. Exocytosis requires the movement of a considerable area of membrane, and it takes energy to move that much membrane, so the answer is (C). The potential across a nerve cell membrane depolarizes channels and keeps them open, allowing ions to move down concentration gradients (B).

6. B

Protein translation is the same as protein synthesis, so the answer to this question should pop out at you right away. The ribosome is responsible for translating the genetic code delivered by mRNA into protein by linking together amino acids delivered by tRNA.

7. D

There is a trick to this question: you might not remember that protists can also have chloroplasts, ruling out (C). Although plant cells have chloroplasts, they still have mitochondria to produce ATP, as do all other eukaryotic cells. This rules out (A). All eukaryotic cells have endoplasmic reticuli and membrane-bound nuclei, ruling out (B) and (E). Even if you didn't know the answer right off, especially with the qualifier of a "mature" plant cell, you should be able to figure out that (D) is the correct answer.

8. D

Lysosomes contain hydrolytic enzymes to break down waste in the cell so the answer can't be either (B) or (E). Both mitochondria and chloroplasts produce ATP from either cellular respiration or photosynthesis.

9. D

This question requires a bit of synthesis from other topics, but there is one key point to remember. If the cell has a lot of mitochondria it's because it needs a lot of ATP to be produced to supply a lot of energy. Which of the cells seems like it needs a lot of energy? When most people think of work, they think about using muscle, so an obvious first choice would be (D). A yeast cell in S phase (A) is duplicating DNA and producing proteins, so it needs many ribosomes. A xylem cell in old wood (B) is probably no longer a living cell, and if it is living it is mostly transporting water to other plant cells that need it. Red blood cells (C) are full of hemoglobin and very little else. Nerve cells (E) require energy to maintain ionic gradients across the cell membrane, but their energy needs don't compare to the energy needs of muscle cells. Choice (D) is correct.

10. A

Step back from this question and think about cell division. When cells divide under all kinds of extreme environmental conditions, it is unlikely that any natural external stimulus will affect cell division, so (C), (D), and (E) can be ruled out. Cells have an incredible range of sizes. They aren't cued to divide when they grow large enough, thus eliminating (B). Two key points define cell division under natural conditions. First, most cells are "programmed" by their DNA with a rate of division, if they divide at all. Second, observations of cells in culture suggest that they have a population-control mechanism that inhibits cell reproduction if there are too many cells around. The answer, therefore, is (A). If cancer cells contained this population-control mechanism, they wouldn't continue to divide and produce tumors.

11. B

Only bacteria have simple flagellae. With questions like this, try to find an answer choice that is very exclusive. Remember that some bacteria have cell walls, even if the walls are different in structure from plant cell walls.

12. C

This question can be answered using only analytical skills because you are expected to answer based only on the observed results. The one thing you know about results of the experiment is that only the inside of the bag turned blue after an hour. The bag may have changed in weight (A), but this can't be known from what is given. Glucose may have diffused into the beaker (B), but you have no way of knowing for certain. Just as with glucose, you don't know what is going on with the water in (D) and (E). You don't know if water is moving or not.

The only correct choice is (C). The bag membrane is isolating a reaction—the change in color—since the reaction is not occurring in the surrounding solution. The only treatment applied to the bag was placing it in a solution of distilled water and KI. Either distilled water or KI is making the solution inside the bag turn blue. Why isn't the surrounding solution turning blue since the same treatments, distilled water and KI, are present there? The membrane isn't letting anything out of the bag. Since the bag is letting something go in, but not letting something go out, it is selectively permeable (C).

13. B

Now you know that KI is a positive indicator for starch, even if you didn't already know that from your laboratory experience, and that there is starch inside the dialysis-tubing bag. In this question, as in the previous question, you need to make a decision based on what you observe or don't observe. Choice (A) can't be correct because the surrounding solution would also be blue if starch were leaving the bag. Choices (C) through (E) are conclusions that can't be made based on given observations; you don't know if water or glucose is moving or what effects either is having on water potential. You know for certain that KI moved into the bag, so the membrane must have pores large enough to allow its passage, giving you (B).

14. B

When the system reaches equilibrium it doesn't mean things stop moving. However, there is no net change in the concentration of any given molecule inside or outside the bag. Any answers that include net movement of something [(A), (C), and (D)] can be excluded, since the question states that the system is in equilibrium. You already know that the membrane isn't permeable to starch, so (E) can't be correct. The correct choice is (B). The water potential inside the bag has equilibrated with that in the surrounding solution, so there is no net movement of water.

15. D

Anaphase is the phase where chromosomes first start pulling apart from the metaphase plate. When you see a diagram of mitosis you should always be able to recognize two phases: interphase, because the chromosomes won't be visible, and metaphase, because the chromosomes will be visible and lined up in the middle of the cell at the metaphase plate. Anaphase occurs right after metaphase.

16. B

Prophase is usually easy to remember because it occurs before any movement of chromosomes takes place. The chromosomes are visible but haven't lined up at the metaphase plate yet, so this phase shouldn't be mistaken for interphase or any other phase of mitosis.

17. A

Duplication of DNA occurs in interphase and is easily recognized by the absence of visible chromosomes.

18. C

Identifying structures in diagrams is usually difficult to get around without simply knowing the structures. The question tells you that the diagram is a plant cell, but you still need to recognize structures. Identify the structures you know right away and go back to the ones you don't.

19. E

Mitochondria are found in both plant and animal cells.

20. A

Plant cells have a large central vacuole (membrane-enclosed sac) used for storage of organic compounds such as proteins and inorganic ions such as potassium and chloride. Water is stored here as well. As a large empty pocket in the cell, the central vacuole should be one of the easiest structures to identify in this drawing.

Free-Response Answer Explanation

After thoroughly reading all parts of the question, note that part (a) questions your basic knowledge of water potential and osmosis, part (b) questions your understanding of the experimental method and your memory of the diffusion and osmosis lab, and part (c) asks you to synthesize the physical concepts of water potential with your knowledge of plant adaptations to severe conditions. To answer part (a) of the question you are asked explicitly to discuss water potential. In addition, a discussion of osmosis and potential short-term and long-term detrimental effects on the plant should be presented in your answer. For part (b), explain a similar methodology to the one you used to measure the water potential of an unknown. Be explicit in explaining treatments and variables. In answer to part (c), you'll need to integrate knowledge of plant ecology or perhaps discuss how you would design a plant to help deal with solute stress. Remember, your free-response answers must be in complete sentences (very important) and formulated as a cohesive essay. You are allowed to use diagrams and figures, but they must be explained and can't stand alone. Feel free to compose a quick outline on scratch paper or in your head, but keep in mind that you only have 90 minutes to complete four essays (22.5 minutes each).

Here is a potential answer:

(a) The effect of soil solute concentration on plants has to do with the process of osmosis, the movement of water across a semipermeable membrane from a hypertonic solution to a hypotonic solution. Water moves to the hypotonic solution because it has a more negative water potential. Acute effects of water loss on plants include plasmolysis or loss of cellular water and loss of turgor pressure, which both reduce the amount of water rising to the top of plants. Overall, these short-term effects reduce the productivity of a plant by reducing its ability to carry out important chemical reactions like photosynthesis and cellular respiration.

(b) A farmer can estimate a safe concentration of fertilizer solution by placing root samples of crop plants in increasingly higher concentrations of solution, then observing the amount of water loss in the roots by weight. The concentration of fertilizer solution would be the independent treatment variable and the percent change in weight would be the dependent variable plotted on a vertical axis. Assuming the relationship between solute concentration and water movement is linear, the point at which a line plotted through all of the data points (see graph) passes through the horizontal axis is where the water potential of the roots is equal to the water potential of the fertilizer solution. The farmer should be able to apply a fertilizer concentration equal to or less than the target concentration indicated on the graph on the next page.

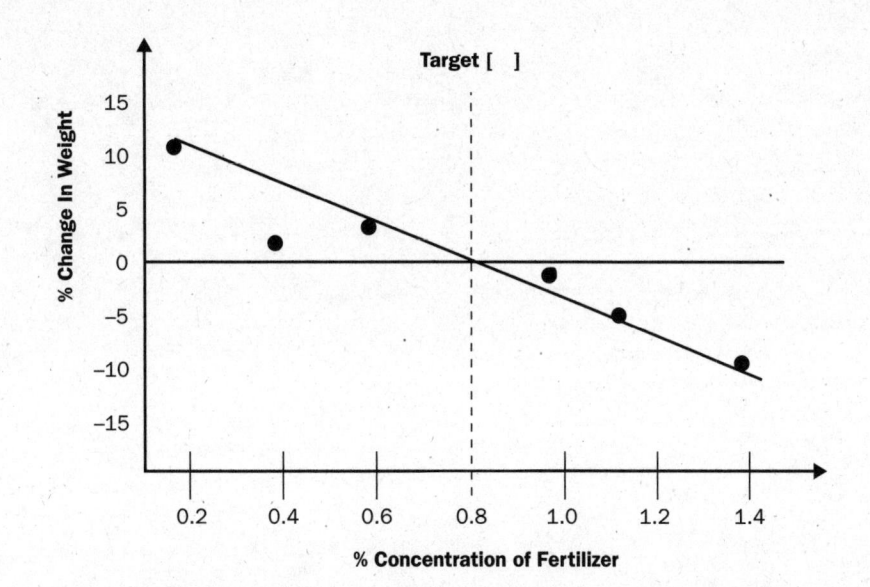

(c) Plants have adapted to high levels of solutes in their environment by lowering their water potential, removing toxic ions, or excluding ion uptake. Some plants express a more negative water potential by increasing the concentration of harmless ions such as K^+ in their cells. A plant will often take up toxic ions like Na^+ with water. These plants exude salts through special structures, store them in leaf structures that fall off, or dilute them with water stored in succulent tissues. Some plants exclude the entry of salt by filtering water at the roots or having a waxy coating on plant surfaces. Other plants reduce their contact with a salty environment by elevating delicate tissues or by reducing the surface area of tissues.

1,2 AP Chem
3,4 AP Physics
5 Pre-calc
6 English
7 Lunch
8 Human Behavior

1,2 AP Chem
3 Lv1 Phisics
4 precalc
5 English
6 Lunch
7 Human Behavior
8 Spanish 4 Lv1

Chapter 5: Cellular Energetics

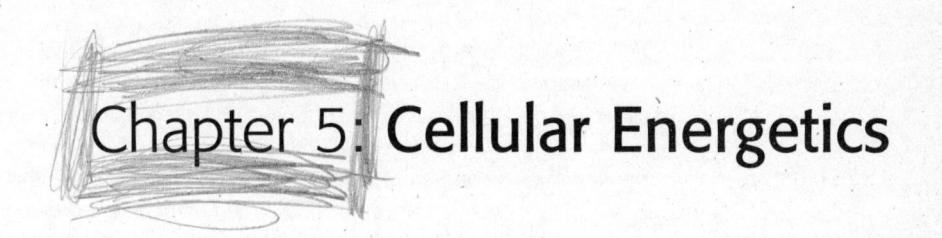

- It's All About the Fuel: How Cells Use and Make Energy
- Cell Respiration Laboratory
- Plant Pigments and Photosynthesis Laboratory
- Applying the Concepts: Questions and Explanations

IT'S ALL ABOUT THE FUEL: HOW CELLS USE AND MAKE ENERGY

The processes covered in this chapter are important to all life on Earth. The energy produced from harnessing sunlight during photosynthesis is transferred through all food webs in nearly every ecosystem. On the other hand, there are a minimal number of questions that cover this material directly on the exam (eight questions out of the 100 multiple-choice questions and perhaps one of the free-response questions). This does not mean, however, that the material is not important. Exam questions place an emphasis on integration and synthesis of concepts in this chapter and how the processes discussed relate to ecology, physiological systems, and form following function. Information is broadly applied in synthesis questions.

In some biology textbooks, extensive diagrams of chains of chemical reactions leading from small building blocks like CO_2 and H_2O to complex products like glucose are shown. Sometimes a textbook shows the reverse, illustrating a diagram of cells breaking down complex molecules to release chemical energy. Diagrams containing such information will not be shown in this chapter. If you wish to view extensive diagrams of these reactions, browse through the "Additional Resources" at the end of chapter 1. In this chapter, you will be presented with information you need to know for the AP exam, and you will find out how your knowledge of that information will be assessed on the exam. It is important to remember that all of the processes discussed in this chapter revolve around the production or utilization of adenosine triphosphate (ATP), the molecule that powers all living systems.

Coupled Reactions and Chemiosmosis

Of all the things to remember from this chapter, there is one that is most important. The energy used by cells to produce ATP in either cellular respiration or photosynthesis typically comes from the movement of H$^+$ ions across a membrane, *down* a concentration gradient. The process of using this energy is called **chemiosmosis**. Glycolysis and some other forms of ATP creation do not involve chemiosmosis. A **coupled reaction** is one in which transport, in this case the transport of an ion, is coupled with a chemical reaction. Special membrane proteins called **ATPases** create proton channels and convert ADP to ATP when a proton passes through.

The mitochondrion moves H$^+$ into the **intermembrane space** via the **electron transport chain**, which creates a proton gradient across the inner membrane. The energy to do this comes from the breakdown of food. The chloroplast moves H$^+$ into the **thylakoid space** in a way very similar to the mitochondrion, but it drives the oxidative phosphorylation process with light energy. This process is called **photophosphorylation**.

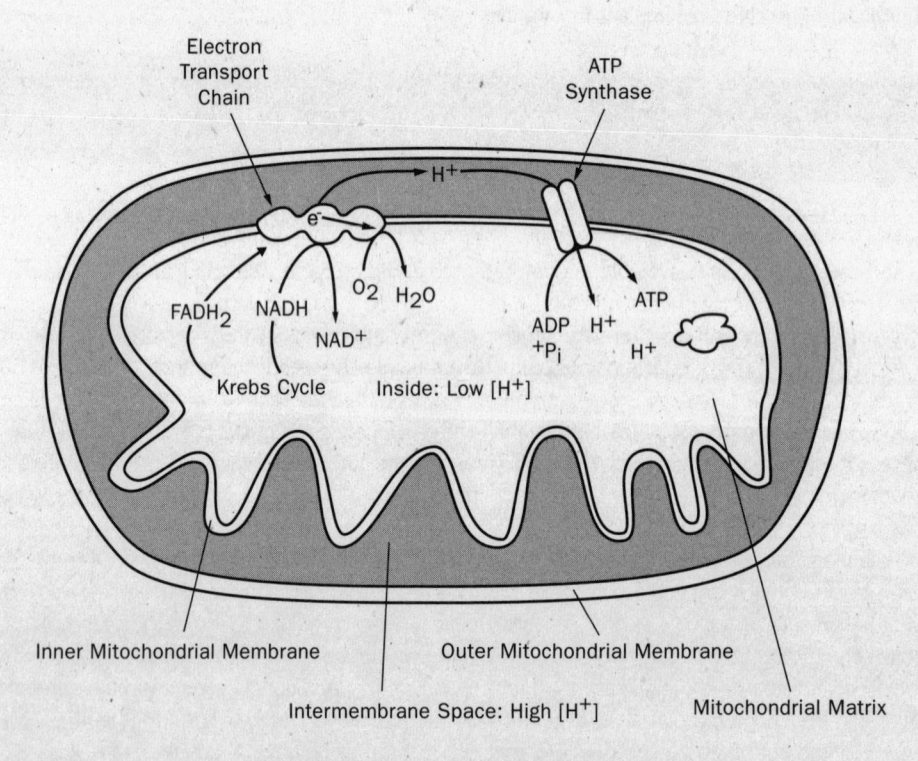

Chemiosmosis in the Mitochondrion

The structures that perform chemiosmosis are examples of form designed for function. The invaginations of the inner membrane (cristae) inside the mitochondrion provide an increased surface area. Even the proteins in the membrane are aligned in a way that spans the membrane width, and their juxtaposition provides the proper architecture to allow chemical reactions to take place. The most important thing to remember about fine protein structures is that their location and form are what allow chemiosmosis to happen.

Anaerobic and Aerobic Catabolism

Catabolism is the breaking down of complex substances into simple substances, making available energy in the process. In **anaerobic catabolism**, there is no electron transport chain or oxygen (O_2) available to carry out a reaction. Anaerobic reactions require almost as much energy to carry out as they yield. From an evolutionary perspective, it is much more advantageous to have the ability to utilize the oxidative properties of oxygen (O_2), as reactions in **aerobic catabolism** do, because more energy can be generated.

The following table includes a few comparisons between the three different components of cellular respiration. In the presence of oxygen (aerobic catabolism or **aerobic respiration**), all three processes can occur. However, in the absence of oxygen, only glycolysis can occur; this is anaerobic catabolism (or **anaerobic respiration**). If you wish to obtain more information on catabolism, a list of further reading can be found at the end of chapter 1.

Anaerobic and Aerobic Catabolism

Reaction	Location	Original Substance	Reaction Product(s)	O_2	Energy Source	ATP Cost	ATP Yield	Net ATP
Glycolysis (yielding ATP directly from NADH via fermentation)	Cytoplasm	Glucose	Pyruvic acid	No	2 NADH	2	4	2
Citric Acid (Krebs) Cycle	Mitochondrial matrix	Citric acid (from pyruvate/ Acetyl CoA)	Oxaloacetic acid (with many intermediates)	Yes	6 NADH 2 $FADH_2$	2	2	0
Electron Transport Chain (oxidative phosphorylation)	ATPases across inner mitochondrial membrane and inner mitochondrial matrix	NADH $FADH_2$	NAD^+ FAD	Yes	Proton motive force	0	34	34
Total yield from one glucose molecule via both anaerobic and aerobic processes								36

Keep in mind that not all energy comes from sugar in the form of glucose. Organisms use food in the form of carbohydrates (starch), protein, and fat. Carbohydrates ultimately break down or are transformed into glucose. The amino acids from proteins ultimately enter the citric acid cycle as pyruvate and acetyl CoA. Fats are broken down into glycerols that can be converted to pyruvate and fatty acids that are converted to acetyl CoA. All of the pyruvate and acetyl CoA from the breakdown of food enter the citric acid cycle and the electron transport chain.

Photosynthesis

First off, remember that although all organisms containing photosynthetic pigments can perform photosynthesis, cellular respiration still takes place in the cells of these organisms. Plant cells have both chloroplasts and mitochondria in order to perform photosynthesis and cellular respiration. Photosynthesis is actually the reverse reaction of cellular respiration.

$$CO_2 + H_2O + energy \rightarrow sugar \rightarrow O_2$$

Photosynthetic organisms are primary **producers**, providing food that supplies the rest of the food web. This "food" starts out in the form of glucose.

The most amazing aspect of photosynthesis is the ability of photosynthetic proteins to split water molecules (H_2O). Once the water molecules have been split, the oxygen atoms are immediately released as O_2 and the hydrogen (H) atoms donate their electrons, which are used to form ATP and NADPH. The two H atoms combine with the C and O atoms from CO_2 to form carbohydrates and more water. Sugars are used for energy storage and O_2 is a waste product that other organisms use for respiration. If the process of photosynthesis had not evolved and produced atmospheric O_2, you wouldn't see most of the organisms that now live on Earth.

Photosynthesis takes place in two stages: the **light reactions** and the **dark (Calvin)** or **light-independent reactions**. In the light reactions, light energy is harnessed to produce chemical energy in the form of ATP and NADPH in a process called **photophosphorylation**. The dark reactions complete **carbon fixation**, the process by which CO_2 from the environment is incorporated into sugars with the help of energy from the reduction of ATP and NADPH. Light reactions produce energy and dark reactions make sugars.

Photosynthesis will be explored further in the lab section of this chapter. Comparisons of C3, C4, and CAM plants will likely be seen in synthesis questions about photosynthesis. Most plants are C3. This means that the initial products of C fixation are two three-carbon molecules (phosphoglycerate or PGA), synthesized through the intermediate enzyme **rubisco**. In C4 plants, CO_2 is initially fixed into a four-carbon molecule (**oxaloacetate**, also found in the Krebs cycle) by the intermediate enzyme **phosphoenol pyruvic acid (PEP)** in a mesophyl cell; this four-carbon molecule later releases a CO_2 molecule when it enters a bundle sheath cell (an instance of form and function). The enzyme PEP is much more likely to bind to CO_2 since it has a higher affinity for CO_2 than rubisco. C4 plants have a physiological advantage in hot, arid environments where they often have to limit the opening of stomata during the day.

Plants that go through C4 photosynthesis are grasses, which include semi-arid to arid crops like corn and sorghum. Many succulent plants like cactus use an alternative method of limiting water loss in arid environments and are called **CAM (crassulacean acid metabolism)** plants because they collect CO_2 at night when it is cooler. CO_2 is then stored in the form of organic acids. C3, C4, and CAM plants all carry out the dark reactions in the Calvin cycle. However, C4 plants complete a carbon fixation step in separate *parts* of the plant, and CAM plants complete a carbon fixation step at separate *times*.

CELL RESPIRATION LABORATORY

This lab is much more about the experimental method than cellular respiration. In this lab, you learn how to set up and use a respirometer to measure the change in volume of a gas, which can be assumed to be O_2 from the germinating seeds placed in water. The effect of increasing temperature on the rate of gas volume change (O_2 utilization) is also measured during the lab. One of the problems with this lab is that gas expands when it is heated. Sometimes the volume increases so much it blows the dye out of the end of the respirometer tube!

This experiment is an excellent example of using control specimens to isolate experimental variables. Put quite simply, the dependent variable being measured is the change in gas volume in the respirometer. The hypothesis is that the change in gas volume is being caused by the utilization of O_2 by the germinating seeds. Other factors can contribute to a change in gas volume (i.e., temperature and pressure) so these variables need to be isolated from the variables you are interested in (i.e., rate of respiration of the seeds). To isolate variables you are interested in from other variables, dormant or dead seeds should be subjected to the same experimental treatment (change in temperature) that the target specimens, the live seeds, are subjected to. The dead seed sample is the control group. Any measurable change in the dependent variable in the control group must be removed from the dependent variable of the target group.

Let's say that when your seeds were heated to 35°C, the volume of gas in the respirometer of the target group (live seeds) decreased by 0.3 ml, but the volume of gas in the respirometer of the control group (dead seeds) increased by 0.1ml. Something occurred in the respirometer of the control group that caused an increase in gas volume. Perhaps the change in volume in the respirometer was created by the expansion of heated gas, or expansion was caused by CO_2 not being absorbed by the soda lime or other CO_2 absorbant. Either way, the dependent variable of temperature must be taken away from the target group because a change in volume should have occurred in the respirometer of the target group as well as the respirometer of the control group. As a result, it can be assumed that an additional 0.1 ml (for a total of 0.4 ml) of O_2 was likely used by the germinating seeds. In the experiment, 0.4 ml of gas were not measured because 0.1 ml of gas was obscured by the expansion of gas from some unknown factor, likely increased temperature. An accurate measure of the target group can never be obtained if the control group isn't included in the experiment.

PLANT PIGMENTS AND PHOTOSYNTHESIS LABORATORY

This is a relatively simple experiment, but it is very equipment intensive. The exact procedure used in the experiment will depend on the availability of specific chemicals, glassware, and an expensive spectrophotometer. The first part of the experiment is a chromatography separation using cellulose paper and a solvent that is more polar than cellulose. The colors in the plant extract separate according to how polar and nonpolar groups interact with the polar solvent and cellulose in the paper. The more polar pigments travel further because they stay in the solvent longer. The purpose of the experiment is to demonstrate that the green color observed in plants is actually composed of several different substances of differing colors. The same experiment can be done with most commercial ink pens, to see the separation of black and blue colors from what looks like black ink.

Remember that photophosphorylation is driven by light energy absorbed by pigments in chloroplasts. White light is composed of many different wavelengths. Plants in particular have developed ways to use light of more than one wavelength. Chloroplasts can only use light energy if the energy is absorbed. Visible color is caused by reflected light. Plants reflect green light so green light is not very useful in photosynthesis.

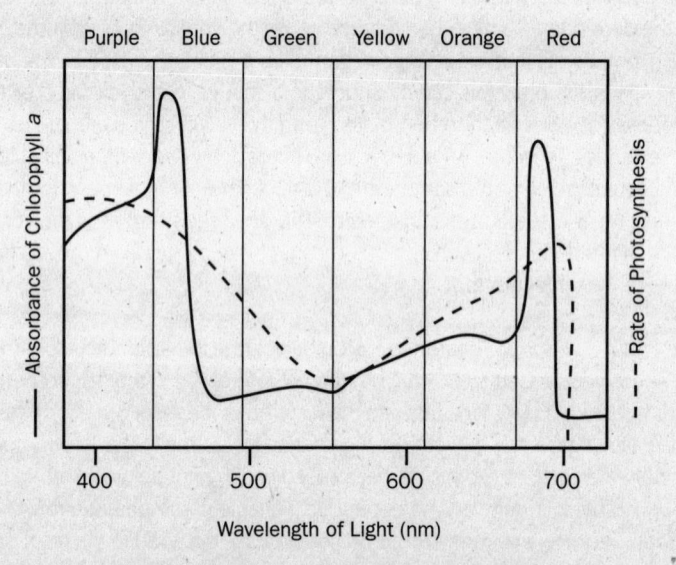

Action Spectrum for Photosynthesis

This image shows action spectra for the rate of photosynthesis (dotted line) and the absorbance of chlorophyll *a*. Note that the two action spectra do not correlate identically, indicating that the rate of photosynthesis depends on the presence of other photopigments as well.

The pigments separated in the lab are chlorophyll *a*, chlorophyll *b*, and carotenoids. Chlorophyll *a* is medium green, chlorophyll *b* is yellow-green, and the carotenoids range in color from yellow to orange. Chlorophyll *a* is the only photopigment that participates in light reactions, but chlorophyll *b* and the carotenoids indirectly supplement photosynthesis by providing energy to chlorophyll *a*. (The carotenoids absorb light the chlorophyll cannot and transfer the energy to the chlorophyll.) The point to keep in mind is that many different photopigments absorb light energy from different wavelengths during photosynthesis. If you want plants to be healthy, give them blue and red light, not green.

The second part of the experiment involves the reduction of a chemical called DPIP by the light-dependent reactions of photosynthesis. The DPIP solution is initially dark blue and loses color as photosynthesis takes place. A spectrophotometer can measure change in color very accurately. If a spectrophotometer is not available, a qualitative judgment of color change must be made visually. More treatment variables may be added to the groups of chloroplasts, such as exposing the chloroplasts to different temperatures, exposing them to different wavelengths of light, and shocking extracted chloroplasts with heat. The main conclusion that can be drawn from this experiment is that photosynthesis is reduced or eliminated in the absence of light. Note that the rate of photosynthesis increases with increased temperature, but decreases if the temperature gets too high (remember the effect of temperature on enzyme activity). If lights of different wavelengths are used in the experiment, the results illustrated in the Action Spectrum for Photosynthesis graph should be obtained.

IF YOU LEARNED ONLY FOUR THINGS IN THIS CHAPTER....

1. Chemiosmosis is energy produced by movement of H^+ ions across a membrane against a concentration gradient.

2. Coupled reactions occur when the active transport of an ion is accompanied by a chemical reaction. The most important example is aerobic respiration, in which cells manufacture their key energy source, ATP.

3. Anaerobic and aerobic catabolism/respiration are methods cells use to create ATP. Processes of anaerobic respiration include glycolysis and fermentation. They yield two ATP for every molecule of glucose. In organisms requiring oxygen, cellular respiration proceeds from glycolysis through the Krebs cycle to the electron transport chain. Hydrogen ions travel across the inner membrane of the mitochondrion to the intermembrane space, creating a proton gradient. This processes yields an additional 34 ATP for every molecule of glucose.

4. In photosynthesis, plants use light energy from the sun to fuel a coupled reaction (photophosphorylation) that produces ATP. In the light-independent (dark) reactions, carbon dioxide is converted to glucose using the ATP produced during photophosphorylation.

APPLYING THE CONCEPTS: QUESTIONS AND EXPLANATIONS

1. The energy needed to form ATP via electron transport phosphorylation comes from the net movement of

    (A) glucose into the cell by facilitated diffusion
    (B) water molecules diffusing into an intracellular hypertonic environment
    (C) oxygen diffusing out of the cell
    (D) light photons into the inner mitochondrial matrix
    (E) protons diffusing across a membrane with the concentration gradient

2. What is the function of O_2 in aerobic metabolism?

    (A) Oxidizes glucose, making it more soluble in water
    (B) Reduces enzymes, limiting glucose synthesis
    (C) Activates enzymes in the citric acid (Krebs) cycle
    (D) Accepts electrons through the electron transport chain
    (E) Transforms ion gates to allow diffusion of K^+

3. CO_2 production can be used as a measure of metabolic rate because

    (A) the heat from metabolism increases its partial pressure
    (B) CO_2 is a waste product of the catabolism of glucose
    (C) a decreased cellular pH liberates gaseous CO_2
    (D) plants need CO_2 to complete photosynthesis
    (E) small, carbon-containing molecules have little energy

4. Plants with C4 photosynthesis are more efficient than C3 plants at

    (A) binding CO_2 and fixing carbon for light-independent reactions
    (B) synthesizing carbon molecules into complex sugars
    (C) utilizing energy from ATPases that perform photophosphorylation
    (D) absorbing incident solar radiation
    (E) removing O_2 waste products

5. From an evolutionary perspective, the most primitive process in cellular energetics is most likely

    (A) fermentation
    (B) the citric acid (Krebs) cycle
    (C) the dark reactions in photosynthesis
    (D) CAM reactions in photosynthesis
    (E) oxidative phosphorylation

6. Which of the following statements is true about both C4 plants and CAM plants?

    (A) They do not need ATP from the light-dependent reactions.
    (B) They have adapted systems for survival in cold, wet habitats.
    (C) They provide alternatives for CO$_2$ fixation.
    (D) They produce more O$_2$ than C3 plants.
    (E) They do not require light to complete photosynthesis.

7. Chemiosmosis is the process by which cells

    (A) obtain food by endocytosis
    (B) use an electrochemical gradient to produce energy
    (C) produce photopigments in plastids
    (D) exchange complex molecules for water across the plasma membrane
    (E) allow the diffusion of any substance other than water across the plasma membrane

8. What is the role of the H atoms in water during photosynthesis?

    (A) They stabilize pH in the thylakoid matrix.
    (B) They reduce the temperature of chloroplasts.
    (C) They magnify solar energy by refracting light waves.
    (D) They split and combine with CO$_2$ to make sugars and free O$_2$.
    (E) They oxidize acetyl CoA into oxaloacetic acid.

9. C4 and CAM plants are more likely to have an advantage over C3 plants

    (A) in habitats with low sunlight and low temperatures
    (B) in temperate climates with high rainfall
    (C) when soil nutrients are scarce
    (D) in alpine habitats with greater snowfall
    (E) in semi-arid to arid habitats

10. The product(s) of the light-dependent reactions of photosynthesis is/are

    (A) pyruvate
    (B) glucose
    (C) ATP and NADPH
    (D) CO$_2$ and H$_2$O
    (E) ribulose biphosphate (RuBP)

11. How many ATP are produced via the chemiosmotic principle for every molecule of NADH that transfers high-energy electrons to the electron transport chain?

    (A) 1
    (B) 2
    (C) 3
    (D) 4
    (E) 5

12. How many ATP are produced for every molecule of FADH$_2$ that transfers high-energy electrons to the electron transport chain?

    (A) 1
    (B) 2
    (C) 3
    (D) 4
    (E) 5

Questions 13–15 refer to the following list.

    (A) Stomata
    (B) Cytoplasm
    (C) Chlorophyll
    (D) Thylakoid
    (E) Stroma

13. Site of the light reactions

14. Regulates entrance of CO$_2$

15. Site of the Calvin cycle

Questions 16–20 refer to the following figure of a mitochondrion.

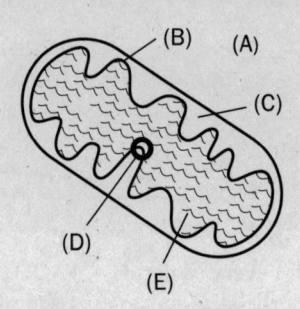

16. Site where glycolysis occurs   A
17. Contains a higher concentration of protons   C
18. Site where citric acid (Krebs) cycle occurs   E
19. Membrane that establishes a gradient of H⁺ ions   B
20. Provides increased surface area for oxidative phosphorylation   D

Free-Response Question

Plants utilize several photopigments to harness light energy of various wavelengths from the sun. The figure illustrates the absorbance spectra of chlorophyll *a*, chlorophyll *b*, the carotenoids, and the action spectrum of photosynthesis. Use the figure to answer the following questions.

(a) **Discuss** the similarities and differences between the peak absorbances of the different photopigments and the peaks of photosynthesis activity at various wavelengths of visible light. Which photopigments seems to have the greatest and smallest effects on the rate of photosynthesis?

(b) **Discuss** the possible evolutionary advantages gained by a plant that has more than one photopigment.

(c) Primitive plants occur in aquatic ecosystems and water allows more blue and red light to penetrate than green light. **Discuss** the frequency with which the color green appears in existing terrestrial plants.

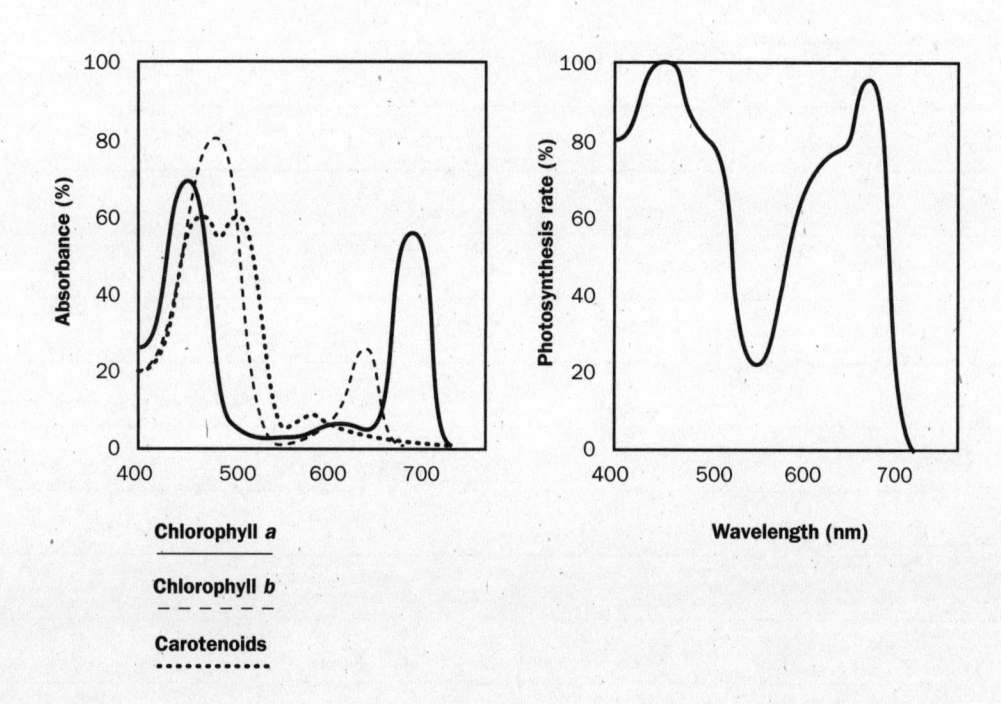

ANSWERS AND EXPLANATIONS

1. E

At least one question about chemiosmosis or coupled reactions will most likely appear on the exam. The correct answer is (E) because the energy to form ATP comes from the proton motive force established by oxidative phosphorylation. Some answers can be eliminated to reduce your chance of error. Diffusion never needs or creates energy, even if it is facilitated, so (A), (B), and (C) can be eliminated. Choice (D) is misleading because light photons don't have an effect on mitochondria; they affect photosynthesis in chloroplasts.

2. D

This is a good question because the need for oxygen is basic to so many organisms on Earth. Recalling that an alternative name for the electron transport chain is oxidative phosphorylation may give you a hint as to the correct answer, but first let's eliminate some other answer choices. Glucose is already fairly soluble in water, so (A) can be eliminated. Oxygen is an oxidizer that doesn't reduce anything, so (B) can be eliminated. When you've shortened the list of options to (C), (D), and (E), you have to use more of your knowledge base and a bit of logic. First of all, the question specifically asks for a function associated with metabolism. Of the choices left, (E) seems least likely to be correct. The correct choice is (D). If you can't remember that O_2 accepts electrons through the electron transport chain in aerobic respiration, you can reasonably reduce the question to a fifty-fifty chance. Choice (C) is associated with cellular respiration, whereas oxygen is a primary component of aerobic respiration. As such, it accepts electrons as part of aerobic respiration, giving you (D).

3. B

If you don't remember that the products of the catabolism of glucose are CO_2 and H_2O, you may have more difficulty with this question. Plants need CO_2 for photosynthesis, but this has little connection to the metabolic rate of other organisms, so (D) can be eliminated. Choices (A), (C), and (E) are all tough because you can't use simple logic to eliminate them as options. You have to use a base of knowledge. To answer the question, you must either know that (B) is correct or know that the other answers are incorrect.

4. A

This is another knowledge-intensive question. Try to remember that C4 plants are grasses and that they live in dry, hot habitats. Also remember that C4 and C3 plants are more similar than dissimilar. Choice (D) is an advantage that a shade plant would capitalize on, not a desert plant. If you consider that most large plants and fruit-bearing plants are C3 plants, (B) can be eliminated. O_2 inhibits photosynthesis and heavily impacts a plant's physiology, so (E) can be eliminated. You are left with (A) and (C), which gives you a fifty-fifty chance at choosing the correct answer.

5. A

This question asks you to consider the world's environment over time and to synthesize this with what you know about cellular energetics. It is a good example of a question the College Board would use to test your knowledge of themes rather than concepts. Over an evolutionary timeline, the organisms that need O_2 to survive (i.e., animals) came later so (B) and (E) can be eliminated. You may be tempted to choose either (C) or (D) because plants are more primitive, but these answers are incorrect. First of all, there was life on earth before plants evolved (e.g., bacteria), which produced energy without photopigments. Second, two answers that pertain to the same subject can usually be eliminated. Both (C) and (D) deal with photosynthesis. You are not likely to be expected to know the phylogeny of plant families to know whether CAM plants are primitive or derived, which eliminates (D).

6. C

Since C4 plants are grasses and CAM plants are succulent plants such as cacti, (B) can be eliminated right away. Choice (E) is incorrect because photosynthesis can't occur without light. C4 plants get their name from the process of creating a four-carbon molecule that has a higher affinity for CO_2, and CAM plants bind CO_2 at night when water loss is reduced, giving you (C).

7. B

Plan to see a question or two about chemiosmosis and/or coupled reactions on the exam. This is something you'll just have to memorize.

8. D

This question goes back to the basics. The simplified formula for the function of water in photosynthesis is $CO_2 + 2H_2O \rightarrow CH_2O$ (sugars) $+ O_2 + H_2O$. Try to memorize this equation—you will find it useful for the exam. In photosynthesis, the water molecule is split and occurs in all products (the underlined atoms).

9. E

Your knowledge of photosynthesis is likely to be tested in the form of synthesis on the exam. Again, you do not need to know the specific reactions involved with C4 and CAM plant photosynthesis. You need to know that these chemical processes are crucial to the physiological adaptation of plants to dry, hot habitats.

10. C

The products of the light reactions of photosynthesis are ATP and NADPH. In glycolysis, a molecule of glucose is split into two molecules of pyruvate. Glucose is the product of the Calvin cycle, the second stage of photosynthesis. CO_2 and H_2O are the starting materials for photosynthesis. Ribulose biphosphate (RuBP) is a five-carbon sugar to which carbon is fixed in the Calvin cycle.

11. C

Three ATP are produced for every molecule of NADH that transfers high-energy electrons to the electron transport chain.

12. E

Most of the ATP produced in aerobic respiration results from oxidative phosphorylation. Oxidative phosphorylation yields a net 34 ATP. Glycolysis yields a net 2 ATP (via substrate-level phosphorylation); substrate-level phosphorylation yields a net 2 ATP in glycolysis and 2 ATP in the Krebs cycle; fermentation yields a net 2 ATP (via substrate-level phosphorylation). Active transport is a form of molecular transport across membranes which moves molecules against their concentration gradient and requires ATP.

13. D

The light reactions occur within the thylakoid.

14. A

Stomata regulate the entrance of CO_2 into the leaves of a plant.

15. E

The Calvin cycle takes place in the stroma.

16. A

All cellular respiration reactions that require O_2 occur inside the mitochondrion. Glycolysis does not require O_2 and occurs in the cytoplasm.

17. C

The inner membrane is the site of chemiosmosis of mitochondria, so there is a proton gradient across this membrane. The electron transport chain is continually placing H^+ into the intermembrane space, so the correct answer is (C).

18. E

The citric acid cycle occurs in the matrix of the mitochondrion.

19. B

If you know the answer to question 17, you know that the membrane that establishes the H^+ gradient is the inner membrane.

20. D

Question 20 is also about the inner membrane because the electron transport chain (oxidative phosphorylation) is what establishes the proton gradient. The inner membrane is invaginated into cristae to provide greater surface area.

Free-Response Answer Explanation

Although this question does not ask you to design an experiment yourself (be prepared to do so on the exam, though), you are expected to discuss the results of experimental observations. This question is asking you to analyze visual data from the two graphs. The first part of the question is testing your ability to understand and relay information from the graphs. Knowledge of photosynthesis is necessary to answer the questions, but you also need to be able to "read" the graphs. Be aware that you may see an action or absorbance spectrum that is different than the standard. For your answer, pick out specific parts of the graph to discuss rather than general aspects. Make sure you talk about the peaks and troughs of particular pigments at particular wavelengths. In your answer, demonstrate that you know the difference between an action spectrum and an absorbance spectrum.

The second and third parts of this question ask you to apply information from the graphs to what you know about evolution, ecology, and physiology. Here is a chance for you to get credit for responses that may not provide a right answer, but are well defined and supported by details. Feel free to be creative, but back up your points with details and factual information.

Here is a potential answer:

(a) These two figures illustrate the difference between the peak activity of photopigments (the absorbance spectrum) and the peak activity of the process of photosynthesis in a plant (the action spectrum) for the wavelengths of visible light. While the absorbance spectrum isolates activity of specific pigments, the action spectrum is the cumulative effect of all the photopigments. There are two peak levels of activity recorded in the action spectrum, at approximately 430 nm, and 675 nm. These levels significantly correspond with the peaks of activity of chlorophyll a at approximately 430 nm and 675 nm and of chlorophyll b at approximately 480 nm and 630 nm. The action spectrum also demonstrates a trough of photosynthetic activity between the wavelengths 500 nm and 600 nm. This trough corresponds to low levels of absorbance for all of the photopigments measured. Photosynthetic activity never reaches 0 percent in the trough despite low levels of absorbance for the photopigments.

(b) The secondary carotenoid photopigments have levels of absorbance with amplitudes nearly as high as those for chlorophyll a, but their effects on photosynthesis are mostly shadowed by the activity of the two chlorophyll photopigments. The curve of the action spectrum is smooth between peaks of chlorophyll a and b activity because the carotenoids supply additional energy to chlorophyll a at the other wavelengths. The highest rates of photosynthesis occur when the absorbance for cholorphyll a is at its highest, suggesting that this photopigment contributes significantly to photosynthesis. Although the percent absorbance of chlorophyll b at 480 nm is higher than any peak absorbance of chlorophyll a, the rate of photosynthesis is not as high as when chlorophyll a has its peaks at 430 nm and 675 nm, the points at which photosynthetic activity is at its highest overall. The secondary carotenoid pigments also do not supply enough energy to chlorophyll a to produce a rate of photosynthesis as high as when chlorophyll a activity is highest at 430 nm and 675 nm.

(c) An organism's ability to maintain reactions of photosynthesis when exposed to light that has more than one wavelength is increased by having more than one photopigment. An example of light with more than one wavelength is visible light from the sun. If plants were black, they would be able to use light from every spectrum. Plants are not black, however, which may be explained by the evolutionary constraint that the ancestors of land plants were aquatic. Water allows less light from green wavelengths to pass through and more light from blue and red wavelengths to pass through. Aquatic plants had the adaptive advantage of being able to use more light from blue and red wavelengths, rather than using light from green wavelengths. Although this would not necessarily drive selection for plants that were green, it would select for plants that were not blue or red. As the photopigments for absorbing blue and red increase, the visible wavelengths that are reflected are more green. Most plants on land are some shade of green because they reflect green wavelengths of light directly from the sun.

Chapter 6: **Heredity**

- The Roots of the Family Tree: Understanding Inheritance
- Mitosis and Meiosis Laboratory
- Genetics of Organisms Laboratory
- Applying the Concepts: Questions and Explanations

THE ROOTS OF THE FAMILY TREE: UNDERSTANDING INHERITANCE

The concepts in this chapter include classical genetics, the principles of inheritance through gamete formation, and sexual reproduction. The College Board suggests that 8 percent of the questions on the exam cover the concepts discussed in this chapter (same percentage as the last chapter). However, you are likely to find more questions on the information presented in this chapter than on cellular energetics. Be prepared. Many of these concepts rely more on sequential logic and are less memory intensive, which is helpful to know when studying for the exam.

Even if you never took a biology class, it's hard not to notice that children look more like their parents than other adults. A chicken hatches from a chicken egg, kangaroos carry around baby kangaroos in their pouches, and apple seeds grow apple trees. The passing on of characteristics or traits is controlled by the genetic information contained within the DNA in cell nuclei. This chapter reviews how genetic information is passed on to offspring.

Meiosis and Gametogenesis

In chapter 4 you reviewed how cells divide and duplicate through the process of mitosis. Mitosis results in the production of identical cells for growth or, in the case of single-celled organisms, **asexual reproduction**. In **sexual reproduction**, two parent organisms each contribute a cell, which combine to form an offspring that shares half of each parent's DNA. Sexual reproduction can also occur when a hermaphroditic organism fertilizes its own eggs to form offspring.

An organism that has two sets of **chromosomes** is said to be **diploid** (designated as 2N). A diploid organism has one set of chromosomes from each parent, for a total of two sets. When a cell has one set of chromosomes, which is half the number of chromosomes that a diploid cell contains, it is said to be **haploid** (1N or N). In some cases cells have several sets of chromosomes, and are called **polyploid** (3N, 4N, etc.). There are a few organisms that exist in a natural state with haploid or polyploid cells, but most organisms are diploid.

When diploid organisms sexually reproduce, one of their own cells is combined with a cell from the other parent. These cells, called **gametes**, are haploid before they combine. When gametes combine, they form a diploid offspring. The formation of these special haploid cells (gametes) is called **gametogenesis**. The process that forms cells with one set of chromosomes (haploid cells) is called **meiosis**.

Here is what you need to remember so far:

Diploid, Haploid, and Polyploid

Ploidy	Number of Chromosomes	Designation	Examples
Diploid	2 of each, one from each parent	2N	Most organisms that reproduce sexually
Haploid	1 of each	N	Most bees, some fungi, some protists
Polyploid	More than 2 of each	3N, 4N, 5N, etc.	Some plants

Remember the goal here is not to rehash everything you've ever learned about biology or even everything you learned in AP Biology. The goal is to prepare you for what's on the exam. When considering meiosis, you are most likely to encounter questions that compare meiosis with mitosis. Meiosis occurs in two steps, **meiosis I** and **meiosis II**. Meiosis I is different from meiosis II; meiosis II is essentially mitosis with half the number of chromosomes. The differences between meiosis I, meiosis II, and mitosis are shown in the following table and the figure opposite. Between the illustration and the information in the table you should have enough repetition to nail down the differences between mitosis and meiosis before going on to the practice questions.

Comparison of Mitosis and Meiosis

Division	Number of Chromosomes in Parent Cell	Prophase/ Prophase I	Metaphase/ Metaphase I	Anaphase/ Anaphase I	Number of Daughter Cells	Number of Chromosomes in Daughter Cells
Mitosis	2N	Chromosomes form into sister chromatids	Individual chromosomes align at metaphase plate	Centromeres separate and sister chromatids travel to opposite poles	2	2N
Meiosis	2N	Chromosomes form tetrads by synapsis; crossing over at chiasmata	Pairs of homologous chromosomes align at metaphase plate	Synapsis ends and homologous chromosomes travel to opposite poles; sister chromatids travel to same pole	4	N

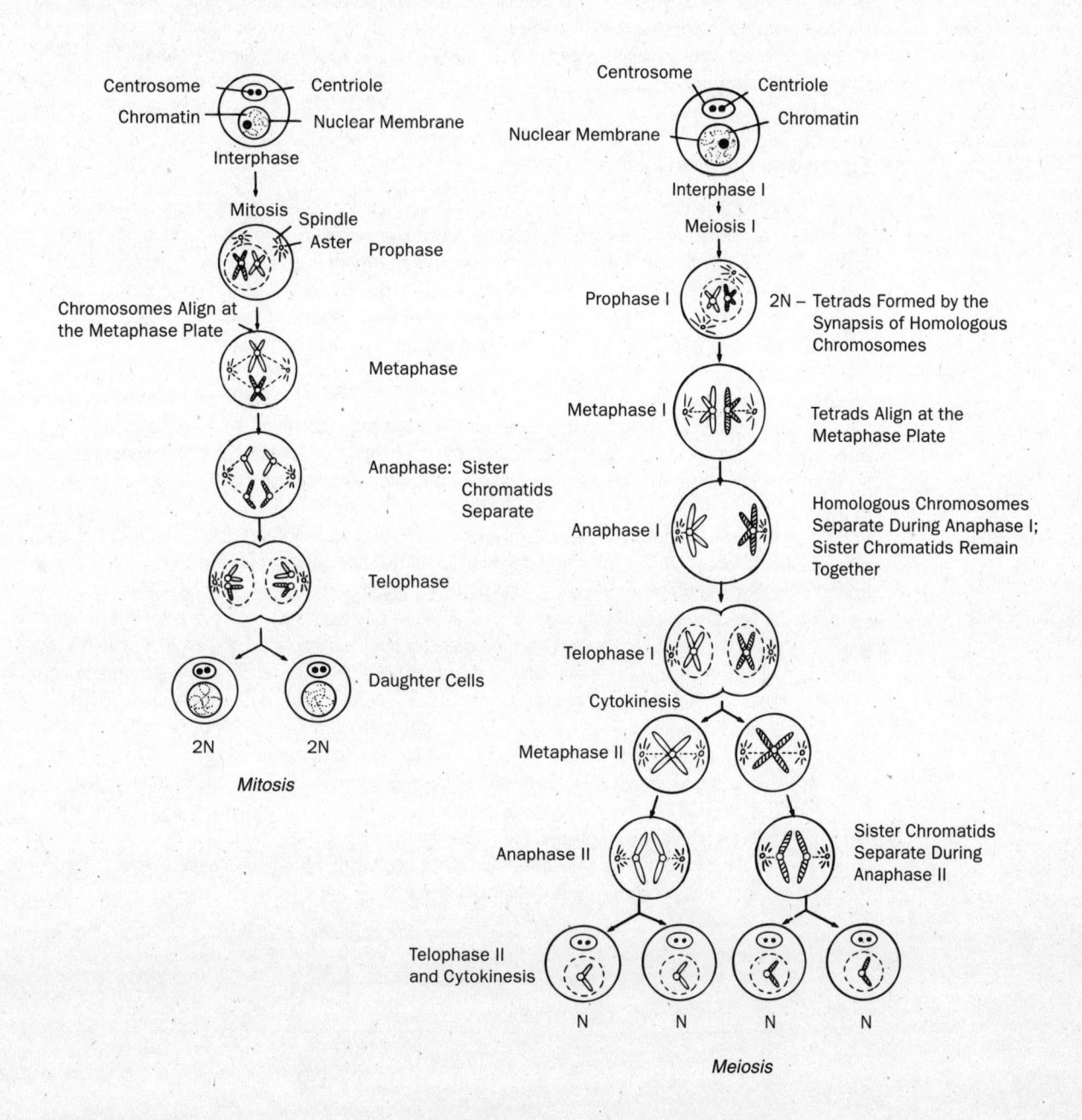

Comparison of Mitosis and Meiosis

The ultimate goal of meiosis is to produce cells with half the number of original chromosomes, so that two cells with half the number of chromosomes can combine to create offspring with a complete set of chromosomes. Later, you will review how chromosomes sort to provide genetic material from each parent to the offspring.

Eukaryotic Chromosomes

Chromosomes are condensed bodies of DNA molecules that store codes for the translation of several different kinds of proteins. These proteins dictate how an organism is put together and functions. In the most simplified way of looking at the **chromosomal theory of inheritance**, each section of DNA that translates a different protein is called a **gene**. A region of DNA may code for a protein that controls eye color, so this region of DNA is called the gene for eye color. Instead of thinking of a gene as the DNA itself, the term "gene" can be associated with a location on the chromosome.

Alleles specifically code for the traits of an organism. In the case of eye color, there might be an allele for blue eye color and an allele for green eye color. The eye color that an organism ends up with (its **phenotype**) is dictated by which allele is placed in the gene location for eye color (its **genotype**). Sometimes an organism might have a genotype for one allele and express the phenotype of another.

In a typical diploid (2N) eukaryotic organism, each cell has two copies of each chromosome. These pairs of chromosomes are a result of the combination of two haploid gametes formed by meiosis, one from the mother and one from the father. The mother and father each provided one allele for each gene, leaving the offspring with either two of the same allele (e.g., two for blue eye color or two for green eye color) or one of each allele (e.g., one allele for blue eye color and one allele for green eye color). In most genes, there is one allele that is **dominant** over the other and hides the expression of that other allele in the phenotype of the offspring. The other allele is called **recessive**. Geneticists record the alleles for genes with uppercase, italicized letters for dominant alleles (*B*) and lowercase, italicized letters for recessive alleles (*b*). If blue eye color is the dominant allele (indicated as *B*) and green eye color is the recessive allele (indicated as *b*) an offspring could have one of three different genotypes (*BB*, *Bb*, and *bb*) when its parents' gametes combine. The phenotype of an offspring with the genotype *BB* is blue eyes. An offspring with the genotype *Bb* will also have blue eyes because the blue eye color allele is dominant. Only offspring with the genotype *bb* will have a green eye phenotype. The following figures clarify this information.

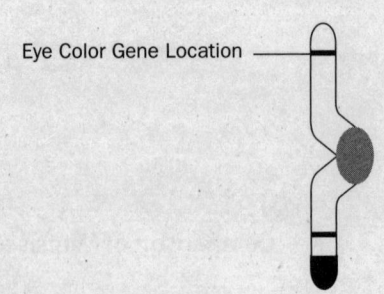

Eye Color Gene Location

The model above indicates the location of the gene for eye color on an animal chromosome. Each animal has two of each of these chromosomes, one from its mother and one from its father.

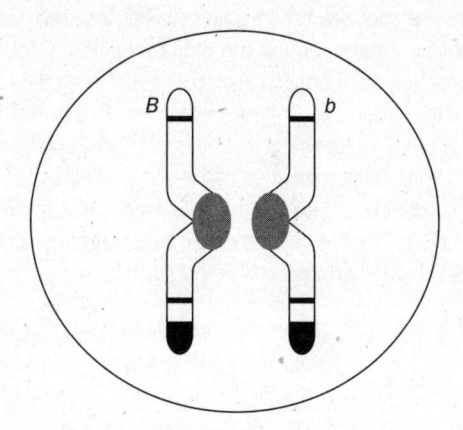

The model above is a cell from an animal with one allele for blue eye color (B) and one allele for green eye color (b). This animal's genotype for eye color is Bb and its phenotype is blue eyes because the blue eye color allele is dominant. The chromosomes are essentially identical in appearance, but the alleles have a slightly different DNA code.

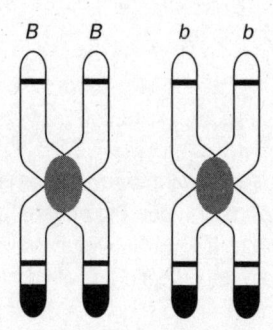

The figure above shows chromosomes after interphase I of meiosis. The DNA strands have been doubled so that there is an exact duplicate of each original chromosome linked by a centromere to sister chromatids. There are now two copies of the B allele and two copies of the b allele.

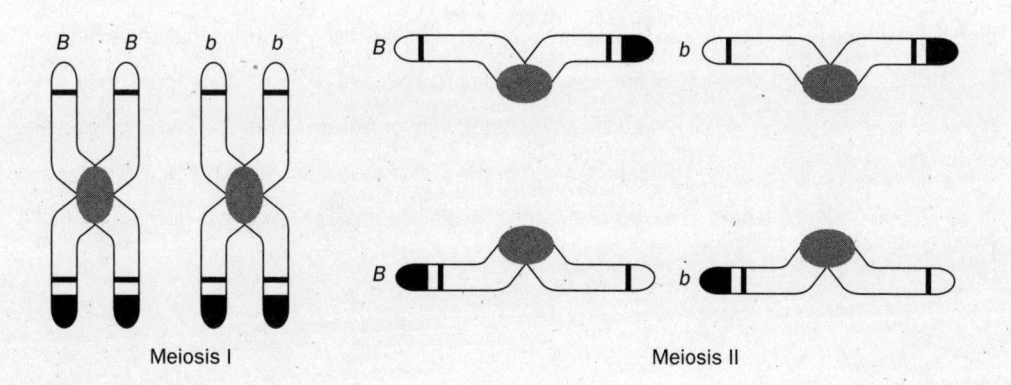

Meiosis I

Meiosis II

After the cell undergoes complete meiosis, the original cell with 2N chromosomes has divided into four daughter cells, each with 1N chromosomes. The daughter cells are four gametes, two with the *B* allele and two with the *b* allele. Since the original animal cell had both a *B* allele and a *b* allele, it can produce gametes with alleles for either blue or green eye color. This occurs in all individuals of the same species. An individual with the recessive green eye color (its genotype is *bb*) can only produce gametes with *b* alleles. When an organism has two of the same alleles (i.e., *BB* or *bb*) it is called **homozygous** for that gene. If both alleles are the dominant allele, the gene is called **homozygous dominant**. If both alleles are recessive, the gene is called **homozygous recessive**. If the organism has different alleles (i.e., *Bb*) it is called **heterozygous** for the gene.

Since the green eye color allele is recessive, it can be discerned that an individual in the population with green eyes has the genotype *bb*. An individual with blue eyes can have either a *BB* genotype or a *Bb* genotype. The genotype of a blue-eyed individual can be determined by mating the blue-eyed individual with a green-eyed individual. This is called a **test cross**. If all of the offspring have blue eyes, then the blue-eyed individual was homozygous dominant. The genotypes of all of the offspring from a mating between a *BB* genotype and a *bb* genotype can only be *Bb*. If any of the offspring have green eyes, the blue-eyed adult must have been heterozygous. The genotypes of the offspring from a mating between a *Bb* genotype and a *bb* genotype are either *Bb* or *bb*. Mating between two heterozygous adults can produce offspring with three different genotypes: *BB*, *Bb*, and *bb*. In this case, two blue-eyed adults can produce a green-eyed offspring.

Inheritance Patterns

After learning the basics of genetics, it is important to review some specific information about inheritance that might appear on the exam. You need to be familiar with mathematical principles of simple probability. Relax; it's not as hard as it sounds. Even if there are more than two alleles for a gene, an individual can only have two alleles on each pair of chromosomes. Ending up with a combination of alleles is like tossing a coin.

Take an organism that is homozygous for a particular gene (*AA*). This organism produces four gametes, each with an *A* allele. Each of the gametes is one out of a possible four gametes, but because all of the gametes have an *A*, four out of four of the gametes have an *A*. Four out of four is a probability of one $\left(\frac{4}{4}=1\right)$. You can look at the question more intuitively by saying that since there are only *A* alleles in the parent cell, all of the daughter cells $\left(100 \text{ percent or } \frac{1}{1}\right)$ will have the *A* allele. If an organism is heterozygous (*Aa*) it produces two gametes with an *A* allele (2 out of 4) and two gametes with an *a* allele (2 out of 4). The probability of either the *A* or *a* allele in the gametes is $\frac{1}{2}\left(\frac{2}{4}=\frac{1}{2}\right)$. Or again, intuitively, if the parent cell has one *A* and one *a* allele, it will produce gametes that are half *A* $\left(50 \text{ percent or } \frac{1}{2}\right)$ and half *a* $\left(50 \text{ percent or } \frac{1}{2}\right)$. Got it?

Now consider two genes and an individual that is homozygous for both (*AABB*). Although there are two genes to consider, the organism still produces four gametes from a single parent cell. Each gamete gets an *A* allele and a *B* allele. What is the probability that a gamete will have BOTH an *A* and *B* allele? Four out of four will have an *A* $\left(\frac{4}{4} \text{ or } 1\right)$ and four out of four will have a *B* $\left(\frac{4}{4} \text{ or } 1\right)$, so four out of four will have *AB* $\left(\frac{4}{4} \times \frac{4}{4} = \frac{16}{16} = 1\right)$. If the individual is heterozygous for one of the genes (*AABb*), the probability of the *A* allele stays the same, but now only half of the gametes get a *B* allele $\left(\frac{2}{4} = \frac{1}{2}\right)$ and the other two get a *b* allele $\left(\frac{2}{4} = \frac{1}{2}\right)$.

What is the probability of producing a gamete with both an *A* and *B* allele from this individual? Four out of four will have an *A* $\left(\frac{4}{4}\right)$ and two out of four will have a *B* $\left(\frac{2}{4}\right)$, so only two out of four will have *AB* $\left(\frac{4}{4} \times \frac{2}{4} = \frac{8}{16} = \frac{1}{2}\right)$. If the individual is heterozygous for both genes (*AaBb*), the probability of each allele in the gametes (*A, a, B,* or *b*) is $\frac{1}{2}$ so the probability of *AB* is $\frac{1}{4}$ $\left(\text{as } \frac{1}{2} \times \frac{1}{2} = \frac{1}{4}\right)$, the probability of *Ab* is $\frac{1}{4}$ $\left(\text{as } \frac{1}{2} \times \frac{1}{2} = \frac{1}{4}\right)$, the probability of *Ba* is $\frac{1}{4}$ $\left(\text{as } \frac{1}{2} \times \frac{1}{2} = \frac{1}{4}\right)$, and the probability of *bb* is $\frac{1}{4}$ $\left(\text{as } \frac{1}{2} \times \frac{1}{2} = \frac{1}{4}\right)$.

Once you can figure out what gametes will be produced from which individuals, you can figure out the genotypes of offspring from your **Punnett square**. The following is a simple cross between two heterozygous individuals for one gene. A mating of this kind is called a **monohybrid cross**.

	A	a
A	AA	Aa
a	aA	aa

Monohybrid Cross

Typically the gametes of the sperm are recorded along the left side of the Punnett square and those of the egg are recorded across the top. The allele from each sperm is paired with the allele from each egg where the column and rows meet, showing the genotype of the offspring. The different indications of *Aa* and *aA* are used only to demonstrate the source of the alleles; they are the same genotype.

The above cross results in three different genotypes (*AA, Aa,* and *aa*), but only two phenotypes because the dominant trait is expressed in both the *AA* and *Aa* individuals. The dominant trait shows up in the offspring in a ratio of 3 to 1.

You can put together a **dihybrid cross** (tracking two genes—*A* and *B*—rather than one) just as easily as a monohybrid cross. Just put the gametes of the male in the left column and the gametes of the female across the top.

	AB	Ab	aB	ab
AB	AABB	AABb	AaBB	AaBb
Ab	AAbB	AAbb	AabB	Aabb
aB	aABB	aABb	aaBB	aaBb
ab	aABb	aAbb	aabB	aabb

Dihybrid Cross

The dihybrid cross produces the famous phenotypic ratio of 9:3:3:1. If you consider a large number of offspring, there will be on average 9 out of every 16 that express the dominant phenotypes for both genes, 3 out of every 16 that express the dominant phenotype of one gene, 3 out of every 16 that express the dominant phenotype of the other gene, and only one that expresses the recessive phenotypes for both genes.

These are the basics of inheritance. However, genes do not operate in isolation from one another; this makes genetics more complex than Mendel's experiments and the Punnett square might suggest. Many different factors can affect phenotype and genotype in offspring. You should be familiar with the following topics.

- **Incomplete dominance**: a form of inheritance in which heterozygous alleles are *both* expressed. This means that the offspring will display a combined phenotype that is distinct from both parent organisms. For example, a plant with purple flowers and a plant with white flowers might produce offspring that have pink flowers.

- **Epistasis**: two or more genes (that are not alleles) interact to control a single phenotype. For example, one gene might act to suppress the expression of another gene; this is what happens in albinism, when recessive albinism alleles governing production of the enzyme that catalyzes melanin production (tyrosinase) prevent expression of the genes that govern the amount of melanin production. In epistatic interactions, a gene's effect on phenotype depends on the presence or absence of other genes elsewhere on the chromosome.

- **Polygenic inheritance**: a type of inheritance in which several interacting genes control a single trait. Many traits result from the additive influences of multiple genes; skin color is one common example of a polygenic trait.

- **Pleiotropy**: a pleiotropic gene controls more than one characteristic in an organism, often controlling aspects of the phenotype that do not seem to be connected to one another. For example, white cats that get their fur color from certain alleles are also generally deaf. Many genes that cause diseases are pleiotropic, including the genes behind sickle cell anemia and phenylketonuria (PKU).

- **Genetic recombination**: molecular process by which an organism's genes are rearranged in its offspring. Through this process, two alleles can be separated and replaced by different alleles, thereby changing the genetic makeup but preserving the structure of the gene. Chromosomal crossing over is an example of a mechanism by which this process takes place.

- **Gene transfer**: vertical gene transfer occurs when an organism receives genetic material (i.e., DNA) from a parent organism or from a predecessor species. Horizontal gene transfer occurs when an organism transfers genetic material to cells that are not its offspring.

More examples of inheritance patterns are shown in the review questions at the end of the chapter. Genetic terminology is further developed in the lab section of this chapter.

MITOSIS AND MEIOSIS LABORATORY

This lab is simple and straightforward. The mitosis portion of the lab involves observing commercial slides of some plants and/or animals fixed at various stages of mitosis. This portion of the lab will help you to recognize, through repetition, the different stages of mitosis. It will also give you some perspective on the sizes of cells and cell structures. The lab will provide you with some experience working with a microscope.

The meiosis portion of the lab is usually completed with cellular models that you can hold in your hand to study Mendel's Laws of Inheritance. The models are usually strings of beads or another kind of model linked to a chain. The two laws of Mendelian inheritance are as follows:

Law of Segregation—This describes the separation of alleles in the parent genotype during the process of gametogenesis. There can only be a maximum of two different alleles in a single parent; half the gametes get one allele and the other half get the other allele.

Law of Independent Assortment—This suggests that different genes sort into different gametes, independently of each other. For example, the sorting of alleles for eye color is not affected by the sorting of alleles for hair color.

The laws above explain the 3:1 and 9:3:3:1 ratios of phenotypes observed in monohybrid and dihybrid crosses. By separating parent genotypes into gametes and recombining them into offspring with the model beads, you can get a hands-on perspective of chromosomal inheritance.

Another observation to make in this lab is that Mendel's laws aren't true for all genes. There are some genes that do not sort independently from others. These genes may be linked to other genes on the same chromosome. It can also be observed that the probability ratios expected are often not quite what is expected based on Punnett squares. This is because some chromosomes exchange genetic information with each other by **crossing over** when the homologous pairs are synapsed into tetrads. The farther away genes are from each other, the more likely they are to exchange DNA with another **chromatid**. The frequency of certain observed phenotypes can be used to estimate a relative rate of crossovers. The higher the frequency of occurrence of a certain phenotype, the more likely there is crossing over between genes, and the further away the genes are from each other.

GENETICS OF ORGANISMS LABORATORY

In this lab you will manipulate several generations of a species of vinegar fly (distinctly different than the true fruit flies in the family Tephritidae) in the genus *Drosophila* (Drosophilidae). Vinegar flies are ideal study organisms, especially for genetics experiments, and have been used for decades to study genetics. The flies are easy to maintain, easy to work with, and have a very short life cycle.

Along with a review of genetics principles, this lab provides some hands-on experience working with diploid organisms; you learn about their life cycle, determine their sex, observe phenotypes directly from individual flies, etc. To some extent, you will be reproducing the work you did in the mitosis and meiosis laboratory, but instead of using plastic models you will be using actual organisms and observing phenotypes instead of genotypes. You will complete monohybrid and dihybrid crosses and look at the ratios of phenotypes in the **first filial (F1)** and **second filial (F2) generations** to see whether the chromosomes of the vinegar flies are following Mendel's two laws or not. At this point in your study of genetics, you should already know what to expect from the crosses, but you will be exposed to two new concepts in this lab.

First, you will perform a **sex-linked cross**. Up to this point you have been studying **autosomal genes**. These are genes that are on any chromosome in the organism that is not a sex-determining chromosome. Most diploid organisms have a chromosome system that determines the sex of an organism. In humans, the 23rd chromosome determines sex. If a human has an *X* and a *Y* pair of the 23rd chromosome, he is a male. If the human has a pair of *X*'s for the 23rd chromosome, she is a female. The presence of the *Y* chromosome is crucial for a human to be male. The system is similar in the *Drosophila*, but not quite the same. Sex is actually determined by the ratio of *X* chromosomes to the number of autosomes (and even this is being questioned). If there are two *X* chromosomes the fly will be a female, and if there is only one *X* chromosome the fly will be a male. The presence of a *Y* chromosome is not crucial for determining sex in vinegar flies, although it is crucial for determining sex in humans.

What is important in understanding sex-linked genes is that *X* and *Y* chromosomes are not homologous. There are many more genes on an *X* chromosome than on a *Y* chromosome. If an *X* chromosome is not paired with another *X* chromosome (in other words, if the fly is male), all of the alleles on the *X* chromosome will be expressed as the phenotype. This includes all alleles on the *X* chromosome, whether they are dominant or recessive. Expression of only genes on the *X* chromosome could be lethal for male flies because many recessive lethal alleles occur in heterozygous parents. If a female is a carrier for a recessive lethal allele on one of her *X* chromosomes and the male that she mates with carries the same gene on the *X* chromosome, half of their male progeny will die!

You will most likely explore a sex-linked, recessive gene in this lab, since it will be easier to spot than a sex-linked, dominant gene. Regardless of which organisms you use in this lab or which genes, you can easily focus on the sex-linked gene by looking at the progeny of the cross between the mutant and wild type adults. In the cross of a wild type female with a mutant male, none of the offspring will show the mutant phenotype. All of the males will receive only a *Y* chromosome from the affected male parent and the females will have one wild type *X* chromosome from the mother to be dominant over the recessive *X* allele from the father. The cross between a mutant female and a wild type male will produce offspring with all females being heterozygous for the recessive allele and exhibiting the wild phenotype. All of the males will be mutant because they only possess one *X* chromosome, which they received from the mutant mother.

The second concept you need to know from this lab is the application of **chi-squared analysis**. The chi-squared analysis measures the difference between the number of observations you make that meet your expectations and the number that don't. You put these values into a formula and compare the answer to a table of standards. The formula is:

$$\chi^2 = \sum \frac{(Observed - Expected)^2}{Expected}$$

As an example, look at the frequencies observed if you toss a coin 100 times. You observe that the coin comes up heads 52 times and comes up tails 48 times. Since there is an equal likelihood that the coin will come up heads or tails, the expected values for heads and tails are 50 and 50. Putting these values into the formula:

$$\chi^2 = \frac{(52-50)^2}{50} + \frac{(48-50)^2}{50} = 0.16$$

Compare the value 0.16 to a chi-squared table at one **degree of freedom (d.f.)**. One degree of freedom is used because two observations were made and d.f. equals number of categories of observations minus one.

Probability (*p*)	Degrees of Freedom (d.f.)				
	1	2	3	4	5
0.05	3.84	5.99	7.82	9.49	11.1

At $p = 0.05$ in the chi-squared table, the value obtained from the above equation would have to be at least 3.84 in order for your observations to be statistically different from what is expected. The value calculated was only 0.16, so it can be concluded that although your observations weren't exactly $\frac{50}{50}$, they are not statistically significantly different from the expected outcome. There are many considerations and assumptions to make when performing statistical analyses, which would be covered in greater depth in a class on statistical analyses. For the chi-squared analysis, there are three key points to remember:

The formula for the chi-squared statistic: $\chi^2 = \sum \frac{(Observed-Expected)^2}{Expected}$

The formula for d.f., d.f. = # of categories of observations − 1

Your calculated χ^2 must be greater than the corresponding *p* value in the table in order for the outcome to be significantly different than what you expected.

IF YOU LEARNED ONLY SIX THINGS IN THIS CHAPTER . . . .

1. Meiosis refers to the process by which sexually reproducing organisms produce sex cells (gametes) with half the chromosomes (haploid) of the rest of the organism's cells (which are diploid). It has two stages, meiosis I and meiosis II, and results in the creation of four gametes. In sexual reproduction, the male and female gametes join to create a new organism with the normal number of chromosomes

2. Different versions of a gene that code for the same trait are called alleles. In classical (Mendelian) genetics, an individual receives one allele from each parent. Individuals with matching alleles are homozygous for that trait while those with different alleles are heterozygous. Usually one version of the allele is dominant (e.g., brown eye color) and the other is recessive (e.g., blue eye color). Heterozygotes are "ruled" by the dominant allele.

3. Geneticists perform test crosses to determine the genetic makeup (genotype) of organisms displaying the dominant phenotype. A Punnett square is used to illustrate a test cross. Mendel's Law of Segregation states that an individual's alleles separate during meiosis, and either may be passed on to the offspring. Mendel's Law of Independent Assortment states that inheritance of a particular allele for one trait does not affect inheritance of other traits.

4. Non-Mendelian inheritance patterns include incomplete dominance, epistasis, polygenic inheritance, pleiotropy, genetic recombination, and gene transfer.

5. Sex-linked genes occur on the *X* or *Y* chromosome. Males who inherit recessive *X*-linked genes from their mothers always express the trait in question, as men have only one *X* chromosome.

6. Chi-squared analysis refers to a means of determining if experimental observations are significantly different from the expected result.

APPLYING THE CONCEPTS: QUESTIONS AND EXPLANATIONS

1. All of the following occur during both mitosis and meiosis I EXCEPT which?

    (A) Chromosomes have sister chromatids.
    (B) Nuclear DNA is duplicated during interphase.
    (C) Chromosomes condense and the nuclear membrane disappears.
    (D) Chromosomes synapse into tetrads.
    (E) Cells have a diploid number of chromosomes at prophase.

2. Assume that genes *A* and *B* are not linked. If an organism is heterozygous for both genes, what is the probability that the organism will produce a gamete with both the *A* and *B* alleles?

    (A) $\frac{1}{4}$

    (B) $\frac{1}{2}$

    (C) 1

    (D) 0

    (E) 50 percent

3. A man and woman have two sons and would like to have a little girl. As the couple's genetic counselor, you explain to the parents that the chance their third child will be a girl is

    (A) 1 in 1
    (B) 1 in 2
    (C) 1 in 3
    (D) 1 in 4
    (E) 0

4. Which of the following defines a test cross?

    (A) A cross between two heterozygous individuals
    (B) A cross between two homozygous recessive individuals
    (C) A cross between an individual with the dominant phenotype and one with the recessive phenotype
    (D) A cross between two individuals with the dominant phenotype
    (E) A cross between two individuals with the recessive phenotype

5. What is the percentage of homozygous yellow individuals in the F2 progeny resulting from the original cross of a homozygous yellow pea plant and a homozygous green pea plant?

   (A) 0 percent
   (B) 25 percent
   (C) 50 percent
   (D) 75 percent
   (E) 100 percent

6. In *Drosophila* the traits for red eye color (*R*) and straight wings (*W*) are dominant and the traits for white eye color (*r*) and curly wings (*w*) are recessive. A cross between two flies produces a progeny with 607 red-eyed flies with straight wings and 202 red-eyed flies with curly wings. Assuming neither of these two genes is sex-linked, which of the following are most likely to be the genotypes of the parents?

   (A) *RRWW* × *RRWW*
   (B) *RRWW* × *RRWw*
   (C) *RrWw* × *RrWw*
   (D) *RrWw* × *RRWw*
   (E) *RrWW* × *Rrww*

7. Red-green color blindness is a sex-linked recessive trait in humans. Which of the following is true if a colorblind woman and a man with normal vision have children?

   (A) None of their children will be colorblind.
   (B) Half of their male children will be colorblind.
   (C) Half of their female children will be colorblind.
   (D) All of their female children will be colorblind.
   (E) All of their male children will be colorblind.

8. Multiple crosses involving genes known to occur on the same chromosome produce frequencies of phenotypes that suggest there is a high rate of crossover between these two genes. Which of the following is the most likely explanation for the phenotypic frequencies observed due to crossing over?

   (A) The two genes are far apart from one another.
   (B) The two genes are both recessive.
   (C) The two genes have incomplete dominance.
   (D) The two genes are both located far from the centromere.
   (E) The two genes are on the same chromatid.

9. Which of the following is true of meiosis II?

    (A) It produces two identical daughter cells.
    (B) It produces haploid daughter cells.
    (C) Chromosomes create chiasmata during crossing over.
    (D) Homologous chromosomes travel to opposite poles.
    (E) It produces four diploid daughter cells.

10. Which of the following phenotypic ratios describes the results of a dihybrid cross between two genes that follow Mendel's Laws of Segregation and Independent Assortment?

    (A) 3:1
    (B) 1:1
    (C) 7:2:2:1
    (D) 5:3:1
    (E) 9:3:3:1

11. Nondisjunction can occur in _____ and may result in _____ .

    (A) anaphase; extra chromosomes
    (B) prophase; extra chromosomes
    (C) anaphase; extra chromosomes or missing chromosomes
    (D) prophase; extra chromosomes or missing chromosomes
    (E) metaphase; extra chromosomes or missing chromosomes

Questions 12–15 refer to the following list.

   (A) Genotype
   (B) Phenotype
   (C) Allele
   (D) Epistasis
   (E) Heterozygous

12. Interaction between two or more genes

13. Observable appearance reflecting the expression of genes

14. Genetic makeup of an individual

15. An organism which has two different forms of a gene

Questions 16–18 refer to the normal ABO human blood type inheritance.

16. A child with blood type O has a mother with blood type B. The father must have which of the following blood types?

    (A) A
    (B) B
    (C) O
    (D) AB
    (E) Any blood type except AB

17. If a mother has blood type O, which of the following blood types could never occur in any of her children?

    (A) A
    (B) B
    (C) O
    (D) AB
    (E) She could have children with any blood type.

18. If the father has blood type A and the mother has blood type B, which of the following blood types could never occur in any of their children?

    (A) A
    (B) B
    (C) O
    (D) AB
    (E) They could have children with any blood type.

Questions 19–20 refer to the pedigree diagrammed below.

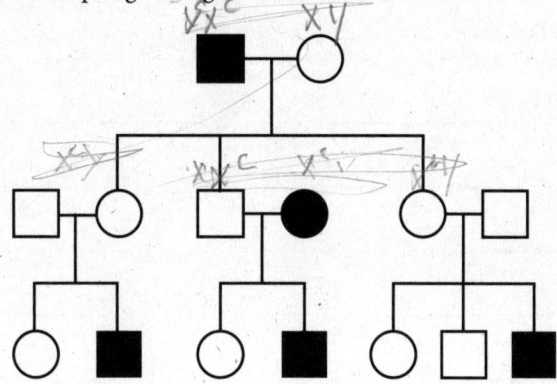

19. This pedigree shows an inheritance pattern that can best be explained by allele(s) that are

    (A) incompletely dominant.
    (B) sex-linked recessive.
    (C) sex-linked dominant.
    (D) autosomal dominant.
    (E) on the same chromosome.

20. What can we conclude about the genotype of the grandmother at the top of the tree for this trait?

    (A) She is homozygous dominant.
    (B) She is homozygous recessive.
    (C) She is a heterozygous carrier.
    (D) Her mother expressed the trait.
    (E) We cannot conclude any of the above from the pedigree.

Free-Response Question

Geneticists can determine which genes will be expressed in offspring by tracking inheritance patterns and using Punnett squares.

(a) **Explain** the role of alleles in determining the genotype and phenotype of offspring.

(b) **Discuss** the purpose of a "test cross." Create a sample Punnett square for the test cross of the alleles *Aa* and *AA*, where *A* is the dominant allele. **Explain** the outcome of the cross.

(c) Suppose the *H* allele codes for Huntington's disease. If one parent has the alleles *HH* and the other parent has the alleles *Hh*, will any of their children have Huntington's disease? **Explain** why and how you came to your conclusion.

ANSWERS AND EXPLANATIONS

1. D

Remember that you are comparing mitosis with the first part of meiosis, which is where the real differences between mitosis and meiosis occur. The two main differences are that in meiosis the chromosomes undergo synapsis as well as crossover (sometimes), and homologous pairs of chromosomes align at metaphase I instead of individual chromosomes. Review these differences again in the Comparison of Mitosis and Meiosis table.

2. A

First off, the probabilities $\frac{1}{2}$ and 50 percent are the same, so the correct answer can't be either (B) or (E). If the organism is heterozygous for both genes, it can't have AB in all of the gametes, so the answer can't be (C). Can the organism produce a gamete that doesn't have an AB genotype or has a probability of $\frac{1}{4}$? Since the organism is heterozygous for both alleles, it has an $AaBb$ genotype. This means that the probability is $\frac{1}{2}$ that any gamete will get an A allele and $\frac{1}{2}$ that any gamete will get a B allele. The probability that a gamete will get both alleles is $\frac{1}{2} \times \frac{1}{2}$ or $\frac{1}{4}$.

3. B

Don't get trapped here. The chance of having either a boy or girl is always 50 percent regardless of the gender of the previous children. The mother is always going to produce eggs with an X chromosome. Half of the father's sperm will carry an X chromosome and half will carry a Y chromosome. This makes it even odds for either a boy or girl.

4. C

A test cross is a way of discovering the genotype of an individual expressing the dominant phenotype. The individual can either be homozygous dominant or heterozygous. Since at least one individual has to express the dominant phenotype, (B) and (E) can be ruled out. The way to discover the unknown genotype is to cross the individual with an individual that is homozygous for the recessive allele. The individual that is homozygous for the recessive allele will express the recessive phenotype.

5. B

This question sounds complicated but it isn't—you don't even need to know which color is dominant. The gametes produced by the homozygous yellow plant are all going to have the yellow allele and the homozygous green plant will only produce gametes with the green allele. This means that all of the F1 progeny will be hybrids for seed color. The phenotypic ratio of a monohybrid cross is 3:1 with one homozygous dominant individual, two heterozygous individuals, and one homozygous recessive individual (on average for every four individuals). It doesn't matter whether yellow is dominant or recessive because both the dominant homozygous and recessive homozygous individuals will show up in the progeny with a frequency of 25 percent.

6. D

Curly-winged individuals would only be expressed with a homozygous recessive gene, in a ratio of 1:3. This rules out (B) and (E). You would get some white-eyed individuals in a dihybrid cross, so this rules out (C). Choice (A) can't be correct because these parents would only produce red-eyed, straight-winged progeny. The correct answer must be (D) by process of elimination. You can test your answer with a Punnett square.

7. E

You are given two crucial pieces of information. One is that the gene is sex-linked recessive and the other is that the mother expresses the trait. This means that the mother must be homozygous for the recessive allele, which is expressed in the male children. All of the children will get a colorblind allele from the mother. The female children will get their second X chromosome from the father, who has a normal allele for seeing colors, and since the colorblind allele is recessive it won't be expressed. Choices (C) and (D) can't be correct. The male children will get a Y chromosome from their father, so they must get a colorblind allele from their mother. All of the male children will be colorblind. This rules out (A) and (B).

8. A

The question specifically asks for the effect of crossing over, which relates to the location of genes on the chromosome. This rules out (B) and (C) because while dominance can affect phenotypic frequencies, it has nothing to do with crossing over or gene locus. Choice (E) is a confusion of terms. If both genes are on the same chromosome, it follows that they are on the same chromatid; in fact, there is a copy of each gene on each duplicate chromatid in meiosis I. Choice (D) sounds appealing because it makes sense that there will be more crossover events farther from the centromere. However, (D) does not address the relative location of the two genes to one another. If both genes are far from the centromere, but very close to each other, there is little chance that the genes will cross in relation to each other. High rates of crossover occur between genes when they are located far from each other on the chromosome, which leads to (A).

9. B

This should be an easy question after you review meiosis. The only tricky choices are (C) and (D), but both of these things occur in meiosis I. Choice (A) is the result of mitosis, not of meiosis. There are four daughter cells produced at the end of meiosis, but they are all haploid, which eliminates (E) and makes (B) the correct answer.

10. E

The question tells you that the genes are autosomal, not linked, and not subject to varying probabilities in their phenotypes due to crossover. Two of the possible answers, (A) and (E), are familiar phenotypic ratios from standard crosses. The other ratios in the list shouldn't look familiar at all. Don't let them confuse you. If you simply can't remember that the results of a dihybrid cross give you the ratio 9:3:3:1 you might try to figure out the correct answer through logic. The question deals with two genes so there are four different gamete genotypes possible. A Punnett square with four different gametes from the parents has 16 boxes for progeny phenotypes. Adding up the numbers in the ratios above gives you $9 + 3 + 3 + 1 = 16$. No other ratio listed gives a cumulative total of 16.

11. C

Nondisjunction can occur in anaphase and may result in extra copies of chromosomes or missing chromosomes.

12. D

Interaction between two or more genes is termed epistasis.

13. B

The observable appearance of an organism that reflects the expression of genes is called the organism's phenotype.

14. A

The genetic makeup of an individual is the individual's genotype.

15. E

An organism that has two different forms of a gene is heterozygous.

16. E

You should acquaint yourself with the human ABO blood phenotype inheritance system. There are three alleles in this system: I^A, I^B, and i. The alleles I^A and I^B are codominant and the i allele is recessive. Although there are three alleles, any one individual can still only have two of the three. In the question above, the child with blood type O must have two recessive alleles, ii, receiving one from his mother and one from his father. The mother has blood type B so must have the genotype $I^B i$. The father can have any genotype that includes a recessive allele: $I^A i$, $I^B i$, or ii. The father cannot have blood type AB because it would not allow him to contribute a recessive allele to the child.

17. D

This question is very similar to the one above and should be approached the same way. The mother must have the genotype ii, meaning that she must contribute a recessive allele to all of her children. Since the children will inherit the recessive allele, none of them can be blood type AB ($I^A I^B$).

18. E

In this question the possibilities increase. The father could be homozygous for I^A or a carrier of the recessive allele. Likewise, the mother could either be homozygous for I^B or be a carrier of the recessive allele. Without knowing the parents' genotypes or the genotype of any of their children beforehand, you only can conclude that they could have a child with any of the blood types.

19. B

You need to be familiar with pedigrees. Pedigrees are family trees with all the males indicated by squares and the females by circles. The shapes that are filled in express the trait in question. If you are given a pedigree, look first at the expression of the trait in the males, especially if sex-linkage is an option among the possible answers. The typical inheritance pattern in sex-linkage is for a trait to be more common in males and to skip generations. Both of these occur in this pedigree. The grandfather at the top of the tree express the trait. If the gene in question were sex-linked, the father would pass on his *X* chromosome to all of his daughters. If the trait were dominant, all of his daughters would express the trait. The daughters do not express the trait, so the answer is (B). Incidentally, this pedigree could also exist for an autosomal recessive allele, but that isn't one of the possible answers.

20. E

If the pedigree is showing the inheritance of the sex-linked recessive trait as decided in the question above, the grandmother CANNOT be homozygous recessive. This is about all you can conclude from the available information, though. The grandmother could be either a heterozygous carrier or homozygous dominant and still produce the pedigree.

Free-Response Answer Explanation

Alleles come in pairs and compose the genetic makeup of a gene. One allele comes from the mother in a sexually reproducing species, and one allele comes from the father. The combination of alleles codes for the phenotype, or gene expression, when placed in the genotype (gene location) for a particular trait.

The second part of the question asks about test crosses. A test cross can be performed with a Punnett square, to determine the possible genetic outcomes for the gene cross of two procreating organisms. In genetic testing and family planning, test crosses are performed to make sure that parental offspring will not receive a lethal or detrimental allelic combination. Test crosses are particularly helpful when a disease runs in a family and the parents want to know the chance they have (which can be expressed as a phenotypic ratio) of bearing a child that expresses the detrimental trait.

The third part is a trick question, since the allele for Huntington's disease is dominant (thus represented with a capital letter). Both parents can have the disease and still breed, unlike disorders that lead to sterility or very low fertility, as in Down's syndrome.

(a) Alleles play an important role in the expression of genes in offspring. Since two alleles make up a gene, the combination of two alleles leads to the expression or repression of genetic outcomes. Since the phenotype is the physical expression of a gene, the phenotype is what you will see when two alleles are combined to form a gene. If, for example, a codes for pink eyes and A codes for black eyes, with a as the recessive allele and A as the dominant allele, an allelic combination of Aa would produce a black-eyed offspring. In terms of genotype, the alleles are found on a specific gene and chromosomal location that codes for the expression of a trait.

Normally, lethal traits or traits that lead to disease are recessive, which means that the alleles that code for a phenotypic trait will only be expressed if both alleles are recessive. For example, Tay-Sachs is a recessive autosomal disease. There are exceptions, however—some alleles that code for disease are dominant, as in Huntington's disease.

Some genes have two different alleles, as in sickle cell anemia. An individual will only express symptoms of sickle cell anemia if he or she is homozygous for the alleles AA. An individual with the alleles AS is resistant to malaria, which is beneficial in certain areas of the world. An individual with an SS gene is neither resistant to malaria, nor will he or she express symptoms of sickle cell anemia. Since alleles produce the phenotype for a gene, some phenotypic outcomes are beneficial, as in resistance to a disease, some alleles have no genetic repercussions (e.g., eye color), and some allelic combinations are detrimental, even lethal, to an organism.

(b) The purpose of a "test cross" is to see what the genetic outcome will be from combining gene(s) of two parents. A Punnett square can be constructed to cross one gene from two parents, termed a "monohybrid cross" or two genes from two parents, termed a "dihybrid cross." The following is a Punnett square showing the test cross of parental genes Aa and AA.

	A	A
A	AA	AA
a	Aa	Aa

The outcome of the Punnett square shows that 50 percent of the offspring will be heterozygous dominant for the trait and 50 percent will be homozygous dominant. That means that the trait, provided it is dominant, will be expressed in all offspring. The inheritance pattern for the Punnett square is 100 percent or $\frac{4}{4}$ for the expression of an A allele and 50 percent or $\frac{2}{4}$ for inheritance of the a allele.

If this were the first filial generation (F1) and if the trait in question were recessive, one offspring of the second filial generation (F2) would express the trait if two heterozygous dominant individuals decided to breed. In the second filial generation, only one of the offspring would express the trait, if it were recessive, since the other offspring would have either a homozygous dominant or heterozygous combination of alleles. A heterozygous individual is a "carrier" of a recessive allele, which can be expressed in the next generation if the individual breeds with another heterozygous individual. The "carrying" of alleles explains why a disease may skip generations or not show up for several generations in a family. Depending upon what outcome you are looking for in a Punnett square or test cross, a homozygous or heterozygous combination of alleles can be beneficial, detrimental, or neutral.

(c) The allele that codes for Huntington's disease is dominant; therefore, as seen in the Punnett square below, all offspring of these parents will have Huntington's disease at some point during their lifetime.

	H	H
H	HH	HH
h	Hh	Hh

The symptoms of Huntington's disease do not usually appear until age 30 to 50, which is when many people are in the process of having children or have already had them. If one or both parents develop the symptoms of Huntington's disease, genetic testing can be conducted to see if the gene for Huntington's disease has been passed on to the offspring.

Since the symptoms of Huntington's disease start out mild and progress to become quite debilitating, individuals may not know they have Huntington's disease in the early stages. Symptoms include slight, uncontrollable muscle movements, depression, short-term memory loss, clumsiness, anger, and antisocial behavior, which happens at some point to most people who don't have Huntington's disease. As the disease progresses, the individual may experience more severe involuntary movements, loss of initiative and organizational skills, speech impediment, difficulty swallowing, difficulty speaking, weight loss, and subsequent mood swings, stubbornness, frustration, and depression.

Fortunately for all of the individuals in the test cross above, there are medications to manage their symptoms. Unfortunately, there is currently no cure for this autosomal dominant disease that affects the nerve tissue of the brain. Age at onset of the disease is different for every individual.

Chapter 7: **Molecular Genetics**

- The Double Helix: DNA and the Mechanisms of Inheritance
- Molecular Biology Laboratory
- Applying the Concepts: Questions and Explanations

THE DOUBLE HELIX: DNA AND THE MECHANISMS OF INHERITANCE

The mechanistic view of looking at biology at the level of molecules and cells took an abrupt historical bend with the study of the topic of heredity. The classical genetics concepts covered in the last chapter were a more simplistic explanation for inheritance than the details of molecular genetics discussed in this chapter. Classical genetics concepts were discovered before more complex molecular genetics topics, as advances in technology allowed more advanced concepts in genetics to be perceived.

There are three overall themes of molecular genetics to remember.

1) Be familiar with the structure of DNA and RNA and how these simple building blocks provide both a simple system for duplication and a wealth of genetic diversity that codes for variation in all existing and extinct organisms.

2) Know that ultimately DNA is a code for the translation of proteins, that these proteins impart function to all of the biological processes that make an organism alive, and that control of expression of the genes for these proteins largely controls bioactivity.

3) Before turning your study towards evolution, you must understand DNA's role as the root of genetic change in the form of mutations.

DNA and RNA Structure

You have undoubtedly learned that **DNA** and **RNA** molecules are long strands of **nucleotides** linked together by their sugar-phosphate backbone between the 5' and 3' carbons of deoxyribose or the ribose ring. DNA has a **deoxyribose** sugar backbone while RNA has a **ribose** backbone. RNA also contains the nitrogenous base uracil instead of thymine. Both complementary strands of DNA serve as templates for duplication during mitosis and meiosis, which relates back to the last chapter. The simple sugar-phosphate backbone allows for exact copies of itself to be reproduced during propagation.

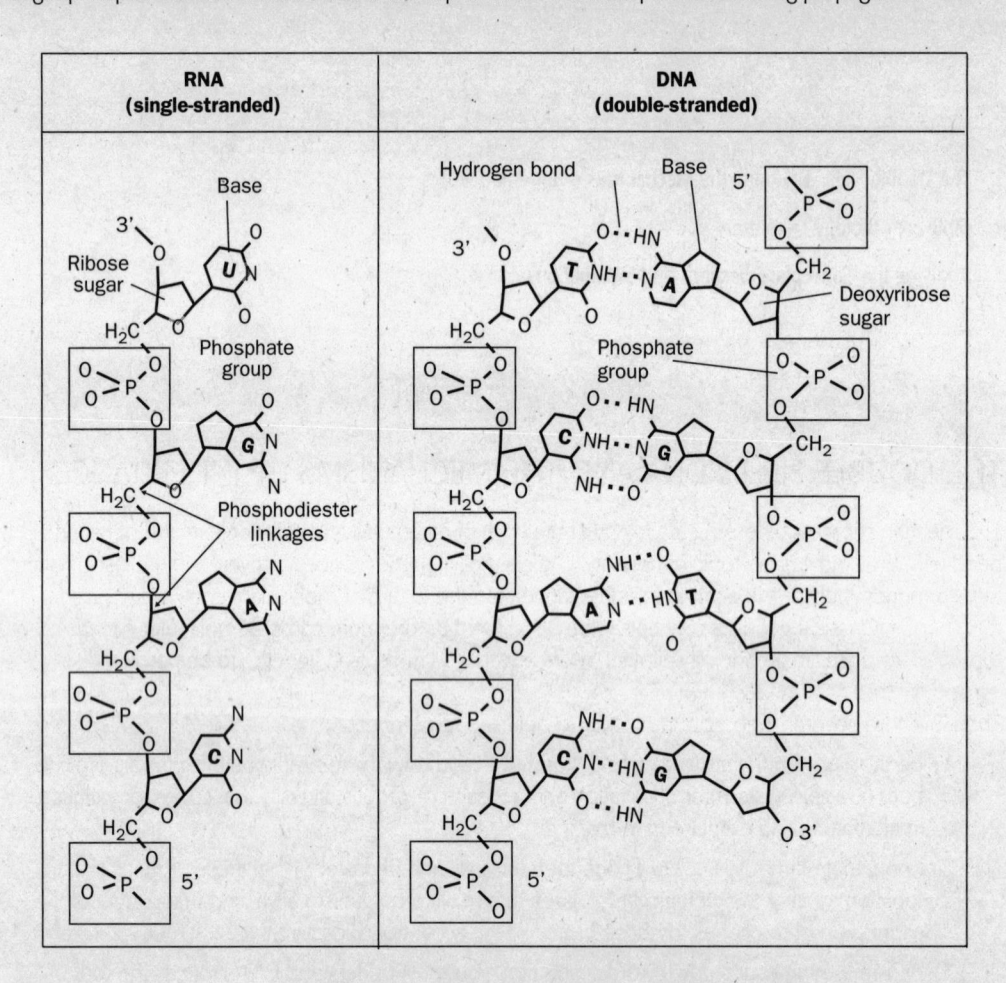

Structural Comparison of DNA and RNA

There are only four different nitrogenous bases that make up the nucleotides in DNA: adenine, cytosine, guanine, and thymine. These bases pair along complementary strands of DNA. Guanine pairs with cytosine and thymine pairs with adenine. In RNA, thymine is replaced with uracil, which pairs with adenine. Four different variables (the combination of nitrogenous bases in the sequence of a nucleotide) code for proteins in all of the organisms that have ever existed on Earth. It is amazing that the combination of nitrogenous bases in the DNA of all organisms on Earth leads to such genetic variety.

Protein Translation

In a eukaryotic organism, DNA is found in the **nucleus** of the cell. **Protein synthesis** takes place in the cytoplasm, so information from the coding regions of DNA in the nucleus must be moved to the cytoplasm. Information is moved to the cytoplasm through **mRNA (messenger RNA)**. The sequence of base pairs on the original strand of DNA is "mirrored" with a strand of **complementary bases** of mRNA (substituting uracil for thymine), starting at a special region called a **promoter** and ending at another special region called a **terminator**. This process is called **transcription**. The mRNA undergoes a sequence of changes before it reaches the cytoplasm. For reasons not fully understood, the original DNA contains many regions that are not coded for in protein synthesis. These regions are called **introns**, and are excised before the mRNA enters the cytoplasm for protein translation. **Exons** are regions that code for protein synthesis in mRNA and are the opposite of introns. The regions on the mRNA that correspond to the introns are excised before the mRNA enters the cytoplasm for protein translation.

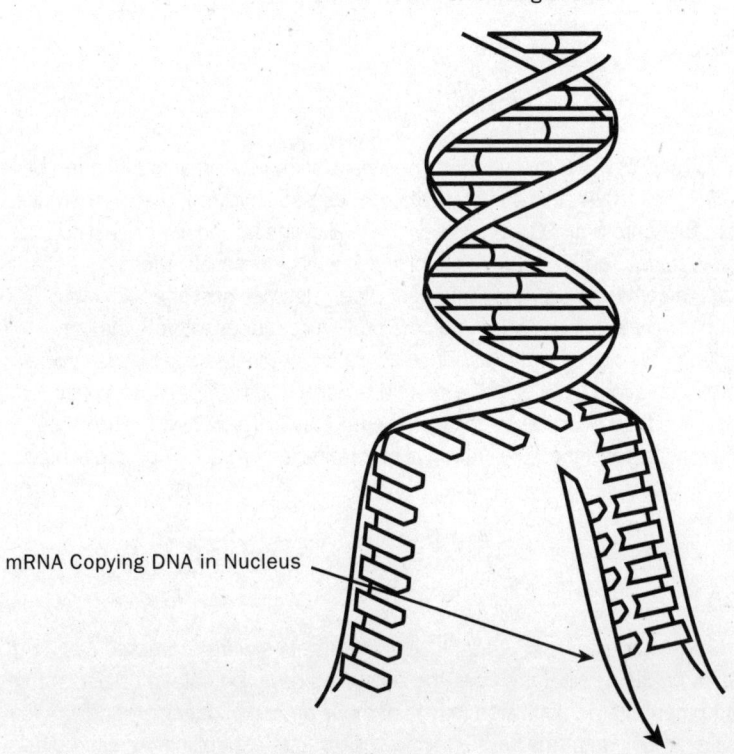

Unwinding of DNA

mRNA Copying DNA in Nucleus

Transcription

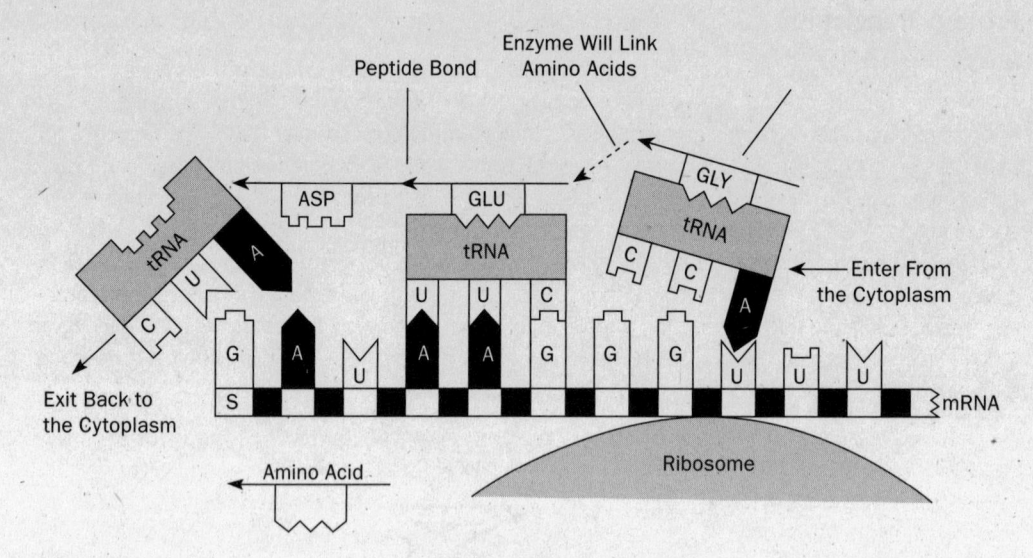

Translation

Once in the cytoplasm, mRNA acts as the template for protein translation. A short series of three bases called a triplet or codon codes for a **tRNA** (transfer RNA) that carries a specific amino acid. There are 64 (4^3) possible triplets, but there are only 20 amino acids. Even allowing for the start signal codon AUG as a site to begin protein translation and several codons acting as stop signals, there is considerable redundancy in the genetic code. This redundancy is sometimes described as being degenerate. Once translation is initiated at an AUG triplet, the protein continues to grow through elongation. The ribosome performs protein synthesis. Once an mRNA enters the cytoplasm, several **ribosomes** attach to it, creating multiple lengths of the protein simultaneously. Elongation continues until a stop codon is reached and the ribosome disengages the mRNA. The newly synthesized polypeptide usually undergoes transformation (i.e., removal of terminal amino acids) before achieving its active state.

Control of Expression

Even a simple bacterium contains thousands of genes. If all of these genes were expressed (transcribed and translated into protein) all of the time, the cell would never be able to obtain enough resources to stay alive, not to mention the confusion that all of those proteins floating around would cause. Prokaryotes and single-celled organisms have a relatively easy time controlling the differential expression of their genomes compared to multicellular organisms that have the same genome in every cell, yet carry out very specific and different functions in different structures. How does a cell in the eye of an eagle express the proper genes to make proteins for eye development and function, while a cell in the eagle's pectoral muscle expresses proteins to build muscle tissue and energy use? Both cells have the same genes. The answer is only partially understood, even with modern technological advances in genomics and proteomics. Furthermore, the intricacies of gene expression are too much to cover in just a few pages of this test-prep book. So what do you need to know?

The majority of what is known about the control of **gene expression** is at the **transcriptional** level. Expression of genes takes place in both prokaryotes and eukaryotes, but eukaryotes have more levels of control because of the separation of DNA within their nuclear membrane. The operon model of transcription is an important system to know. Operons appear in certain bacterial genomes.

An **operon** is a series of genes that include a promoter and a terminator that synchronize to perform a biological function. The operon consists of the structural genes that code for proteins for other functions, but may also contain genes that code for **repressor proteins** that inhibit the transcription of the operon by blocking a region of the operon called the operator. The repressor protein usually is removed from its blocking position by the presence of a **substrate** (a sugar), which allows the structural genes to be transcribed and translated. The structural genes then perform some action on the substrate, making it unavailable. The repressor again blocks the expression of that operon. This is a classic example of a **negative feedback loop**. After all, there is no need for the proteins until the substrate is available. This is also an example of the control of gene expression by **induction** because the expression of the genes is induced by the presence of some substrate. The important thing to remember is that genes are always present in the DNA, but certain environmental variables affect their expression. Some other forms of gene regulation are those that are based on an internal clock or synchronized with other cellular events.

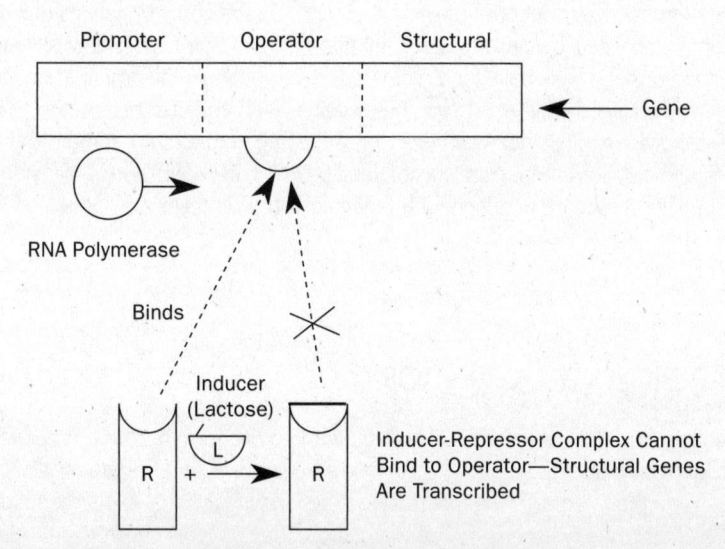

The *Lac* Operon in *E. coli*

In this illustration, a protein repressor binds to the operator in the absence of lactose in the growth medium. When lactose (the inducer) is present it binds to the repressor and disengages from the operator. The transcription of the operon by RNA polymerase then commences. The transcribed mRNA codes for the proteins that metabolize lactose, a process which eventually leads to the removal of lactose from the medium. When lactose is gone, the repressor binds to the operator again and the operon is back to square one.

Another important concept to appreciate from this model of gene expression is that the promoter and operator of an operon are "unaware" of what happens at another part of the DNA strand, unless the DNA performs a specific activity on them. For example, if the gene to produce insulin were placed within the operon to produce the enzymes that metabolize glucose in a bacterium, the insulin gene would be translated as well as the glucose-metabolizing genes if the bacterium were placed in a medium with glucose. This type of genetic manipulation has revolutionized the discipline of molecular genetics.

Kinds of Genetic Change

The goal of a cell is usually to maintain the same code in its DNA, base pair by base pair. Sometimes changes occur spontaneously during replication or are due to environmental factors such as irradiation. These changes in the DNA code are called **mutations**. Mutations can involve a change in only one base or several bases. Mutations are also placed in one of two categories, base-pair substitutions or insertions and deletions. **Base-pair substitutions** occur when one base pair is incorrectly reproduced and exchanged with a different base pair. An **insertion** is when any number of extra base pairs are added to the code and a **deletion** is when any number are removed from the code.

The effect of mutations is dependent upon where they occur in the code. Mutations of introns often have no effect on an individual because they are not translated into proteins. Likewise, base-pair substitutions in third position base pairs for a codon often do not affect the amino acid it codes for (remember the code is degenerate), therefore the protein is unchanged. Most mutations in structural proteins are deleterious, meaning that they negatively affect the nature of the protein and usually produce a non-viable cell. Some mutations in structural proteins are viable and, if the mutation is in a gamete, are potentially passed on to offspring. The next chapter will cover the passing on of mutations to offspring.

MOLECULAR BIOLOGY LABORATORY

This is a difficult laboratory to conceptualize. It is also difficult to perform because it requires a considerable amount of specialized materials, expensive equipment, and aseptic conditions. Some companies have even prepared software that replaces the lab with a virtual experiment. Regardless, you will almost assuredly see questions on the exam about the concepts covered in this laboratory.

There are two parts to the laboratory. The first part deals with the transformation of bacteria using plasmids that contain known genes. Bacteria can incorporate loose DNA in a medium into their own genome after being shocked by a chemical or with temperature. The shock causes breaks in the bacterial genome that recombine with the plasmid genes inserted. As long as the gene from the plasmid inserts after a promoter region, it will be expressed as part of the new, recombined bacterial genome. The common example is to take plasmids that contain an ampicillin-resistant gene and transform bacteria that are affected by ampicillin. After applying the treatment, you can test for transformation by looking for bacterial growth on a medium that contains ampicillin. You start with a stock of ampicillin-sensitive bacteria and a stock of plasmid containing the gene for ampicillin resistance. You apply the shock treatment to the bacteria and add the plasmids, then apply the bacteria to growth media with and without ampicillin. Only bacteria that have incorporated the ampicillin-resistant gene from the plasmid into their genome will survive in the medium containing ampicillin. Your goal in this part of the laboratory is to see firsthand the products of recombination and transformation that you have already learned on paper. You will see some examples of how this knowledge will be tested in the sample questions.

The second part of the experiment involves splicing DNA using restriction enzymes and visualizing these fragments using gel electrophoresis. A gel is a matrix composed of a polymer that acts like filter paper. Think of the gel as a thick, tangled jungle. DNA fragments are negatively charged, so will move away from the negative pole of an applied current and move toward a positive pole. As the DNA fragments move away from the negative terminal of the charge applied, they get tangled in the jungle. The smaller the fragments of DNA, the faster they can move through the tangle. The movement of the fragments can be adjusted by changing the density of the gel (the tangles in the jungle) or by changing the quantity of charge (the pushing/pulling power) applied to the gel.

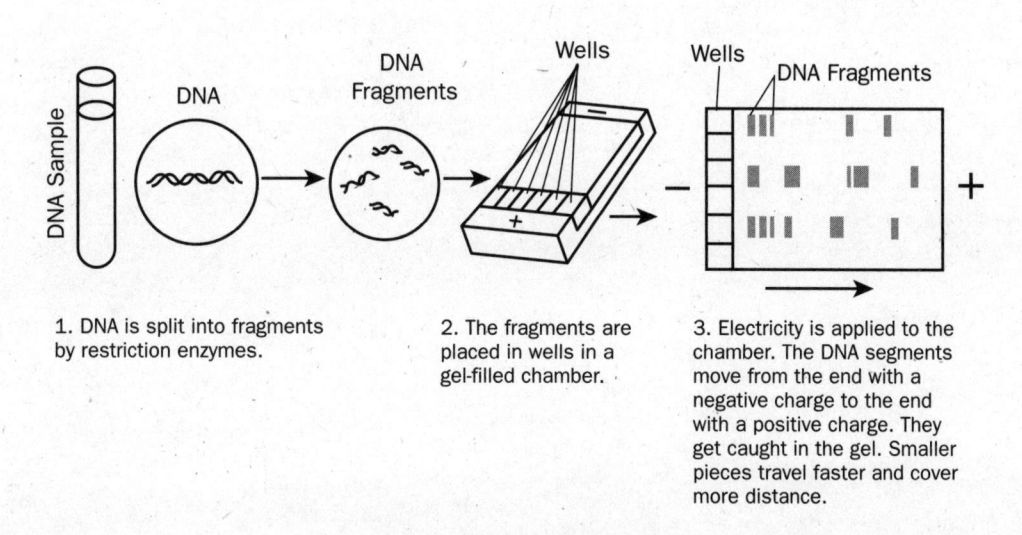

1. DNA is split into fragments by restriction enzymes.

2. The fragments are placed in wells in a gel-filled chamber.

3. Electricity is applied to the chamber. The DNA segments move from the end with a negative charge to the end with a positive charge. They get caught in the gel. Smaller pieces travel faster and cover more distance.

Diagram of Gel Electrophoresis

Restriction enzymes are manufactured by molecular genetics laboratories to have a high specificity to a series of base pairs on a strand of DNA or RNA. When the enzyme finds and binds to these base pairs, it snips the DNA. The more this series of base pairs appears on the DNA or RNA strand, the more cuts the restriction enzyme makes in the strand. After the strand is "digested" by the restriction enzyme, the fragments with different lengths can be separated using gel electrophoresis. This technique is commonly used to determine genetic differences between organisms, as long as the appropriate region of DNA can be isolated and the right combination of restriction enzymes can be determined. The most important thing to remember from this lab is that smaller fragments move farther on the gel. They move farther *because* they are smaller. All of the fragments move for the same reason: they are charged molecules moving away from an applied charge of the same kind (positive or negative).

IF YOU LEARNED ONLY FIVE THINGS IN THIS CHAPTER....

1. Nitrogenous base pairs make up DNA and RNA: adenine pairs with thymine (DNA only) or uracil (RNA only), and cytosine pairs with guanine.

2. In transcription, the DNA strands separate and mRNA copies one side. The mRNA takes the information to the ribosome, where protein synthesis occurs. In translation, tRNA carries amino acids to the mRNA and assembles them into proteins based on the mRNA code.

3. Genetic expression is controlled by many factors. Operons occur in bacterial genomes. They are sets of genes that perform a function. Genes in the operon code for repressor proteins that inhibit the function of the operon. Their function is controlled by a negative feedback loop (they bind to a substrate so the operon can do its task, which includes depleting the substrate so that repressor proteins inhibit the operon once again).

4. Mutations are the source of genetic change. Types include base-pair substitutions, which affect one amino acid, as well as insertions and deletions, which shift the genetic code and affect many amino acids.

5. Scientists can modify a bacterium's DNA by adding new genes.

APPLYING THE CONCEPTS: QUESTIONS AND EXPLANATIONS

1. Perhaps the simplest and oldest form of life on Earth, this organism exists as a protein capsule surrounding nucleic acids.

   (A) Virus
   (B) Bacterium
   (C) Protist
   (D) Fungal spore
   (E) Yeast

2. A scientist places free strands of DNA, which contain a gene that codes for the protein allowing the metabolism of glucose, in a medium containing bacteria that can only survive on the sugar lactose. The scientist heat shocks the bacteria in $CaCl_2$ and lets them recover before plating them in several petri dishes with only glucose as a nutrient source. After several days there are no signs of bacterial growth in the glucose medium. All of the following are possible explanations for the results EXCEPT which?

   (A) All of the bacteria died from the heat shock treatment.
   (B) The gene for glucose metabolism was not incorporated into any of the bacteria.
   (C) The gene for glucose metabolism was incorporated, but not within an expressed operon.
   (D) The free strands of DNA were from a sheep so will not function in bacteria.
   (E) The free strands of DNA contained only a structural gene without its own promoter.

3. The complementary strand of DNA to the DNA fragment 5'-GGC ATA CAT-3' is

   (A) 3'-CCG UAU GUA-5'
   (B) 3'-GTA TAT CCG-5'
   (C) 3'-ATG TAT GCC-5'
   (D) 3'-CCG TAT GTA-5'
   (E) 3'-CCG AUA GUA-5'

4. An operon in a bacterial genome is composed of

   (A) several structural genes that express proteins with similar function
   (B) one or more structural genes, a promoter, an operator, and a terminator
   (C) incorporated DNA from plasmids
   (D) viral DNA from infection and reverse transcription
   (E) foreign RNA from a phage

5. The following diagram shows the results of a gel electrophoresis of several fragments of DNA obtained from a restriction enzyme digestion. What can be concluded from the diagram?

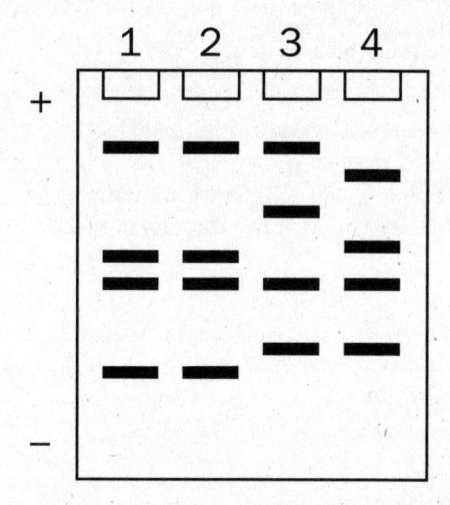

   (A) The DNA in wells 1, 2, 3, and 4 are likely from the same organism.
   (B) The DNA in wells 1 and 2 are likely from the same organism.
   (C) The DNA in wells 1 and 3 are likely from the same organism.
   (D) The DNA in wells 1 and 4 are likely from the same organism.
   (E) The DNA in each well is likely from a different organism.

6. Which of the following types of mutation involves only one base pair?

   (A) An inversion
   (B) A deletion
   (C) An addition
   (D) Triploidy
   (E) A substitution

7. What type of bonds joins one strand of DNA to another strand of DNA, forming a double helix?

   (A) Ionic bonds
   (B) Hydrogen bonds
   (C) Polar covalent bonds
   (D) Hydrophobic interactions
   (E) Nonpolar covalent bonds

8. Which of the following statements is false?

    (A) Mutations are important in evolution.
    (B) Errors in DNA replication can cause mutations.
    (C) Mutagens are substances which cause cancer.
    (D) Products of cellular metabolism can cause mutations.
    (E) DNA polymerase proofreads newly synthesized DNA.

9. In gel electrophoresis, DNA fragments migrate toward the _____ electrode; the _____ the fragment, the faster it moves through the gel.

    (A) negative; smaller
    (B) negative; larger
    (C) positive; smaller
    (D) positive; larger
    (E) None of the above

Questions 10–13 refer to the following list:

    (A) Uracil nucleotide
    (B) Guanine nucleotide
    (C) Translation
    (D) Transcription
    (E) Splicing

10. Takes place only in eukaryotes

11. Forms three hydrogen bonds when linked with a cytosine nucleotide

12. DNA → RNA

13. Found only in RNA

Questions 14–17 refer to the following list.

(A) Conjugation
(B) Replication
(C) Transcription
(D) Transformation
(E) Translation

14. Process in which the mRNA template is used to construct amino acids

15. Process in which free DNA is incorporated into the genome of a bacterium

16. Process that creates duplicate strands of DNA

17. Process in which bacteria engage in sexual exchange of DNA

Questions 18–20 refer to the following scenario. A scientist conducts an experiment with penicillin-sensitive bacteria in which he adds a plasmid containing a gene that confers penicillin resistance. Following a protocol that elicits normal growth and uptake of the plasmid DNA, the scientist then adds bacteria to four new plates as shown below.

	Glucose Medium With No Antibiotic	Glucose Medium Penicillin Added
Bacterial Strain Without Plasmid	#1	#2
Bacterial Strain With Plasmid Added	#3	#4

18. The scientist should expect to see the most bacterial growth on plate(s)

(A) 1, 2, 3, and 4
(B) 1 and 2
(C) 3 and 4
(D) 1 and 3
(E) 2 and 4

19. The scientist is likely to see no bacterial growth on plate(s)
    (A) 1
    (B) 2
    (C) 3
    (D) 4
    (E) 1, 2, 3, and 4

20. How did the scientist likely promote the uptake of plasmid DNA by the bacteria?
    (A) The scientist added lactose to the growth medium.
    (B) The scientist irradiated the bacteria.
    (C) The scientist increased the heat in the medium.
    (D) The scientist exposed the plasmids to UV light.
    (E) The scientist digested the bacterial cell wall with restriction enzymes.

Free-Response Question

DNA makes up the genetic code for all living organisms. Although it is omnipresent in all living organisms, DNA cannot transmit this code on its own. A series of processes and functions must occur within the cell to carry out the genetic instructions encoded in DNA.

(a) **Describe** the process of transcription in protein synthesis. Include in your answer: mRNA, DNA, complementary base pairs, terminator, promoter, nucleus, introns, exons, and cytoplasm.

(b) **Describe** the process of protein translation in the cell. Include in your answer: mRNA, codon, tRNA, amino acid(s), elongation, ribosome, stop codon, and polypeptide.

(c) **Explain** the concept and provide an example of a negative feedback loop.

ANSWERS AND EXPLANATIONS

1. A

Because viral and bacterial research has played such a major part in our understanding of molecular genetics, the College Board has determined that these concepts should be part of the AP Biology curriculum. You have already learned about bacteria and the relative sizes of organisms as seen under a microscope. The smallest of the organisms studied are viruses. Smaller doesn't always indicate simpler or older, but in this case it does. Viruses are very simple structures and scientists still argue as to whether they are alive or not. Viral infections of bacteria and the actions of retroviral RNA have been crucial to developing new techniques for studying the DNA molecule and have led to advances in genomic research.

2. D

The correct answer to this problem is elusive because of the slim possibilities of some of the choices. Both (A) and (B) are very unlikely, but still possible explanations. Choice (C) is not as unlikely as the first two answer choices, but still possible. Choice (E) is very likely, but is essentially a repeat of (C). The incorporated gene does not have to have its own promoter if it is downstream from a promoter in the bacterial genome. The correct answer is (D). All DNA is composed of base pairs; it doesn't matter what organism it comes from. DNA can even be manufactured from scratch and never exist in an organism.

3. D

All you have to do here is remember that bases pair up between G and C and A and T in DNA. Remember this is different with RNA where there are no Ts. The sample fragment is given from the 5' to the 3' end, so your complementary strand will be from 3' to 5' and be in the same order but with the appropriate base pair. Choices (B) and (C) have the code in reverse, individually (C) or by triplets (B). Both (A) and (E) contain uracil, so they can't be DNA complements. The correct answer is (D). If the question had asked for the complementary strand of RNA the answer would be (A), but it asked for the complementary strand of DNA.

4. B

An operon has several different types of DNA regions. You may consider answer choice (A) because an operon has the structural genes that code for proteins for similar function, but there is more to an operon. Choices (C), (D), and (E) can be part of an operon, but do not define it completely. The most correct answer is (B) because it lists all of the parts of the operon.

5. B

You can approach this question two ways. First, you can guess that since the possible responses are about similarity and dissimilarity, (B) is most likely correct because the banding pattern is the same in wells 1 and 2. This is the correct answer. The other approach is to remember that restriction enzymes are very specific so will cut DNA only at sites with the same series of base pairs. The only way the base pairing can be identical is if the DNA code that is being digested is identical.

6. E

First eliminate the answer choices that are obviously wrong. Triploidy is the condition of having three sets of a particular chromosome, so (D) cannot be the correct answer. An inversion by nature must involve more than one base pair, otherwise it would not mean anything to reverse it. Both (B) and (C) can involve only one base pair and are the tricky choices to eliminate, but they can also involve any number of base pairs. Choice (E) is a direct exchange of only one base pair, so it is the correct answer.

7. B

In forming the double helix, one strand of DNA is joined to another strand via hydrogen bonds.

8. C

Mutagens are substances that cause mutations; carcinogens are substances that cause cancer. Not all mutations cause cancer. Also, if a mutation occurs in a gene that is not expressed, the mutation may not result in cancer. The other statements are true: Mutations are important in evolution, since without mutations, there would be no new alleles that could be selected for; errors in DNA replication can cause mutations; products of cellular metabolism, such as superoxide, can cause mutations; and DNA polymerase proofreads newly synthesized DNA, correcting incorrectly placed nucleotides.

9. C

DNA fragments are negatively charged, so in gel electrophoresis, DNA fragments migrate toward the positive electrode. The gel acts as a sieve, sorting out the fragments by size; the smaller the fragment, the faster it moves through the gel, since larger fragments will encounter more resistance.

10. E

Splicing is the removal of introns, or sections of DNA that are not expressed. Splicing takes place only in eukaryotes.

11. B

A guanine nucleotide forms three hydrogen bonds when linked with a cytosine nucleotide.

12. D

Transcription is the process of transferring genetic information in DNA to an RNA message that is decoded to produce protein.

13. A

Uracil nucleotides are found only in RNA.

14. E

In translation, tRNA assembles nucleotide bases to form amino acids, in the order given by mRNA.

15. D

Scientists have used this phenomenon in genetic engineering research.

16. B

DNA replication involves the unwinding of DNA and production of a complementary strand for each strand.

17. A

You may be able to remember this because the word "conjugal" is used to refer to marriage.

18. D

The first thing you should realize is that there will be no growth of the penicillin-sensitive bacteria that weren't transformed on any plates that contain penicillin. This means that there will be no growth on plate 2, so (A), (B), and (E) cannot be correct. The next step in getting this answer correct is to understand the question. It is asking which plates will get the MOST growth. There will likely be bacterial growth on plate 4 because some of the bacteria will be transformed with the plasmid and will acquire penicillin resistance, but there should be considerably more growth on the plates where penicillin resistance is not necessary. There should be more growth on plates 1 and 3, which is (D).

19. B

You already figured out the answer to this question above. There will be no growth on the plate that has penicillin and bacteria that have not acquired resistance from the plasmid.

20. C

There are several ways scientists treat nucleic acids to cause an effect, but here we are looking for the specific response of bacteria being transformed. Adding lactose to the growth medium might induce a *lac* operon, but shouldn't affect the bacteria's tendency to transform. Both radiation and UV light can cause harmful effects to all organisms and scientists use radiation to cause mutations in DNA, but these factors don't increase transformation. Choice (E) is a trick answer. Restriction enzymes digest DNA and RNA and should do nothing to the cell wall of a bacterium. There are two ways scientists promote transformation in bacteria: increasing temperature and adding chemicals such as $CaCl_2$. Increasing temperature is represented by (C).

Free-Response Answer Explanation

Transcription of DNA to RNA must take place before protein synthesis can occur. DNA needs to be transcribed before it is transferred to the cytoplasm of the cell.

Here is a possible response:

(a) Since DNA is found in the nucleus of a cell and protein synthesis takes place in the cytoplasm, base-pair coding information in the DNA has to be moved to the cytoplasm before protein synthesis can commence. The DNA is "transcribed" by mRNA, which mirrors the original strand of DNA with complementary base pairs, to be translated later on into proteins in the cytoplasm. In mRNA, uracil replaces thymine as a complementary base pair for adenine.

Transcription of the DNA starts at a region called the "promoter" and ends at a region called the "terminator." All base pairs within this region will be coded for with complementary base pairs in the mRNA. For an unknown reason, areas of the original strand of DNA, called introns, are not coded for in protein synthesis. Since these areas are not necessary for protein synthesis, they are cut out before the mRNA enters the cytoplasm. The areas that do code for protein synthesis, called exons, are transcribed to the mRNA to be moved to the cytoplasm.

(b) Once DNA is transcribed and the mRNA (messenger RNA) enters the cytoplasm, protein translation is able to commence. Ribosomes, the organelles that perform protein synthesis, attach to the mRNA when it enters the cytoplasm, allowing multiple strands of protein to be created simultaneously. Before protein can be synthesized, the ribosomes must know where to start translating the mRNA.

A codon, or three-base triplet, codes for tRNA (transfer RNA) that carries a specific amino acid. The codon that codes for the commencement of protein translation by the ribosomes is AUG. There are several three-base triplets that serve as "stop codons" to end protein translation, but AUG is the only "start codon." There are 64 possible combinations of triplets, but only 20 amino acids that are composed of these triplets; this leads to redundancy in the genetic code and several of the same amino acids being produced. The chains of amino acids, which form polypeptides, become longer as they are translated by the ribosomes, termed "elongation." Elongation of the protein continues until a "stop codon" is reached, at which point the ribosomes disengage and the formation of polypeptide chains of proteins is complete.

(c) A negative feedback loop slows down a process; it is a type of feedback in which a system, such as the circulatory system, responds to reverse the direction of change, to regulate homeostasis. An example of a negative feedback loop is the circulatory system. Another, more simplified example of a negative feedback loop is the Krebs cycle, or citric acid cycle.

Repressor proteins, which stop the transcription of the operon (a series of genes including the promoter and terminator, which synchronize to perform a biological function), are found in a negative feedback loop. The operon actually contains structural genes that code for repressor proteins to inhibit transcription of the operon. These inhibitory proteins can be removed by a substrate (sugar) that allows the structural genes of the operon to be transcribed and translated. To reverse the process, the genes perform an action upon the substrate, rendering it unavailable to continue allowing the transcription of the operon. The repressor proteins are again free to block transcription of the operon, which regulates homeostasis within a negative feedback loop.

Part Three: AP Biology Review | 175

Chapter 8: **Evolutionary Biology**

- A Little Change Can Add Up: Origins of Life and Evolution
- Population Genetics and Evolution Laboratory
- Applying the Concepts: Questions and Explanations

A LITTLE CHANGE CAN ADD UP: ORIGINS OF LIFE AND EVOLUTION

There is no biological concept more controversial and more misunderstood than biological evolution. The concept can be simply defined with the statement "descent with modification." This means that over time, populations of organisms exhibit changes in characteristics that are passed on through inheritable (i.e., genetic) means. The important distinction between this simple definition and what is commonly accepted as biological evolution is mechanism. Mechanisms will be covered on the exam. You should also clarify the semantic distinction between evolution and biological evolution. Personalities evolve. Societies evolve. But only populations of organisms undergo biological evolution by means of a modification of inheritable characteristics. In this chapter, the term evolution will always refer to biological evolution.

Early Origins of Life

Scientists estimate that the Earth is about 4.5 billion years old and that in the beginning, it was a very inhospitable place. When Earth first came into existence, there was no or very little atmospheric oxygen and the surface of the Earth was bombarded by intense ultraviolet radiation. Eventually there were heavy rains and violent storms, which led to the production of basic inorganic chemical building blocks from the soil and the energy needed to drive reactions for producing simple organic molecules. The prevailing theory of the origin of life is that these organic molecules became more and more complex until amino acids and nucleic acids were formed. Once strings of nucleic acids formed, they could self-replicate within the organic "soup." These self-replicating structures organized into **protobionts**, which were droplets of segregated chemicals. Chemicals continued to organize, until the first identifiable organism came into being. The origins of life have been theorized based on biochemistry, some actual laboratory experiments, and models of early Earth environments. It cannot be substantiated with direct evidence at this time that this theory of the origin of life is true. Fossil evidence of early life is the only way the origins of life can be directly substantiated. The first fossil records of prokaryotes exist from geological deposits thought to be 3.5 billion years old. The oldest eukaryotic cells are from deposits that are about two billion years old. The oldest multicellular fossils are from deposits that are 1.25 billion years old.

Early Evolutionary Ideas and Darwinism

Although some early scientists believed that species are immutable and do not evolve or change, others believed that evolution occurs. Theories of evolution developed long before the time of **Charles Robert Darwin** (1809–1882). Prior to Darwin, even the scientists who accepted the idea that species change over time did not have a solid idea for a mechanism that could explain the changes they observed. The most well-known hypothesis prior to Darwin was that of **Jean-Baptiste de Lamarck** (1744–1829) who suggested that organisms pass on acquired traits in an attempt to reach a more perfect form. In other words, if a mother works out at the gym and becomes strong and healthy, she will pass her acquired health to her children and so on. We know today that parents only pass on genetically controlled traits and not those they have acquired from the environment they live in.

Darwin presented his postulates for his **theory of evolution** by **natural selection** in his work *The Origin of Species* (1859). There have been slight modifications to the theory based on more up-to-date knowledge about genetics and molecular biology, but for the most part, Darwin's theories are accepted today. Many scientists consider evolution by means of natural selection to be established fact.

Darwin had several original postulates for his theory of evolution by natural selection.

1. Individuals vary in their characteristics within a population. This means that all giraffes have long necks, but their necks aren't all the exact same length.
2. The variations observed in populations are inherited. When a big dog has puppies they tend to be big, and a little dog's puppies tend to be little.
3. A considerable number of individuals in a population seem to die as they compete for limited resources in the environment. This is where the term "survival of the fittest" emerged; it simply means that some individuals have characteristics that make them more likely to survive. These characteristics could include being bigger or stronger, but they also could include being smaller and smarter. It simply means that the characteristic(s) suit(s) the environment better.
4. Individuals that have more resources because of their particular characteristics tend to produce more offspring that survive. For example, if a bird with a long beak can get more food from holes in trees because of its long beak, it will be more likely to survive and provide more food for its offspring. If beak length is a result of genetics, that bird's offspring are more likely to have long beaks, and the following generations of offspring are more likely to have long beaks, until every bird in the population has a long beak.

The selection for more "adaptive" traits tends to narrow a population of individuals down to those that are best suited for a particular environment. If changes occur in the environment, selection favors individuals best suited to the new environment. The theory of natural selection seemed to explain the differences Darwin observed in species he studied and helps to explain the biodiversity in organisms today.

Evidence Used to Support Evolution

The Fossil Record

The geological layers in the Earth's crust stack on top of each other, with the oldest layers deeper and the youngest layers closer to the surface. Older geological layers hold more "primitive" fossils.

The Fossil Record Over Geologic Time

Period	Millions of Years Ago	Evidence
Precambrian	> 3,500	Definite fossils of prokaryotes
Precambrian	> 1,000	Earliest fossil eukaryotes
Cambrian	540–505	All extant and some extinct animal phyla
Silurian	438–408	First terrestrial organisms
Devonian	408–360	Diversification of bony fishes; first insects
Permian	286–248	Diversification of reptiles
Jurassic	213–144	Age of dinosaurs
Cretaceous	144–65	First modern birds
Tertiary	65–1.8	Diversification of all major groups of land animals, including hominids
Quaternary	1.8–Present	Extinction of large land mammals

Biogeography

Organisms are more like other organisms in their geographic vicinity. Organisms in adjacent dissimilar environments are more similar than organisms in similar environments on opposite sides of the Earth. This suggests that organisms in adjacent dissimilar environments are descended from recent common ancestors and have not magically come about in the environment in which they currently live.

Comparative Anatomy

Organisms have very different structures that are composed of the same basic components. For example, the human arm has the same bones as the wing of a bat. These structures are called homologous structures because they are considered to have arisen from a common ancestor. Analogous structures are structures that may perform a similar function, but have not arisen from the same ancestral condition. The wings of a bat and the wings of a butterfly are analogous.

Embryology or Ontogeny

Organisms that share a more recent common ancestor have similar modes of development. A classic example is that all vertebrate embryos have a stage of development in which they possess gills, whether they are aquatic or terrestrial. The presence of these more "primitive" characters in the embryos of "advanced" organisms suggests that they share genetically controlled developmental physiologies that have been passed on from their common ancestors. The process through which an organism develops from an embryo to an adult is called ontogeny.

Taxonomy

Organisms are classified into smaller and smaller subgroups based on similar and dissimilar characters. This hierarchy is an implicit illustration of the tree of life, leading to common ancestry by linkage to a superceding group. For example, a plant in the family Euphorbiaceae is more closely related to other plants in Euphorbiaceae than it is to plants in the family Cactaceae. The housefly, *Musca domestica*, is more closely related to other flies in the genus *Musca* than it is to flies in the genus *Stomoxys*.

Molecular Biology

Siblings share more similar DNA with each other than they do with other members of the same species. Similar species have more similar DNA than do more distantly related species (in different genera, for example). Related genera share more similar DNA with one another than they do with genera in another family. A domain is the broadest classification of organisms; thereby organisms in different domains will be the most distantly related.

Rates of Evolution

Prior to Darwin and even for scientists today, there is an underlying assumption that evolution takes a long time. The reasoning behind this theory is that because mutation is the ultimate source of variation and mutations that allow viable offspring are extremely rare, the probability of accumulating enough mutations to cause considerable change in organismal form requires a lot of time. Due to this assumption, prior to chemical dating techniques, geological layers were considered to be very old. Chemical dating has allowed modern scientists to accurately assess the age of geological layers.

Two more hypotheses have been proposed by scientists that are related to rate of evolution. The first hypothesis is called **punctuated equilibrium**. The hypothesis of punctuated equilibrium suggests that changes in organismal form did not take millions of years. Instead, very large changes in form happened relatively quickly (i.e., over thousands or tens of thousands of years) and were maintained thereafter over long periods of time. There is some evidence to suggest this hypothesis may be true, such as the Cambrian explosion in the fossil record, the discovery of cascading developmental genes, and the observation of large changes in phenotypic expression caused by single base-pair mutations. The second hypothesis is the notion that genetic mutations occur in a genome at a linear rate. This is called the **molecular clock hypothesis**. Assuming the molecular clock hypothesis is true, one could extrapolate the age of divergence of two organisms by counting the number of genetic differences in their genomes. This latter hypothesis assumes that mutation rates are constant over time and between species; a likely incorrect pair of assumptions.

Variation, Modes of Natural Selection, and Sexual Selection

Natural selection is the differential survival and reproduction of individuals based on variation in genetically controlled traits. These differing rates of survival and reproduction are due to forces in the environment and/or to forces exhibited by other species. In order to understand natural selection, it is necessary to understand variation.

Ultimately all variation originates in the mutation of DNA in an individual's genome. In order for a mutation to have an impact on evolution, it must occur in a gamete and be passed on to offspring. If mutation occurs in a gamete that forms a zygote, the offspring will inherit the allele for that trait and pass it on to its offspring. Genetic variation can also occur during recombination in meiosis I. As stated in the last chapter, most mutations are harmful, but occasionally a mutation exists in viable offspring. These offspring may then exhibit a phenotype that differs from the rest of the population.

Variation that occurs in a population will have a distribution based upon the kind of natural selection that is taking place in the population. The three types of variation that can occur in a population are stabilizing selection, directional selection, and disruptive (diversifying) selection. If a population is subject to **stabilizing selection**, extremes at both ends of a phenotype are eliminated, resulting in less genetic variability. For example, if the variation in color of a bird species ranges from dark gray to white and the population is subject to stabilizing selection, the medium-gray phenotype will be most common. If the population is under **directional selection**, one extreme is selected against but not the other (e.g., against white), so that the average in the population moves in one direction. **Disruptive (diversifying) selection** favors both extremes but selects against the average(s), which would be medium gray in the case of the birds.

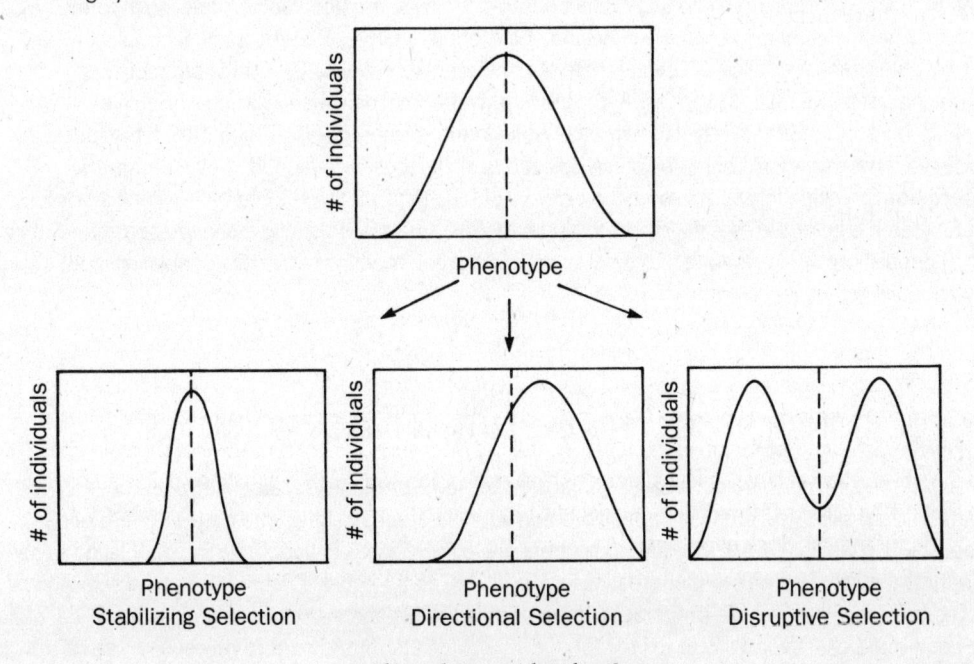

Modes of Natural Selection

Sexual selection is a force exhibited by a member of the same species of the opposite sex. This is most commonly seen when a female selects particular mating characteristics in a male. Many male birds have bright mating plumage that makes them conspicuous in a forest and more prone to attack from predators. Male ungulates often have huge antlers that grow every year and make it more difficult to travel through dense woods. The bright feathers on the male bird and the large antlers on the male deer are characteristics that are **selective disadvantages** in the animals' natural environment, but are **selective advantages** when it comes to courtship and mating. The balance between survival and the ability to mate dictates how a species evolves in terms of natural selection.

Speciation

Over a dozen different concepts of speciation have been promoted in scientific literature, but the most prevalent by far is the **biological species concept (BSC)**. This concept states that a species is defined by a naturally interbreeding population of organisms that produces viable, fertile offspring. In other words, different species are organisms that can't breed with each other or don't naturally breed with each other due to certain barriers. There are two kinds of barriers to interbreeding: prezygotic and postzygotic.

Prezygotic barriers to interbreeding include isolation of species due to ecological, temporal, behavioral, or mechanical factors, or physiological incompatability of gametes. **Postzygotic barriers** include ultimate inviability or sterility of hybrid organisms from the interbreeding of two species. Hybrid organisms may not die off in one generation, but ultimately the offspring of the mating of two species will die without producing offspring of their own.

Geographic isolation is not the only factor that can cause speciation, so additional terms have been defined to describe how species evolve due to isolating mechanisms. **Allopatric speciation** is when one population is separated into two distinct populations by some geographic barrier such as the movement of a tectonic plate or the elevation of a mountain range. After the original population is no longer able to share its alleles, it evolves into distinct populations that have a high probability of acquiring distinctive traits. In contrast, **sympatric speciation** occurs when individuals within a population acquire distinctively different traits while in the same geographic area. Sympatric speciation requires some other form of reproductive isolation, such as those mentioned above. **Parapatric speciation** is less definitive. This occurs when two populations are able to interbreed along a border, but the exchange of alleles is negligible compared to the amount of genetic exchange occurring within each population. A narrow zone of hybridization exists at the meeting of the two populations, but the two populations never coalesce into one.

POPULATION GENETICS AND EVOLUTION LABORATORY

This lab is easy to perform, but it causes confusion for some students because it involves quantitative and analytical skills. The premise of the lab is that your classroom will become a virtual interbreeding population in which you will measure the current allele frequencies in your breeding population to see if or how they change with random mating and, perhaps, after imposing natural or artificial selection. You can use any measurable phenotypic trait in your virtual interbreeding population, such as attached or unattached earlobes, ability to roll the tongue, eye color, etc., but test kits to measure a person's ability to taste the chemical PTC (phenylthiocarbamide) are readily available and commonly used. A person who is either homozygous dominant or heterozygous can taste the bitter PTC, while a homozygous recessive person cannot. You can ascribe the letters T and t to the dominant and recessive alleles respectively for the ability to taste PTC.

Now comes the hard part: putting information into equations. The way to determine the frequencies of the alleles in your breeding population is to assume the population is in **Hardy-Weinberg equilibrium**. If your population is in Hardy-Weinberg equilibrium, two things will be true. First, the addition of the frequencies of the alleles will equal one. The simple formula is

$$p + q = 1$$

where the frequency of the dominant allele is indicated by p and the frequency of the recessive allele is indicated by q. As indicated above, for this experiment, $p = T$ and $q = t$. The frequencies of the phenotypes in a Hardy-Weinberg population follow the equation

$$p^2 + 2pq + q^2 = 1$$

The genotypes in a Hardy-Weinberg population are indicated by each term on the left side of the equation above. The frequency of the homozygous dominant phenotype in the population above is

$$T \times T = T^2 = p^2$$

The frequency of the heterozygote phenotype is

$$2 \times T \times t = 2Tt = 2pq$$

and the frequency of the homozygous recessive phenotype is

$$t \times t = t^2 = q^2$$

There are five assumptions that must be met before assuming a population is under Hardy-Weinberg equilibrium.

1. The population is very large and not subject to small perturbations in the frequencies of alleles. There are no bottleneck effects.
2. The population is isolated from both immigration and emigration. There is no gene flow.
3. There is no mutation.
4. There is no selective breeding and mating is random between individuals.
5. There is no genetic drift. All genotypes code for phenotypes that have equal chance of viability and reproduction. There is no selection of phenotypes.

Keep in mind that the theoretical Hardy-Weinberg population is the benchmark to which all naturally occurring populations are compared. There are probably few if any populations in complete equilibrium. Comparing the expected frequencies under the above assumptions to what actually occurs in a population gives insight into which of the above assumptions is being violated. In this way a scientist can determine which evolutionary or environmental forces prevent a population from maintaining equilibrium.

For this experiment, everyone in the lab (or perhaps several labs) tastes a test strip to see if he or she can detect bitterness. Let's say that 5 percent of the class cannot taste bitterness, leaving 95 percent that can. The students who cannot taste bitterness are homozygous recessive for the trait. This means that the frequency of the recessive allele can be determined by letting

$q^2 = 0.05$, which means

The frequency of the dominant allele can be determined with the equation

$p = 1 - q = 1 - 0.22 = 0.78$

The frequencies of the genotypes that can taste the bitterness are

$p^2 = 0.78 \times 0.78 = 0.61$ for the homozygous dominant genotype and

$2pq = 2 \times 0.78 \times 0.22 = 0.34$ for the heterozygous genotype.

You can check your work by adding up all the genotypic frequencies and making sure that they add up to 1.

$0.05 + 0.61 + 0.34 = 1$

For the next step of the experiment, everyone in the mating population produces gametes, usually by writing down, on four index cards, the four possible alleles that he can produce based on his genotype. Each person is then assigned to someone else in the laboratory to randomly exchange gametes. Gametes are exchanged for several generations and the genotypes of each generation of offspring are recorded and added to a cumulative tally of gametes within the entire population of the classroom. Calculations made in the first part of the experiment are repeated with the new generation, to see if the allele frequencies have changed. Some teachers have students apply a selective advantage/disadvantage to one or more of the genotypes and remove those individuals from the population. Calculations are completed on the cumulative population to demonstrate that equilibrium is lost because assumption five of Hardy-Weinberg equilibrium is violated.

IF YOU LEARNED ONLY SIX THINGS IN THIS CHAPTER....

1. It is thought that chemical components of life on Earth originated through radiation and storms. These compounds became increasingly complex, forming protobionts and ultimately living organisms.

2. According to Darwin, species evolve via natural selection, where animals with certain traits are more likely to survive and reproduce, passing on those traits. Selection can be stabilizing (median is encouraged), directional (the norm shifts toward an extreme), or disruptive (extremes are favored over the norm).

3. Evidence for evolution comes from comparative anatomy (homologous and analogous structures), biogeography, embryology, the fossil record, biological classification, and molecular biology (relatives share DNA).

4. Biological species concept: A species is a reproductively isolated population able to interbreed and produce fertile offspring.

5. In allopatric speciation, geographically separated populations develop into different species. Sympatric speciation occurs when populations in the same environment adapt to fill different niches. Parapatric speciation occurs with limited interbreeding between two groups.

6. Hardy-Weinberg equilibrium states that genetic distribution remains constant in large, isolated, randomly mating populations with no mutation and no natural selection. These conditions rarely (if ever) occur together.

APPLYING THE CONCEPTS: QUESTIONS AND EXPLANATIONS

1. Which of the following is the likely source of energy for the synthesis of the small organic molecules that presumably predated the first forms of life on Earth?
   (A) Fermentation by bacteria
   (B) Photosynthesis by microscopic algae
   (C) Lightning from constant storms
   (D) Shifts in ocean currents
   (E) Nuclear fission of water molecules

2. The earliest forms of life on Earth were recorded from fossil deposits dating from about 3.5 billion years ago. Which of the following were likely found in these deposits?
   (A) Microscopic fungi
   (B) Microscopic protozoa
   (C) Bacteria
   (D) Viruses
   (E) Microscopic algae

3. A population of birds on the coast range in size from 9 cm to 15 cm, with the majority of the birds being 12 cm. Another population of birds exists on a distant island off the same coastline, but all of these birds are 15 cm. What is the most likely explanation for the difference in sizes of the members of these populations?
   (A) Sexual selection
   (B) Mutation
   (C) Disruptive selection
   (D) Recessive genes
   (E) Genetic bottleneck (founder effect)

4. Which of the following is NOT an assumption made about a Hardy-Weinberg population in equilibrium?
   (A) There are an equal number of males and females in the population.
   (B) The population is large.
   (C) There is no immigration into the population.
   (D) There is no mutation.
   (E) All phenotypes have the same fitness.

5. Which of the following would be a statement most likely supported by Lamarck?

    (A) Four out of five hyena pups die because of a lack of resources in the environment.

    (B) The occurrence of new forms is a random event produced by a change in heritable factors.

    (C) The wings of a butterfly and the wings of a bat are analogous structures.

    (D) The long neck of a giraffe is a result of its ancestors continually trying to reach higher branches.

    (E) A species is a population of individuals that are reproductively isolated from other individuals in an area.

6. Which of the following is the ultimate source of all variation within and among populations?

    (A) Natural selection in different environments

    (B) Viable genetic mutations in gametes

    (C) Nonrandom mating between individuals with different traits

    (D) The diversity of habitats that exist on Earth

    (E) Environmental factors that affect development

7. All of the following provide evidence for evolution EXCEPT which?

    (A) The fossil record

    (B) Comparative anatomy

    (C) Embryology

    (D) Food webs

    (E) Molecular biology

8. The assumption that genetic mutations occur at a regular rate over time is part of which hypothesis?

    (A) Recombination

    (B) Natural selection

    (C) Molecular clock

    (D) Speciation

    (E) Hardy-Weinberg equilibrium

9. All of the following pairs of structures are homologous EXCEPT which?

   (A) The wing of a bat and the wing of a bird
   (B) The leg of a human and the leg of a centipede
   (C) The scales on a bird's legs and the scales on a lizard
   (D) The shell of a clam and the shell of a snail
   (E) The pouch of a female opossum and the pouch of a female kangaroo

Questions 10–13 refer to the following list.

   (A) Allopatric speciation
   (B) Sympatric speciation
   (C) Parapatric speciation
   (D) Stabilizing selection
   (E) Sexual selection

10. This type of evolution favors the average members of a population, while eliminating the extremes.

11. A male peacock has such outrageously colored plumage during mating season that its survival is reduced.

12. The isthmus of Panama separates two species of fish that are more closely related to each other than they are to the other species of fish in the bodies of water in which they exist.

13. Two closely related but reproductively isolated species of bees live in the same forest; one bee gathers nectar from a flower that peaks in early spring and the other gathers nectar from a flower that peaks in early summer.

Questions 14–16 refer to the following paragraph.

A large population of flour beetles is allowed to breed randomly for several generations under laboratory conditions. Initially, 36 percent of the population had short elytra, a homozygous recessive trait, but after several generations of breeding, the number of short-elytra beetles decreased to 25 percent of the population. The rest of the beetles in the population have long elytra, which is the phenotype for both the homozygous dominant and heterozygous genotypes.

14. Which of the following can be said about the population of flour beetles based upon the data given above?

    (A) The short-elytra beetles have a selective advantage over the long-elytra beetles.
    (B) There is strong sexual selection for beetles with long elytra.
    (C) The population of beetles is not in Hardy-Weinberg equilibrium.
    (D) The population of beetles is subject to genetic drift.
    (E) The population of beetles is subject to gene flow.

15. What is the frequency of the homozygous dominant genotype in the initial population of beetles?

    (A) 0.62
    (B) 0.6
    (C) 0.4
    (D) 0.16
    (E) 0.04

16. What is the frequency of the heterozygous genotype in the final population of beetles?

    (A) 0.75
    (B) 0.50
    (C) 0.25
    (D) 0.05
    (E) 0.01

Questions 17–19 refer to the following list.

(A) Cambrian
(B) Devonian
(C) Jurassic
(D) Cretaceous
(E) Quaternary

17. Mammals evolved during this geologic period.

18. Gymnosperms developed during this geologic period.

19. Chordates developed during this geologic period.

20. One method of dating organic material involved finding the ratio of

   (A) carbon dioxide to carbon monoxide
   (B) ^{18}O to ^{16}O
   (C) water to carbon dioxide
   (D) ^{14}C to ^{12}C
   (E) carbon monoxide to oxygen

Free-Response Question

Some scientists believe that in order for ecosystems to maintain a balanced or steady state, biodiversity must be conserved. Using both classical Darwinian thought and the modern synthesis of evolutionary theory, answer the following questions as they relate to biodiversity.

(a) **Compare** current biodiversity with the biodiversity that existed during the late Triassic and early Jurassic periods.

(b) Which group of organisms is the most diverse on Earth? Give some possible **explanations** for why this diversity exists.

(c) **Discuss** an example in which human interaction significantly affected the selective forces acting on an organism in a natural system as well as the results of that interaction.

ANSWERS AND EXPLANATIONS

1. C

It should be obvious that the correct answer will not include any energy that comes from an organism. The energy to create the first forms of life could not come from the first forms of life. This rules out (A) and (B) right away. Ocean currents and nuclear fission were never part of the discussion for the origins of early life, so (D) and (E) can be ruled out. Early Earth was a place of continual thunderstorms, which correlates with (C).

2. C

The first thing that stands out in this question is the age of the fossil deposits, dated 3.5 billion years ago, which means the fossils should be very, very "primitive." All of the possible answers are microscopic organisms and small often means less complex, so you might pick a choice based on this key word, but the answer is not based on the organism being small. Also, the smallest organism among the choices is viruses, but there are no fossil records for viruses. The oldest fossil records are from stromatolite deposits containing prokaryotes.

3. E

There is no apparent genetic difference between mainland and island populations. The birds on the island don't exhibit the same range of variation. The birds on the island could be an example of a very strong stabilizing selection, but this isn't one of the choices and is not likely. The occurrence of only large birds on the island could be an example of sexual selection (A) in which only large birds mate successfully, but this is much less likely than the hypothesis that only large birds can make it to the distant island. This question is tricky because when discussing evolutionary scenarios, one can imagine all sorts of possibilities and all of the answers become possible, if not probable. Since it is unclear which alleles are recessive in bird size, (D) can be eliminated. When a new population is formed from a narrow range of the phenotypes in an original population, this is called a genetic bottleneck. The analogy is that the range of phenotypes is narrowed to a select few as if they were being forced through the narrow neck of a bottle.

4. A

This is another tricky question because the wording you learned in association with the assumptions of the Hardy-Weinberg population is most likely different from the wording given in the answer choices. Choices (B), (C), and (D) can be ruled out because they contain key words about assumptions made about a Hardy-Weinberg population at equilibrium. Choice (E) is a different way of saying that all of the individuals have the same chance of viability and reproduction. Fitness is a measure of an organism's ability to reproduce. The correct answer choice is (A), but it still isn't completely obvious after the other choices are ruled out. The assumption of Hardy-Weinberg is that all individuals have an equal chance of mating and that mating is random. If the number of males and females in a population were heavily skewed to only a few individuals of either sex, there would be nonrandom mating, but in most cases the population is large enough to prevent this. Random mating is still likely to occur if there are 10,010 males and 9,990 females, for example.

5. D

Lamarck believed that acquired traits are passed on to offspring. Some of the statements above are part of the developing history of the perceptions of evolution, but are not attributable to Lamarck. A giraffe striving to reach higher branches during its lifetime is an act that is not preserved in the genome. Should this action lead any given individual to develop a longer neck, increased neck size would be an acquired trait that cannot be passed on to offspring.

6. B

All evolutionary change ultimately comes from genetic mutation. Genetic mutations in gametes are passed on from generation to generation.

7. D

This is one of those knowledge-based questions that are hard to reason through without the facts. The best possible approach if you don't immediately know the answer is to consider all of the possible choices within the context of the question. Fossils (A) and the genetics of molecular biology (E) can be associated with evolution. Comparative anatomy (B) and embryology (C) are parts of form and function, another topic in AP Biology. Food webs (D) are part of another topic as well, ecology. However, you might remember that Darwin used comparative anatomy and embryology in his studies and that these techniques are still relied upon today. The correct answer is (D).

8. C

Even if you don't know the answer right off, you can still figure out the answer to this question. A rate is always a change in a quantity over a given period of time. The correct answer is (C), molecular clock, a "clock" being a tool to measure time.

9. B

Okay, quick review. Analogous structures are those that usually share a similar function but do not have a common ancestral origin. Homologous structures might have completely different functions, but are composed of the same building blocks. There is only one obvious answer choice. The leg of a human and the leg of a centipede both allow the organism to walk, but they are completely different structures. A human leg has an internal skeleton and a completely different developmental origin than the leg of any arthropod.

10. D

Always remember stabilizing selection as a bell curve in which the middle trait is most prevalent.

11. E

Sexual selection affects factors that help an organism obtain a mate and often acts against the survival of the organism.

12. A

This is an example of geographic isolation. An initial population of fish was separated by the isthmus and acquired new traits after being isolated.

13. B

This is an example of prezygotic isolation due to ecological and, perhaps, behavioral mechanisms. Because the two bees exist in the same space, they are sympatric.

14. C

Since the frequency of short-elytra beetles decreases, (A) does not make sense. These beetles are raised under laboratory conditions, so are not likely to be subject to gene flow, changes in the population's genetic makeup based on immigration and emigration (E). It is possible that the long-elytra beetles are under selection pressure or that there is genetic drift, but it is impossible to tell from the data provided. The only conclusion that can be made is that the population is not stable and not in Hardy-Weinberg equilibrium, which is (C). Notice that you do not need to know what elytra are to answer this question.

15. D

This question should be easy to answer. The crucial bit of information we have is that the frequency of the homozygous recessive individuals (call them ee) is 0.36. We can use this number to calculate the frequency of the e allele.

$$ee = q^2 = 0.36 \text{ so } q = \sqrt{0.36} = 0.60$$

We then use the equality

$$1 - q = 1 - 0.6 = 0.40 = p$$

and then calculate the frequency of the homozygous genotype

$$EE = p^2 = 0.40^2 = 0.16$$

The correct answer is (D).

16. B

Again, this is just another example requiring simple arithmetic. The frequency of ee is 0.25.

$$ee = q^2 = 0.25 \text{ so } q = \sqrt{0.25} = 0.50$$

We then use the equality

$$1 - q = 1 - 0.5 = 0.50 = p$$

and calculate the frequency of the heterozygous genotype

$$Ee = 2pq = 2 \times 0.50 \times 0.50 = 0.50$$

The correct answer is (B).

17. C

Mammals evolved during the Jurassic period.

18. B

Gymnosperms developed during the Devonian period.

19. A

Chordates developed during the Cambrian period.

20. D

Finding the ratio of ^{14}C to ^{12}C is one method used to date organic material.

Free-Response Answer Explanation

This question, rather than asking you to recall particular facts or even to apply the facts you know about evolution, asks you to apply an appreciation and synthesis of the principles of evolution. There are additional biological concepts from other course topics that you will have to recall in order to answer parts of the question as well.

The first part of the question asks you to compare the conditions that exist today to those of an easily recognizable geological time span, the age of the dinosaur. In order to answer the question, you need to recognize the period, the dominant organisms that supposedly lived during that time, and the environmental conditions of the period. You then need to make some comparisons between the organisms and environmental conditions that existed during the late Triassic and early Jurassic period, and the organisms and environmental conditions that exist today. There is no right or wrong answer—just be explicit without being wordy.

Part (b) requires a knowledge of the patterns of biodiversity that exist today. The obvious answer is that either arthropods or insects are the most diverse group of organisms. Either answer should be acceptable as long as you support it. You then need to make some statements that address the high amount of adaptive radiation that took place in the case of arthropods/insects.

The last part allows you to apply a particular piece of information that you provide. Providing your own example makes it easier for you to answer the question, but more difficult for the grader to give you points. The key to answering this part of the question is to be very explicit and pay close attention to the question you are answering with your explanation. You will not want to "wax poetic" or be political about an environmental disaster caused by humans for part (c). Give your example and explain the cause-and-effect relationship within the context of evolutionary biology.

Here is a possible response:

(a) The transition between the Triassic and Jurassic periods marks the time in Earth's history when life on Earth was dominated by dinosaurs. The major continental plates were still massed into the super-continent Pangaea and were just beginning to separate. The major limitation to the distribution of organisms was environmental conditions that existed at the time, as most of the major landmasses were contiguous and only local geographic isolations caused speciation. The majority of the terrestrial organisms that existed on Earth were centered around a warm, dry climate with a centralized region near the equator. There was no glaciation or polar ice caps present. The marine systems were warm and full of both simple and complex organisms.

The limited variation in climate during the late Triassic and early Jurassic periods made a great impact on diversity. There were fewer niches available for an organism to take advantage of, so there were fewer organisms. Flowering plants, mammals, and birds were just starting to evolve and the land was dominated by reptiles. The Earth today is much more diverse in terms of climate and species. After the Jurassic period, the major tectonic plates separated, isolating several large lines of organismal groups. Modern flowering plants, insects, mammals, and birds all have major adaptive radiations that are much more recent than the late Jurassic period. Geographic isolation and the availability of a large number of new ecological niches allowed the diversification of organisms that could take advantage of these niches.

(b) The most diverse group of organisms by far is the Class Insecta. Their diversity is explained by a combination of certain morphological advantages and ecological associations. First of all, insects evolved wings that allowed them to escape predation and take advantage of several new ecological niches. Insects are closely associated with flowering plants and their fast radiation is closely linked to the radiation of the angiosperms as they colonized a changing land climate. Insects also have short generations, allowing them to produce many offspring in short periods of time. These factors combined in traits that conferred selective advantages and provided insects with higher fitness. Also, the higher rate of reproduction and gamete production allowed more chances for viable mutations to occur. Since mutations are the ultimate source of genetic variation, the greater the rate of maintaining a mutation, the higher the rate of accruing variation.

(c) Both biotic and abiotic (nonliving) factors in the local environment act as selective agents on the fitness of each species of organism. When these selective factors are stable, population numbers are stable. When human intervention impacts the selective forces affecting an organism, it changes the pressure for a population to remain constant. A good example of this is the effect of global warming on the seasonal activity of migrating birds. The northern temperate regions are becoming warmer sooner in the spring, but the migratory birds are arriving at the same time. The insect communities the birds use to feed their nestlings are emerging earlier in the year, so that the food resource is becoming limited for the birds that had previously adapted to the later arrival time. Some birds have a genetic predisposition to arrive a little sooner, so they will have a selective advantage over the later-arriving birds. This should cause directional selection away from the birds that arrive at the previously adapted time toward birds that arrive earlier. After several generations, the bird populations with earlier arrival times should be the dominant phenotype. This likely will not cause speciation, but will probably change the behavioral phenotype of some bird species. The bird species will either have to adapt to the new ecological niche or suffer extinction because of extremely low fitness.

Chapter 9: **Diversity of Organisms**

- Cataloging Biodiversity: The Linnean System
- Applying the Concepts: Questions and Explanations

CATALOGING BIODIVERSITY: THE LINNEAN SYSTEM

In 1758, the Swedish naturalist Carl von Linné (a.k.a. Carolus Linnaeus, 1707–1778) published a version of the *Systema Naturae*, which used a consistent method of naming organisms called **binomial nomenclature** in an attempt to construct a "natural classification" that would reveal order in the universe. The study of biology has never been the same since. Now there are multiple codes for naming organisms that follow the method put forth by Linnaeus. The concepts presented within this chapter and topic should acquaint you with the difficulty of creating a language with which scientists from all over the world can communicate about the forms of life they observe.

Classification

Scientists need a way not only to identify the kinds of organisms they observe, but a way to communicate to other scientists which animal(s) they study. If, for example, a scientist refers to an organism by a regional name in a native language, it is difficult for a scientist on the other side of the world to identify the organism in question or to correlate any associative information by relationship with other organisms. The way the scientific community has addressed this need has been to create a hierarchical system of naming called the **Linnaean system of classification**. On the next page are the Linnaean classifications of the common housefly, the vinegar fly, the human, and cultivated corn.

Linnean Classification System

Linnaean Hierarchy	Housefly	Vinegar Fly	Human	Corn
Kingdom	Animalia	Animalia	Animalia	Plantae
Phylum (Division)	Arthropoda	Arthropoda	Chordata	Angiosperms
Class	Insecta	Insecta	Mammalia	Liliopsida
Order	Diptera	Diptera	Primates	Poales
Family	Muscidae	Drosophilidae	Hominidae	Poaceae
Genus	Musca	Drosophila	Homo	Zea
Species	domestica	melanogaster	sapiens	mays

The Latin binomial for an organism, which includes the genus and the species (e.g., *Musca domestica*, *Homo sapiens*, etc.), provides an internationally applied standard to which all people can refer. In addition, the collective group names to which the organism belongs imply an evolutionary relationship if the classification system follows phylogeny (evolutionary events involved with the origin of a group of organisms). The two species of flies above belong in the same order, humans belong in the same kingdom, and corn is in a completely different kingdom. The point of common origin of these organisms can be extrapolated by simply knowing the binomial.

The Diversity of Life and Distinguishing Features

In addition to naming organisms, scientists are always trying to simplify things. These simplifications usually take the form of generalities. In general, there are few generalities that apply to biology (irony intentionally applied here). Classification, however, has always relied on the similarities and differences that occur between species. You cannot review all of microbiology, cell biology, zoology, and botany in one chapter, but this section presents a summary of some of the major groups and their characteristics.

Until recently, the first level of classification of organisms was the **kingdom**. There were five recognized kingdoms (**Monera**, **Protoctista**, **Fungi**, **Plantae**, and **Animalia**), since too many organisms defied the simple classification of being either a plant or animal. Now all of life is first divided into three **domains**. You should be familiar with the domains for the AP Biology exam. The first two domains, **Archaea** and **Bacteria**, include all of the prokaryotes. Remember that **prokaryotes** are unicellular organisms with cell walls and no membranes on any of their subcellular structures. The Archaea are special organisms that have adaptations for extreme environmental conditions such as high temperature or high salt concentrations. The Archaea are separated into two groups, the **Halophiles** (salt loving) and the **Thermophiles** (heat loving). New species of Archaea are often discovered in deep-sea thermal vents or hot springs. Bacteria compose the rest of the prokaryotes. There are distinct chemical differences between the Archaea and Bacteria, but you should know that the Archaea have adapted to extreme conditions. The last domain is the **Eukarya**, which encompasses the rest of life on Earth and the groups that you should spend the majority of your time reviewing.

The Eukarya are split into four groups, the Protoctista, Fungi, Plantae, and Animalia. Simply put, the Eukarya are all of the **eukaryotes**. They can be divided into four kingdoms by the information in the next table. Keep in mind that the table contains general information and there are always exceptions in biology.

The Kingdoms of the Eukarya

Kingdom	Structure	Reproduction	Examples
Protoctista	Microscopic; mostly unicellular; 9 + 2 cilia or flagella	Asexual; no embryo	Amoeba, paramecia, diatoms, slime molds
Fungi	Multicellular; no cilia or flagella; chitinous cell wall	Haploid or dikaryotic; haploid spores	Yeasts, molds, mushrooms
Plantae	Large, sessile, multicellular; cellulose cell wall	Fertilization of female by male gamete; alternation of generations	Mosses, gymnosperms, angiosperms
Animalia	Large, motile, multicellular; use muscles for movement; no cell walls	Sperm and egg form zygote	Sea stars, worms, insects, mammals

Be familiar with the distinguishing features and some representative members of the major divisions of Plantae. There are at least ten divisions of Plantae. You should be familiar with at least four: the **Mosses**, **Ferns**, **Conifers**, and **Angiosperms**. The table below summarizes some of the major differences between each division of Plantae.

The Major Plant Divisions

Division	Structure	Reproduction
Mosses	Nonvascular; produce a mat consisting of many plants and rootlike structures called rhizoids	Water necessary for swimming sperm; haploid gametophyte is dominant generation; homosporous
Ferns	Vascular; compound leaves called fronds develop from coiled fiddlehead; true roots	Water necessary for swimming sperm; diploid sporophyte is dominant generation; homosporous
Conifers	Vascular; produce cones with internal gametophyte; evergreen, needlelike leaves	Wind pollinated; diploid sporophyte is dominant generation; heterosporous
Angiosperms	Vascular; produce flowers and fruit	Wind or animal pollinated; diploid sporophyte is dominant generation; heterosporous

Be familiar with the hypothesis that terrestrial plants evolved from green algae (Protoctista) because they share several features.

1. Chlorophyll *a* and accessory pigments
2. Thylakoid membranes stacked locally into grana
3. Cell wall made of cellulose
4. Carbohydrate stores made of starch
5. Cell plate dividing cytoplasm in cytokinesis formed from Golgi complex vesicles

Whereas the major determinate of plant divisions is mode of reproduction, the major determinates in animal phyla are body plans and development. The first branch in the animal tree has to do with body symmetry. The most primitive animals have either no body symmetry, as with the sponges that grow in amorphous masses, or **radial symmetry**. The rest of the animals have **bilateral symmetry**. Animals with radial symmetry have a side with a mouth (oral) and one without (aboral) but no front or back end or left or right side. Animals with bilateral symmetry have a head and tail and a right and left side in addition to a front and back or top and bottom.

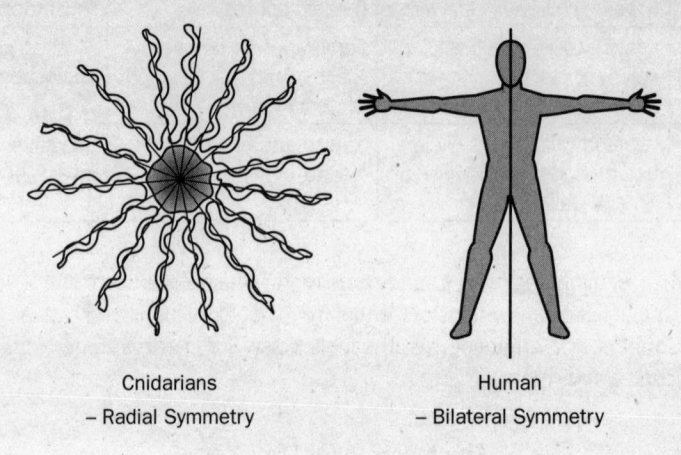

Cnidarians
– Radial Symmetry

Human
– Bilateral Symmetry

Radial and Bilateral Symmetry

Sponges are sessile filter feeders. They actively draw water through their bodies with specialized cells that have flagella. Different sponges are characterized by the composition of their internal **spicules**, which can be made of calcium carbonate, silica, or a soft protein called spongin.

The animals with radial symmetry include the Cnidarians, which encompass jellies, corals, and sea anemones. Simply put, Cnidaria are bags with a mouth. Their mouths are lined with stinging tentacles containing cnidocytes (*cnide* means *nettle*).

The remaining animals are further subdivided by the way they develop internally. There are two characteristics to consider. One is whether the animal has a true **coelom** (a mesoderm-lined cavity in the body between the gut and the outer body wall). This separates animals into **acoelomates** (no coelom), **pseudocoelomates** (possessing a body cavity not completely lined by mesoderm), and **coelomates**. The other characteristic to consider is whether during development the zygote has determinate or indeterminate cleavage. Animals with **determinate cleavage** have early developmental cells with a predetermined fate. If you separate a cell after initial cleavage it will not develop into a complete animal, it will be missing parts. Animals with **indeterminate cleavage** have early developmental cells that can go on to produce a whole animal. This is what makes identical twins possible. Animals with determinate cleavage are called **protostomes,** and animals with indeterminate cleavage are called **deuterostomes**. Like that of most groups of organisms, the current classification of animals is constantly changing, as new characteristics are discovered and analyzed. For example, the use of DNA analysis has led to recent advances in genetic research.

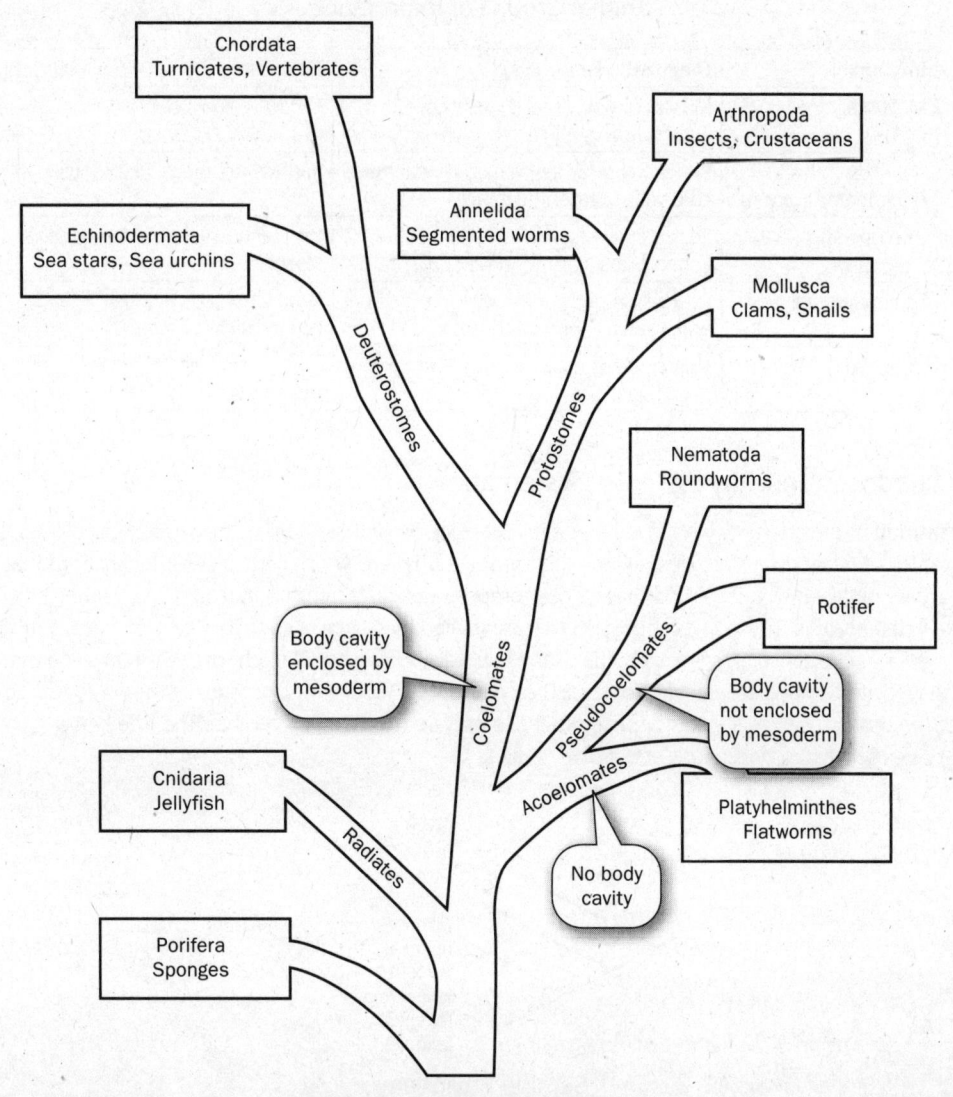

Phylogeny of Major Animal Groups

Use the next table to review some of the diagnostic features of some of the more important animal groups.

Animal Groups of Importance

Phylum	Diagnostic Feature(s)	Examples
Mollusca	Muscular foot, visceral mass containing gut, mantle (shell)	Octopus, snail, clam
Annelida (segmented worms)	Repeating body segments, ventral nerve cord with segmented ganglia	Earthworm, leech, sea mouse
Arthropoda	Segmented exoskeleton and appendages, compound eyes	Lobster, horseshoe crab, spider, insects
Echinodermata	Radial symmetry, water vascular system, endoskeleton of calcerous plates	Sea star, sea urchin, sea cucumber
Chordata	Notochord; dorsal, hollow nerve cord, pharyngeal slits	Fish, amphibians, reptiles, mammals

Relatedness and Phylogenetic Systematics

Systematists are scientists who study and formulate classifications as well as the relationships among **taxa**. These relationships are illustrated in the form of a **phylogeny** (branching evolutionary tree) that is usually obtained by analyzing data matrices composed of characteristics that are shared among the taxa in the analysis. These trees show taxa that are related by sharing a common evolved characteristic or perhaps several characteristics. On the exam, you will likely be asked to choose which taxa are most closely related among several pairs of taxa. The organisms on the same branches of the phylogenetic tree are more closely related, as opposed to those that are closest to each other. The tree below illustrates some misconceptions about phylogenies.

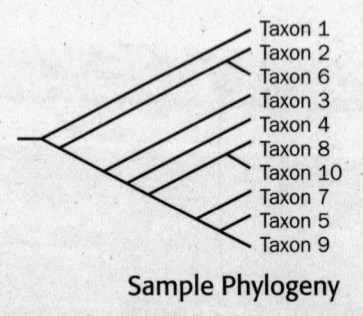

Taxon 1
Taxon 2
Taxon 6
Taxon 3
Taxon 4
Taxon 8
Taxon 10
Taxon 7
Taxon 5
Taxon 9

Sample Phylogeny

This hypothetical phylogeny shows the evolutionary relationships among ten taxa. A phylogenetic tree does not show a timescale. Taxon 7 shares a common ancestor, which isn't shown on the tree, with taxa 5 and 9. Taxon 7 is not the ancestor of taxa 5 and 9, nor does it necessarily have more characteristics in common with the ancestral taxon of taxa 7, 5, and 9 than taxa 5 or 9 do. It could have just as many **anagenetic** (evolutionary changes without speciation) changes along its branch as **cladogenetic** (evolutionary events that lead to speciation) changes that lead to taxon 10. Taxon 4 is more closely related to taxa 8, 10, 7, 5, and 9 than it is to taxon 3. Even though taxon 3 is close to taxon 4 on the tree, it is not on the same branch that includes taxa 4, 8, 10, 7, 5, and 9. Likewise, taxon 2 is more closely related to taxon 6 than it is to taxon 1. Taxa 2 and 6 belong to the same branch that does not include taxon 1.

IF YOU LEARNED ONLY FOUR THINGS IN THIS CHAPTER....

1. Scientists use the Linnaean system of binomial nomenclature to identify organisms based on their relationships to one another.

2. The divisions of classification are, from largest to smallest: domain, kingdom, phylum, class, order, family, genus, species. Organisms that share a phylum are more closely related than those sharing only a kingdom, and so on. Commonly, organisms are identified by genus and species (e.g., *Felix domesticus*).

3. Domains include Archae, Bacteria, and Eukarya (eukaryotes). Kingdoms within the Eukarya include Plantae (plants), Animalia (animals), Fungi, Monera, and Protoctista.

4. A phylogeny is a diagram indicating evolutionary relationships between specimens. It depicts an ancestral species with other species branching off from it. More closely related taxa are nearer to one another in the diagram. The diagram shows anagenetic evolutionary changes (which do not lead to speciation), as well as cladogenetic evolutionary changes (which do).

APPLYING THE CONCEPTS: QUESTIONS AND EXPLANATIONS

1. The Cnidaria are unique among animals because they possess which of the following characteristics?

    (A) Radial symmetry
    (B) Tentacles with stinging cells
    (C) Water vascular system
    (D) Muscular foot
    (E) Ventral nerve cord

2. Which group of organisms is the most prolific of all the organisms that have ever existed on Earth, with over a million existing species known and perhaps several million more to discover?

    (A) Bacteria
    (B) Vertebrates
    (C) Fungi
    (D) Arthropods
    (E) Plants

3. A phylogenetic tree constructed from data about characteristics of taxa provides all of the following information EXCEPT what?

    (A) The time of divergence between two taxa
    (B) A framework with which to classify organisms
    (C) A hypothesis of the way characters evolved among the taxa
    (D) A hypothesis of whether similar characteristics have a common origin (homologous) or are convergent (analogous)
    (E) An objective way to estimate the evolutionary relationships among the taxa

4. Based on the current scientific consensus of evolutionary relationships among living organisms, which of the following pairs of taxa share the closest relationship with each other?

    (A) Mushroom and moss
    (B) Sea star and human
    (C) Bacterium and amoeba
    (D) Snail and jelly
    (E) Leech and lamprey

5. The prokaryotic group Archaea differs from the other prokaryotes by being able to withstand extreme environmental conditions such as

    (A) high pressure
    (B) low pH
    (C) high O_2 concentration
    (D) high altitude
    (E) high temperature

6. This heterotrophic group of organisms can be microscopic and has chitinous cell walls.

    (A) Actinopoda (radiolarians)
    (B) Ciliophora (ciliates)
    (C) Chlorophyta (green algae)
    (D) Fungi
    (E) Cyanobacteria (blue-green algae)

7. A scientist returns from an expedition in the Sargasso Sea with a peculiar organism that is apparently unknown to science, but has a water vascular system. The scientist can begin to classify the organism by placing it in which of the following groups?

    (A) Annelida
    (B) Arthropoda
    (C) Echinodermata
    (D) Cnidaria
    (E) Mollusca

8. Linnaeus provided the modern synthesis of classification with his consistent application of a system of

    (A) binomial nomenclature
    (B) cladistics
    (C) phylogenetic systematics
    (D) natural selection
    (E) evolutionary hypotheses

9. Which of the following terrestrial plant groups fulfills a life cycle in which a haploid gametophyte is the dominant generation?

    (A) Conifers
    (B) Mosses
    (C) Cycads
    (D) Ferns
    (E) Grasses

10. All of the following groups of organisms are members of the Protoctista EXCEPT which?

    (A) Diatoms
    (B) Amoebas
    (C) Rotifers
    (D) Paramecia
    (E) Euglenoids

11. Organisms classified in which category show radial cleavage during early development and have a dorsal nerve cord?

    (A) Protostomes
    (B) Deuterostomes
    (C) Radial symmetry
    (D) Bilateral symmetry
    (E) Coelom

12. Which organism has radial symmetry?

    (A) Frog
    (B) Sea urchin
    (C) Rabbit
    (D) Kangaroo
    (E) Crab

Questions 13–15 refer to the following list.

    (A) Endosperm
    (B) Endoderm
    (C) Gymnosperm
    (D) Angiosperm
    (E) Tracheids

13. This group of plants includes the pine.

14. This tissue nurtures the developing plant zygote.

15. These cells produce tubes and fibers which function in fluid transport in plants.

Questions 16–20 are based on the following phylogenetic tree. Pick the letter on the branch that matches the most likely placement of the taxon.

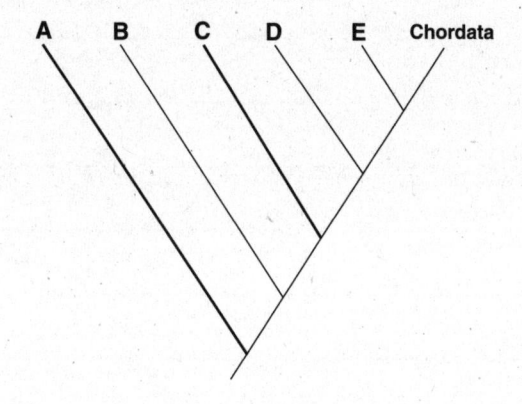

```
   A   B   C   D   E  Chordata
```

16. Echinodermata

17. Archaea

18. Protoctista

19. Plantae

20. Fungi

Free-Response Question

An incredibly diverse biota exists today, with species exhibiting many similar and different features when compared to each other. Evolutionary relationships have always been inferred from observations of existing organisms and the diagnostic characteristics they possess.

(a) **Describe** and **discuss** two diagnostic features and their adaptive significance for any two existing phyla.

(b) **Discuss** the similarities and differences between the phyla you chose and the evidence these characteristics lend toward support of their evolutionary relationship.

ANSWERS AND EXPLANATIONS

1. B

Your first impulse might be to choose (A) since one of the features that separate taxa at the base of the tree of Animalia is general body plan, but remember that echinoderms also have radial symmetry and aren't closely related to Cnidaria. Scientists often use a diagnostic feature to name a group of organisms, which is true in this case. Cnidaria are named for their "cnide" (Greek for *nettle*), the stinging tentacles they possess.

2. D

Both bacteria and fungi have many species with many new ones to discover, but whenever you encounter a question asking about the group with "over a million species," "the most diversity," "the most diverse group," etc., think either arthropods or insects. There are more species of arthropods than all the rest of the species combined. The more familiar groups (vertebrates and plants) do not compare in terms of numbers of species.

3. A

The question says that the phylogenetic tree was constructed from character data, which is important for developing hypotheses of character evolution (C). It also means that the tree could be constructed based on more objective methodologies and provides an objective way to compare different individual hypotheses of characteristics from total evidence [(D) and (E)]. Not all scientists use phylogenetic information to classify organisms, however. A more pragmatic approach based on similarity (e.g., most people refuse to call birds reptiles) is sometimes used, but in most cases, a phylogenetic tree is a good basis for classification (B). The correct answer is (A). A tree can show relative evolutionary positions of taxa, but does not identify exactly when the evolutionary events happened. Even a strong fossil record can only suggest minimum and maximum ages of divergence, not exact dates.

4. B

According to evolutionary theory, all of these taxa are related at some point on the Tree of Life. To answer this question correctly you need to choose the pair of taxa that are highest on the same branch of the tree. Another way to look at this question is to find the two taxa that are in the most inclusive taxonomic group. Two species in the same genus would be very closely related. The correct answer is (B). Chordates and echinoderms both share the characteristic of being deuterostomes among all of Animalia. Mushrooms are not even in the same kingdom as mosses (A). Bacteria and amoebas are both microscopic, but are not in the same domain (C). Snails are protostomal coelomates with bilateral symmetry and jellies are acoelomates with radial symmetry (D). Both leeches and lampreys latch onto other organisms and suck their blood, but a leech is a segmented worm and a lamprey is a vertebrate (E).

5. E

There are three groups of Archaea: methanogens (they produce methane), halophiles [they live in environments with very high concentrations of salt (NaCl)], and thermophiles (they live in environments with very high temperatures). The characteristic that all three groups share is the ability to exist in extreme environments. From the list of possible answers, the only one that is applicable is high temperature (E).

6. D

The term heterotrophic refers to an organism's need to obtain energy from sources other than itself. In other words, the organism isn't photosynthetic. Animals are heterotrophs because they eat other organisms. The correct answer is not one of the photosynthetic organisms in the list [(C) or (E)], since these organisms are autotrophic. Actinopoda and Ciolophora are heterotrophic, but Fungi are organisms that are characterized as having cells with chitinous cell walls (D).

7. C

This question is rather straightforward if you remember which group of organisms has a water vascular system. Unfortunately, of the possible answers, all can be found in marine ecosystems so you cannot rule out possibilities based on ecology. One way to remember this system is to think of the movement of sea stars. Sea stars move by pumping fluid in and out of their tubelike feet. The correct answer is (C), since sea stars belong in the group Echinodermata.

8. A

All of the possible answers are actual terms used in classification, but (B), (C), and (D) are methodologies and theories that postdate Linnaeus. Although the concept of changing forms was around at the time of Linnaeus, he did not believe in evolution and thought that species were immutable. Linnaeus is credited for providing a framework of consistent classification that was based on a binomen composed of a genus and species name.

9. B

To get this answer correct, think of the most primitive plants. In most terrestrial plants, it is the sporophyte that is the dominant form and this is considered an "advanced" characteristic for survival on land. The primitive terrestrial plants are the ones that still retain the gametophyte as the dominant form. Of the possible answers, mosses are the most primitive.

10. C

With a question like this, if you don't know the answer right away, you are left with little recourse but to eliminate the possible answers you are sure are incorrect and then guess which answer is correct. The obvious incorrect answers are amoebas, paramecia, and euglenoids because these are common protozoa examined under the microscope in pond samples. Diatoms are very small protoctists, often with beautiful shells that look sculpted from glass. Rotifers are multicellular pseudocoelomates closely related to roundworms, so they do not belong in kingdom Protoctista.

11. B

Organisms classified as deuterostomes show radial cleavage during early development and have a dorsal nerve cord.

12. B

The sea urchin has radial symmetry, while the frog, rabbit, kangaroo, and crab have bilateral symmetry. In radial symmetry, lines of symmetry meet at a point at the center of the organism. In bilateral symmetry, an organism has two equal parts; a plane divides the body into two halves which are mirror images of each other.

13. C

Gymnosperms is the group of plants which includes the pine.

14. A

The endosperm is the tissue which nurtures the developing plant zygote.

15. E

Tracheid cells in plants produce tubes and fibers which function in fluid transport.

16. E

You need to approach this question by first looking for the organism(s) that is/are most closely related to the Chordata. Also note that all of the questions ask for mutually exclusive categories, so you can solve this cluster using process of elimination. After reading this chapter, you can discern that echinoderms, despite the way they look, are more closely related to chordates (the group that contains humans) than any other animal group. Since there are no other animals in the list, this group should occupy the place on the tree closest to the chordates. The correct answer is (E). After obtaining this answer, the best course of action would be to find the next closest relative to the collective grouping that includes the echinoderms and the chordates. The next closest relative would be one of the "higher" eukaryotes, the Fungi.

17. A

Choice (A) makes sense as a possible taxon to be at the base of any phylogenetic tree.

18. B

Now your goal is to determine which, of the remaining two listed taxa, is the one most closely related to the "higher" eukaryotes. Since the Archaea are prokaryotes, the obvious choice is the Protoctista.

19. C

Among the remaining answers, there is only one "higher" eukaryote left, the Plantae.

20. D

The only remaining taxon is Fungi. You should be able to approach any question like the one above with the same methodology. First, look for the taxon that is most closely related to the one listed, then work your way backward. Be careful to make sure the taxa are mutually exclusive.

Free-Response Answer Explanation

You get to choose the phyla to discuss in your response to this question, which gives you a great deal of freedom in choosing how to answer the question. The question allows you to pick two phyla with which you are most familiar, instead of needing to come up with information about organisms the College Board has picked. You may very well see a question on the exam in this format, but it may ask you about kingdoms, classes, or orders.

If possible, you should pick two phyla in the same kingdom, as they are more likely to provide a longer list of characteristics to discuss. Alternatively, you could pick two phyla that are similar in structure, behavior, ecology, and such, but in different kingdoms, comparing, for example, the photosynthetic properties of Chlorophyta with Bryophyta. Regardless of your decision, you should explicitly answer the questions asked and include extra information that is specific to the question, including names of any inclusive groups that are representative of the diagnostic features you discuss. After picking two diagnostic features for each group, be sure to discuss how those features probably led to the evolutionary divergence of the group. When discussing part (b), see if you can include terms such as homologous, analogous, convergence, or coevolution, for example.

Here is a possible response:

(a) The phylum Arthropoda is characterized by an exoskeleton composed of hardened cuticle made of chitin. Because of this armored coating, arthropods have jointed appendages, another diagnostic feature of the group. Both of these morphological attributes have contributed to the prolific radiation of species that exist in both marine and terrestrial biomes. The Crustacea are very diverse in both marine and freshwater ecosystems and are able to withstand water pressure because of their strong exoskeleton. Their jointed appendages provide several modes of locomotion including walking legs and swimming tails. On land, insects are the most successful group. The cuticle of terrestrial arthropods is resistant to desiccation, allowing them to remain small with limited water loss. The mechanical advantage of the muscular attachments to their jointed appendages gives terrestrial arthropods disproportionate strength and endurance, including the adaptation for flight with appendages that do not derive from other limbs.

The phylum Annelida is characterized by having a multi-segmented body with repeating homologous units (metamerism). Each unit has its own nervous tissue in the form of a ganglion and excretory organs called nephridia. One of the evolutionary advantages of this segmentation is the ability to bend at extreme angles. This allows the animal to occupy extreme spaces and multiple forms of mobility (e.g., swimming or burrowing) without the need of complex appendages. Some annelids also can survive with a number of their segments missing, as the earthworm does when it is cut in half. Annelids possess specialized structures along their lateral surface called chaetae. These structures can be simple spines used for locomotion while burrowing, or expansive fans in some of the marine polychaetes that increase surface area for gas exchange.

(b) Annelids were considered for a long time to be the evolutionary ancestors to the arthropods because some of the segmentation in annelids is retained in arthropods. The arthropods combined several of these segments into body regions, but the segmentation is evident in the millipedes and centipedes as well as in the abdominal regions of some arachnids and insects. The chaetae on annelids are also derived from chitinous proteins, which are similar in composition to the exoskeleton of arthropods. Arthropods show the distinct characteristic of compound eyes, which seems to be a novel structure in their evolution, as annelids do not have compound eyes and only some species have simple eyes. Arthropods also have taken segmented appendages to an extreme, with any number of pairs possible along similar body segments or differentiated body regions. Annelids perform gas exchange across a constantly moist body surface, so are limited to certain habitats. Arthropods exchange gases through small spiracles, which limits water loss, or with lungs in aquatic systems. These adaptations have allowed arthropods to colonize almost every ecosystem that exists on Earth.

Chapter 10: Structure and Function of Plants and Animals

- Amazing Biological Machines: Physiological Processes and Responses
- Transpiration Laboratory
- Physiology of the Circulatory System Laboratory
- Applying the Concepts: Questions and Explanations

AMAZING BIOLOGICAL MACHINES: PHYSIOLOGICAL PROCESSES AND RESPONSES

It is a wonder that organisms function at all. Most organisms are complex machines with many intricate parts and processes. It is perhaps easier to fathom the workings of a simple prokaryote cell—coordinating activity with other organisms and the liquid matrix in which it lives—than to study complex organisms. The synchronicity of the cells, tissues, and organs in a multicellular organism involves a highly complex set of processes that are not easy to simplify. All physiological processes have an underlying means of chemical or electrochemical control, which will be reviewed in this chapter as a foundation for further investigation into specific systems in different groups of organisms. It is important to learn the processes covered in this chapter within the context of their evolutionary significance and their interactions with other body systems and other organisms. In chapter 8, fitness was a measure of reproductive success. Reproduction will continue as a dominant theme in this chapter, along with the processes involved in early development.

This is the most fact-filled section of the entire course. The last chapter detailed the diversity of life and the likelihood that millions of species still remain to be discovered. Every species has its own unique genetic identity and its own unique biology. Every species has a set of structural, behavioral, and physiological adaptations that contributed to the successful evolution of the organism through the process of natural selection. It would be impossible for any one person to know the adaptations and features of every organism on Earth. Due to the amount of information that questions are gleaned from for the exam, synthesis is even more crucial to successfully reviewing these concepts. In addition to synthesis of information, the College Board most strongly emphasizes form and function in the course outline and for this topic on the exam.

Physiological Response Mechanisms

There is no cellular specialization in single-celled organisms because they are so small and are only composed of one cell. The entire organism lives within a homogenous environment with a uniform set of environmental cues for behavior and physiological processes. As multicellular organisms evolved, there was an associated increase in body size as well as a specialization of structures that performed different functions in different parts of the organism with varying environmental stimuli. The key to coordinating these structures was the evolution of successful communication from one area of the organism to another. Messages between structures in an organism had to be mobile and somewhat specific. **Organic molecules** are ideally suited for this task because they are relatively small, can move fairly freely, and can be highly specific. An organism could produce a different chemical for each message it needed to send as long as there was a way for the message to be received on the other end. This is a description of the **hormone-receptor system**.

Hormones are defined as chemical signals that:
- are synthesized by specialized cells,
- travel throughout a multicellular organism by some kind of bodily fluid, and
- coordinate systemic (total body) responses by activating specific **receptor cells**.

For example, if an animal were injured it would release the hormone epinephrine (adrenalin) from the adrenal glands. The epinephrine would travel in the bloodstream throughout the entire body. Any cell in the body that is accessible by blood flow and has **receptor proteins** for epinephrine responds to the "message" of the injury. Blood vessels respond by constricting blood flow, the heart and lungs increase their rate of activity, and the liver releases sugar to the bloodstream. The key issues are the widespread release of hormones, specificity controlled by the receptor's ability to recognize the signal, and the location of the cells that contain the receptor. Some specific hormones, their target tissues, and their activity are discussed later in the chapter.

Hormonal responses take time and are indirect. This is a reasonable form of communication for a plant that doesn't move, but for a bird in flight, a fish swimming from a predator, or a gibbon brachiating through the trees, a quicker, more intricately coordinated form of communication inside the body is necessary. Coordinating environmental signals with movement responses and complex processing is the job of the **central nervous system**. The **endocrine (hormone) system** interacts with the **nervous system**, but the mechanism by which nerves communicate messages is entirely different.

Nerves are like electrical wiring throughout the body, connecting stimuli from both the external and internal environment with the central processing units, the brain and spinal cord (if the animal has them). A **neuron** (nerve cell) is made up of a **soma** (cell body), **dendrites**, and an **axon**. Neurons are closely associated with each other at gaps called **synapses** or **synaptic junctions**. Electrochemical signals travel down the length of the nerve cell in a process called an **action potential**, and it is the action potential that is the basis for all nervous activity.

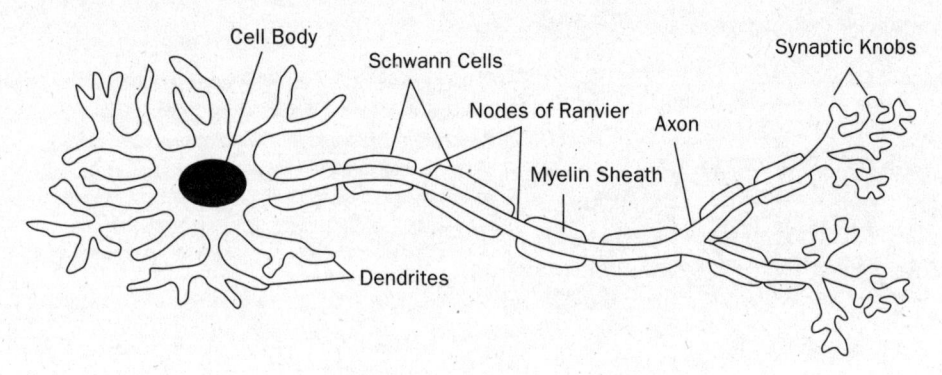

The Neuron and Its Components

An action potential is a temporary discharging of the battery power stored in neurons. It is also the up-and-down signal transmitted on a heart monitor that is commonly seen in movies and on television shows. Almost all cells in the body are constantly pumping Na^+ ions out and K^+ ions in, creating an electrical potential between the more positively charged outside matrix and the more negatively charged cytoplasm. The **resting membrane potential** of most neurons is about −70mV (the minus sign indicates a negative charge). When some kind of stimulus depolarizes the membrane potential to a specific, predetermined threshold, a gate opens and lets all of the Na^+ into the cell, down the **electrochemical gradient** (a combination of the attraction of unlike charges and the diffusion gradient). After the neuron completes its depolarization, the Na gates close and the K gates open, reestablishing the resting membrane potential. The **action potential** takes place locally across the cell membrane, but propagates down the axon as the original depolarization causes Na gates in an adjacent part of the membrane to depolarize as well. The depolarization travels down the length of the axon like the current in an electrical wire.

Outside Cell: Na⁺ > K⁺

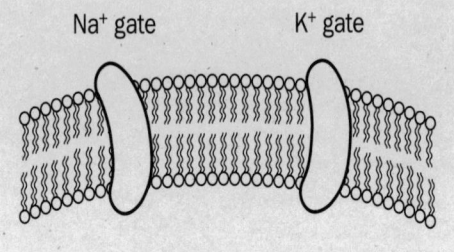

Na⁺ gate   K⁺ gate

Inside Cell: K⁺ > Na⁺

1) Resting Potential

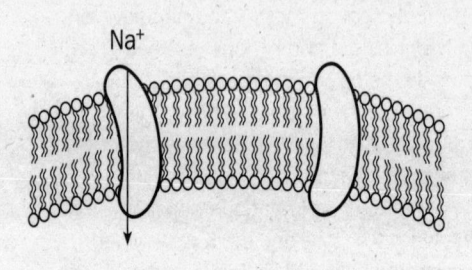

Na⁺

2) Depolarization, Na⁺ Gate Open

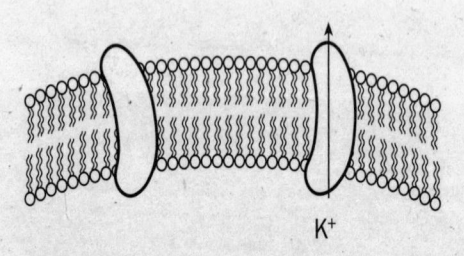

K⁺

3) Repolarization, K⁺ Gate Open

1) The neuron is at rest, with a higher concentration of sodium outside the cell and potassium inside. 2) The membrane is depolarized and the Na+ gate is open, allowing sodium to enter the cell as indicated by the increasingly positive charge on the graph to the right. 3) The K+ gate is open and potassium is leaving the cell, bringing the membrane's charge back to its original level as indicated on the graph.

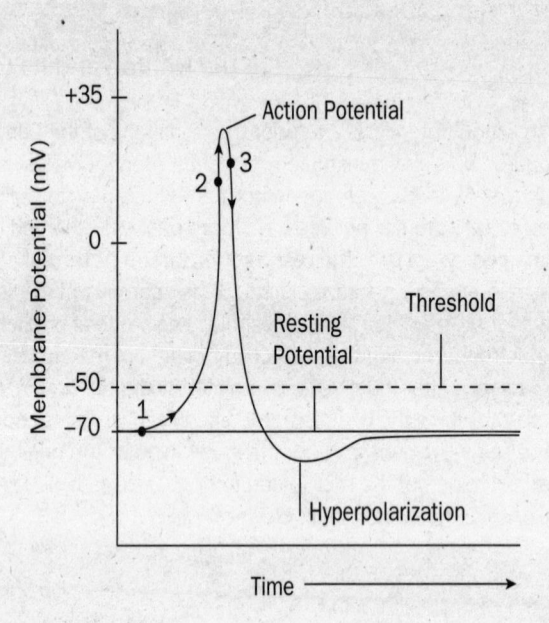

Action Potential

The depolarization that occurs across the membrane of a neuron is controlled by special gates that control the diffusion of Na^+ and K^+ ions. The previous figure shows the timing of the ion gates during the course of an action potential.

When the action potential reaches the end of the cell, it initiates the release of special chemical messengers called **neurotransmitters** that travel across the synaptic gap between two neurons. When the neurotransmitter binds to receptor proteins on the other neuron, it causes an action potential in that neuron. The process repeats along several neurons until the original signal reaches a processing area where a signal is often sent back to the point of origin, causing some kind of reaction. To enhance conductivity, some neurons are surrounded by an insulating sheath called **myelin**, which speeds up the movement of the action potential. Action potentials can travel very quickly through a neuron and are directed down the length of an axon to a specific location, like the soma of another neuron, for instance.

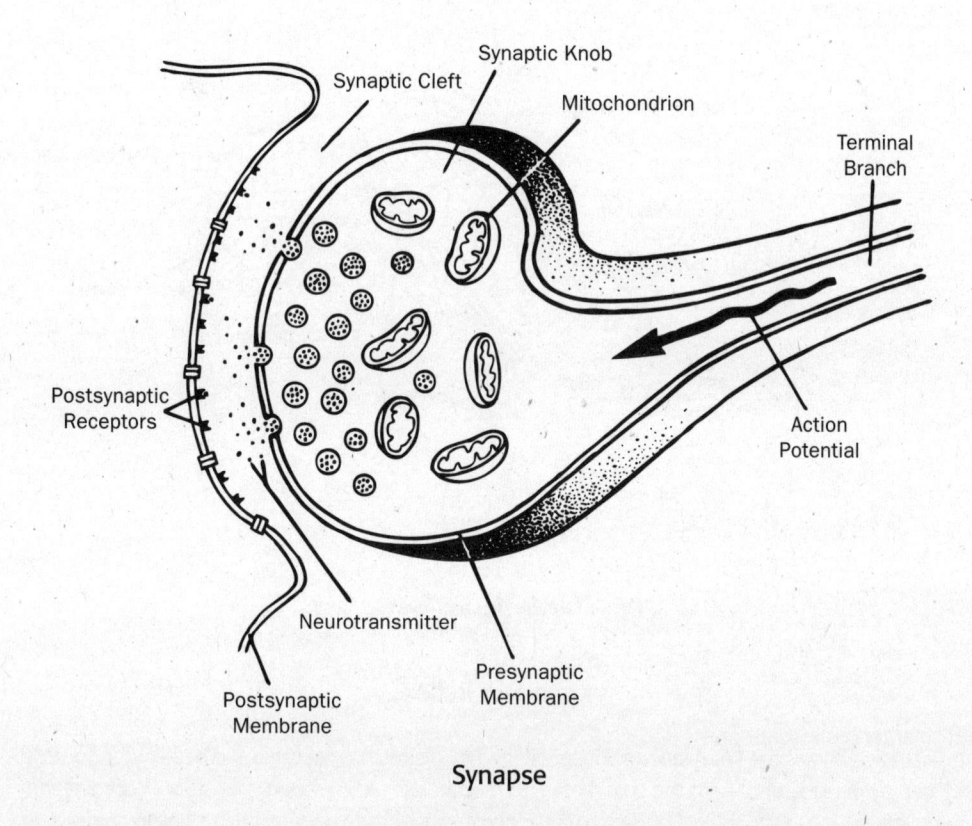

Synapse

The previous image shows a chemical signal sent across a synaptic gap. Neurotransmitters are released from vesicles in the presynaptic neuron (upstream neuron) and bind to receptors on the membrane of the postsynaptic neuron (downstream neuron). The binding of the neurotransmitters on the postsynaptic neuron is the stimulus for another action potential.

A stretch receptor in the anterior muscle of the thigh (quadriceps) provides a good example of how information is sent from one part of the body to another. Stimulus is provided when a doctor softly hammers a specific point below a person's kneecap, causing a special kind of cell called a stretch receptor to lengthen and initiate an action potential in a sensory neuron. The sensory neuron propagates its action potential to the spinal cord, where the action potential synapses with a motor neuron for the quadriceps muscles. An interneuron then synapses with a motor neuron for the knee flexors (hamstrings), initiating action potentials in all of these neurons. The signals sent back to the leg cause muscles of the quadriceps to contract quickly and the muscles of the hamstrings to relax. The result is a quick extension of the knee joint. The reflex response mediated through the spinal cord happens so fast that the brain, which is farther away, doesn't even know that the reflex response occurred until afterward.

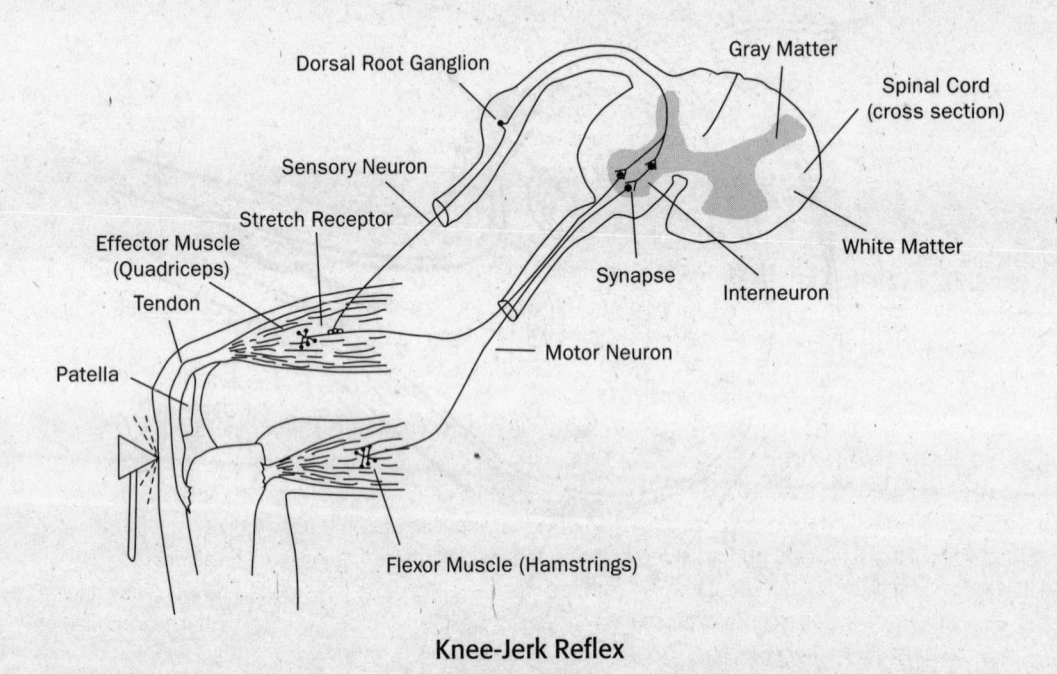

Knee-Jerk Reflex

The illustration shows the knee-jerk response initiated by a stretch receptor in the patellar tendon and mediated by direct control from the spinal cord. Remember, the knee-jerk response is a coordination of two different muscular controlling signals; one signal causes the quadriceps muscles to contract, the other causes the knee flexor muscles to relax.

Plant Structure and Systems

Terrestrial plants left an aquatic environment during their evolution, losing, in the process, some of the benefits that a liquid medium had previously provided. **Cell walls**, which provided simple rigidity for structures in small plants, became aligned to allow for the growth of trees over 100 meters in height and to create channels for the delivery of important soil nutrients. Water loss became a constant challenge in environments where rainfall could be unpredictable or nearly nonexistent. The sun, formerly relied upon solely for light-giving energy, became a source of dehydration and overheating. The terrestrial environment also provided a great degree of variation, in contrast to the constant temperatures and cycles of aquatic systems. Seasonal changes and the variability of weather required plants to synthesize new chemical messengers that would allow plant tissues to respond to both acute and chronic environmental cues. **Novel structures**, primarily for reproduction, required new forms of control mechanisms to function in time with a changing environment.

One of a plant's many challenges is providing all of its structures with the water needed to complete photosynthesis and to maintain an aqueous solution for all biological reactions. Water is heavy and a plant cannot rely on muscular contractions to move materials, as animals do. Instead, a plant has narrow, lifeless channels in its **xylem** tissue that take advantage of the cohesive properties of water. A plant loses the majority of its water during **transpiration** while its **stomata** are open for the exchange of CO_2 and O_2 in **photosynthesis**. As water is lost through the leaves, it creates a negative pressure, just like sucking on a straw. The cohesive force of water keeps a steady flow of water connected to the negative pressure developed by transpiration occurring in the leaves. The force exerted in the water column due to cohesion is as strong as steel wire of the same diameter. The plant also uses water flow to deliver minerals that have been actively taken up from the soil by the roots.

Plants need a mechanism to deliver the stored energy created during photosynthesis to all of their structures. While water travels upward from the roots through the xylem via the pull of transpiration and the cohesion of water, nutrients flow both up and down through the **phloem** via the pull of **osmotic pressure**. **Osmotic pressure** is built up between areas that have high concentrations of nutrients and those that have low concentrations. **Translocation** of materials in the phloem is like long-range diffusion. Similar to the cohesion of water, lower solute pressure on one end of a vessel of phloem is translated along narrow vessels to an area where the solute pressure is greater. The adaptive significance of these transport systems is the colonization of a terrestrial environment. Plants also harnessed a physical disadvantage, water loss, and turned it into a benefit, the force powering water transport through xylem.

Terrestrial plants were able to grow much taller than aquatic plants, since xylem and phloem provided structure to the plant. Plant height necessitated transport of information throughout the plant. Plants transport energy through chemical messengers, organic molecules usually in the form of proteins. There are a variety of plant compounds that act on different plant tissues to elicit general responses. For example, a plant can respond to the direction of light by turning its leaves to maximize incident solar radiation, a process called **phototropism**. Plants also respond to the pull of Earth's gravity by orienting their growth perpendicular to the force of gravity. This is called **gravitropism**. Plants can change which of their parts grow faster and when, can cause the complete **senescence** of all of their leaves in the fall, and create the branching pattern of their stems. The next table lists some of the more important plant nutrients and plant messengers, as well as their effects on plant tissues.

Plant Chemical Compounds

Compound	Type	Origin	Action
Phytochrome	Photopigment	Systemic	Detection of light to control photoperiodism
Potassium	Nutrient	Root uptake in soil	Protein synthesis, operation of stomata
Calcium	Nutrient	Root uptake in soil	Cell wall stability, enzyme activation
Magnesium	Nutrient	Root uptake in soil	Chlorophyll synthesis, enzyme activation
Phosphorous	Nutrient	Root uptake in soil	Nucleic acid and ATP synthesis
Abscisic acid	Hormone	Leaves, stems, green fruit	Stomata closing (water stress), growth inhibition
Auxins	Hormone	Seed, meristems	Stem elongation, tissue differentiation and branching, apical dominance, photo- and gravitropism
Cytokinins	Hormone	Roots	Root growth and differentiation, germination and flowering
Ethylene	Hormone	Older leaves, stems, fruit	Fruit ripening, growth inhibition
Gibberellins	Hormone	Meristems, embryo	Germination, flowering and fruit development

Animal Structure and Systems

Although some botanists and plant physiologists may disagree, animals are much more complex than plants in almost all respects. A vast amount of cell and tissue specialization and organ systems have evolved to support animal life in a variety of habitats. For example, animals in arctic climates have evolved to produce antifreeze compounds to prevent cellular crystallization and breakage; other animals have countercurrent gas exchange to support the high metabolism needed for flight, reduced body temperature in arid habitats, or complicated neural networks that allow abstract thought and high-level language. Many of these adaptations have separate origins (they are **convergent**). It would take a book thicker than this one to discuss the biology of just one family of animals (this is also true for any family of plants, by the way), let alone attempt to explain all of animal biology in one section of this chapter. There are, however, a few animal systems that are more prominently covered by the College Board. These include the **endocrine**, **respiratory**, **circulatory**, and **digestive systems**. Some key points for each system will be covered below. The **reproductive system** will be covered separately.

Before embarking on the study of selected animal systems, general terminology for bodily systems needs to be understood. In animals, a collection of specialized cells performing the same or similar function(s) is called a **tissue**. There are **muscle tissues**, **epidermal tissues**, **nerve tissues**, and **connective tissues**. A collection of segregated tissues is called an **organ**. Organs are such structures as the heart, lungs, and brain. The hair and skin covering an animal are considered to be organs. Organs work together to perform regulatory functions within an animal's body. A collection of organs is called an **organ system** or **system** for short. The skin, hair, nails, sebaceous glands, and sweat glands on an animal's body are collectively referred to as the **integumentary system**. The endoskeleton of vertebrates is called the **skeletal system**.

Animals have a myriad of chemical messengers as part of their endocrine systems, with the associated structures to produce these messengers. The College Board frequently asks questions about some of the more common hormones, which are reviewed in the next table. Keep in mind that hormones are delivered systemically and the specific action of each hormone is controlled by specialized receptors in different tissues. This allows systemic responses that are less finely tuned than nervous responses.

Important Hormones in Vertebrates

Compound	Type	Origin	Action
Androgens	Steroid	Testis	Secondary sex characteristics
Epinephrine (adrenalin)	Catecholamine	Adrenal gland	Fight-or-flight response
Estrogens	Steroid	Ovary	Secondary sex characteristics
Follicle-stimulating hormone (FSH)	Glycoprotein	Anterior pituitary	Regulate gonads
Glucagon	Polypeptide	Pancreas	Increase blood glucose
Insulin	Polypeptide	Pancreas	Decrease blood glucose
Melatonin	Catecholamine	Pineal gland	Circadian rhythm
T_3 and T_4	Amino acids	Thyroid	Stabilize metabolic rate
Thyroid-stimulating hormone (TSH)	Glycoprotein	Anterior pituitary	Regulate thyroid

Animals left an aquatic environment and their body sizes increased as terrestrial life diversified. In the past animals could simply exchange O_2 and CO_2 required and produced from respiration across their skin, but adaptation to larger bodies and life on land necessitated more efficient means of moving gases across specialized tissues in respiratory organs. **Gas exchange** requires that molecules be dissolved in liquid. When animals moved onto land, they had to adapt with structures that worked internally, so that moist membranes could be maintained without excessive water loss. **Gills** exposed to the outside environment were replaced by **tracheal openings** in smaller animals and **lungs** in larger animals.

Oxygen is transported in vertebrate animals attached to hemoglobin molecules in red blood cells. **Hemoglobin** is a molecule with a high affinity for O_2, meaning it takes very low partial pressures of O_2 to cause it to disassociate from hemoglobin molecules. When blood travels across the respiratory tissues, it passes an area of high partial pressure of O_2 and a low partial pressure of CO_2. The blood picks up O_2 molecules and releases CO_2 molecules. When the blood comes in contact with tissues that are low in O_2 and high in CO_2 (due to metabolism and cellular respiration) it releases O_2 and picks up CO_2. As long as there is circulation, gas is delivered where it is needed. Smaller terrestrial animals like some invertebrates still rely on diffusion of gases once they enter the body, a severe limitation to body size. While O_2 rides along with hemoglobin, CO_2 molecules travel as bicarbonate ions because free CO_2 would cause too much of a reduction in blood pH.

After an animal takes care of its body regulation with hormones and nervous coordination and supplies its cells with the gases necessary for cellular respiration, the next most immediate concern is getting nutrients. **Nutrients** include not only the energy required to fuel cellular activity, but the **micronutrients** (**vitamins** and **minerals**) necessary for chemical function, and, most importantly, **water**. The specifics of the digestive and excretory systems cannot be covered in this section. Instead, it is important to synthesize the information into an appreciation for how life has evolved to produce these systems to perform tasks and relate form to function.

First of all, as in respiration, animals gradually evolved from single-celled organisms that exchanged molecules with the outside environment across their entire bodies to complex organisms with tissue and organ specialization. Even though the exchange of molecules occurs inside the body cavity of these organisms, the actual surface where this exchange takes place in the respiratory and digestive systems is the interface between the inside and outside of the animal. This means that the contents of the stomach, although inside the abdomen of an animal, are actually part of the outside environment (think of the entire digestive system as being a hole that passes through an organism). The internal compartments of the alveoli are also part of the outside environment. These systems occur externally in more primitive animals.

The digestive tract is simply a long tube that travels through the animal's body allowing the exchange of nutrients. The lumen (the liquid matrix inside the tube) carries particles along like a conveyor belt, while the different tissues the tube passes pull in what nutrients they need and get rid of what they don't. By the time the digestive tract leaves the animal through the anus, the tissues have taken what they want and gotten rid of what they don't, leaving only waste. The different tissues of the digestive tract are localized into organs like the stomach and intestines that coordinate with each other for the uptake of carbohydrates, proteins, fats, vitamins, and any other nutrients the body needs. Just as with gas exchange and the circulatory system, the digestive tract simply provides a mode of transportation for these specialized tissues to transport goods. Again, don't think of the digestive tract as being inside the body. Think of it as an external surface that has been moved inside so it can imitate the liquid environment from which all animals originated.

Reproduction and Development

These concepts are mentioned separately from the other processes in this chapter because they are so important to the evolutionary process. If organisms didn't have a way to propagate there wouldn't be any species, nor would there be the great diversity of species seen today. Plants possess a reproductive advantage in that they can reproduce vegetatively while very few groups of animals can reproduce without some form of sexual reproduction. Plants also live dual lives in that they have specific structures for reproduction (flowers, pollen, and fruit), while the rest of the plant continues life as usual. Female animals' entire bodies are affected by the reproductive process so reproduction has to be an entire stage of existence for them. Some animals' adult life stage is solely for the purpose of reproduction; they don't even have mouthparts for the ingestion of nutrients. As seen in chapter 9, developmental strategies play an important part in defining groups of animals; some of these intricacies of vertebrate development will be discussed below. Concepts of vertebrate development seem to be included often on the exam.

Plants can **reproduce vegetatively** as a function of their **indeterminate growth**. Plant cells can differentiate when isolated from the rest of the plant, so one could place the stem of a rose in the soil and it would grow **adventitious roots** and become a whole plant. This by default produces a series of plants that are all genetically identical and reduces variation in the population, but is advantageous for success when **sexual reproduction** is not suitable for any given environmental situation.

Most terrestrial plants live their lives as **diploid sporophytes** and produce haploid tissue in the form of **ovules** and **pollen** (**gametophytes**), sometimes within the same flower. Pollen is delivered to the female gametophyte either by an animal, such as a bird, or by the wind, and fertilization takes place. The flower functions as the entire reproductive organ, completely dependent on the sporophyte plant. This is a departure from the aquatic condition in which spores travel independently from the plant that produced them. Flowering plants seem to have evolved a specialized region where a plant can be "pregnant." These "seed babies" are housed within a seed coat and often a fruiting body. The seed coat and fruiting body aid in plant propagation by either acting as bait to an animal or providing a source of nutrition to the developing plant embryo.

Animals reproduce sexually by a meeting of **gametes**, the male **sperm** and the female **eggs**. There are several strategies for uniting these cells, including both external and internal means and the use of a variety of direct and indirect intromittent organs. Animals can produce developing embryos by encasing them in a covering (egg) and placing them into the external environment, delivering underdeveloped young and protecting them in a pouch (as marsupials do), or delivering a completely developed juvenile offspring that can either provide for itself or survive with the behavioral support of its parents and other members of the population. Some invertebrate animals are **hermaphrodites**, organisms that can produce both egg and sperm, but the remainder of the animal kingdom is almost entirely made up of species that are one sex or another during their adult lifetime.

After fertilization, the **zygote** undergoes several fast **mitotic divisions** in stages called **cleavage**, **gastrulation**, and **organogenesis**. This process takes a very short time, but in the end, tissues of the developing animal have been predetermined to grow into layers called the **endoderm**, **mesoderm**, and **ectoderm**.

Cleavage

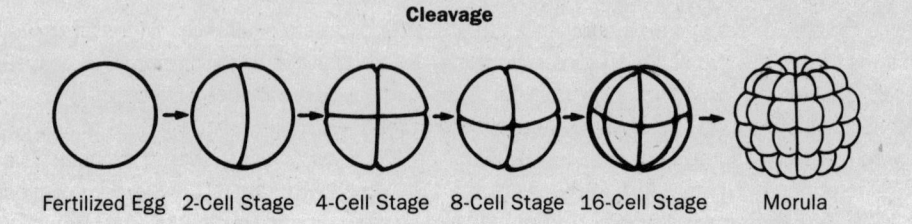

Fertilized Egg → 2-Cell Stage → 4-Cell Stage → 8-Cell Stage → 16-Cell Stage → Morula

Gastrulation

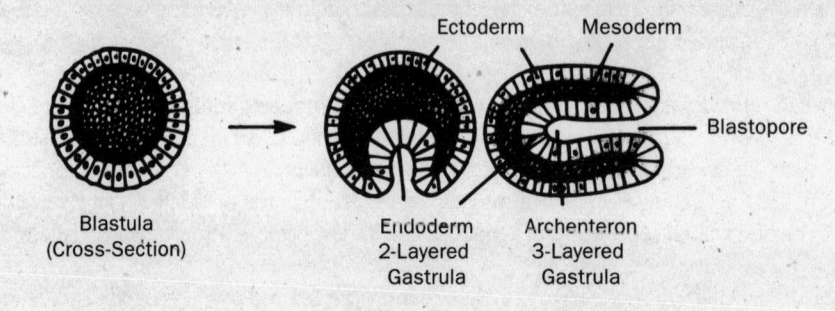

Blastula (Cross-Section) → Endoderm 2-Layered Gastrula → Archenteron 3-Layered Gastrula

Ectoderm, Mesoderm, Blastopore

Organogenesis

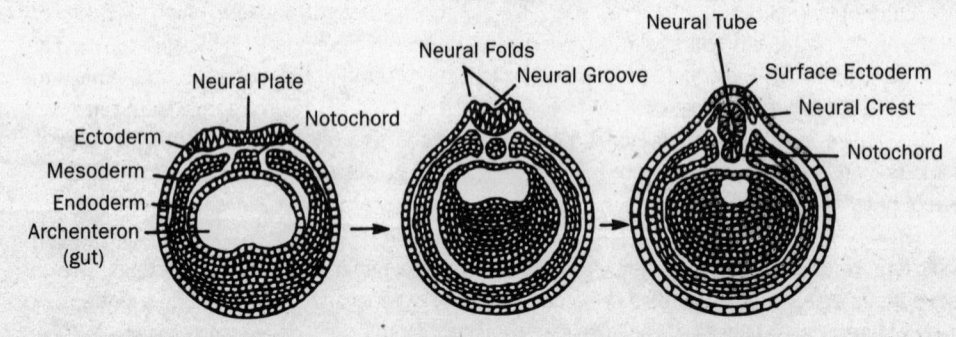

Ectoderm, Mesoderm, Endoderm, Archenteron (gut), Neural Plate, Notochord → Neural Folds, Neural Groove → Neural Tube, Surface Ectoderm, Neural Crest, Notochord

Early Animal Embryonic Development

The following table lists the tissue layer origins of some of the important vertebrate organs, rounding out the review of concepts in this chapter.

Tissue Layer Origins of Organs

Tissue Layer	Organs
Endoderm	Lining of organs of digestive tract, lungs, liver, pancreas, etc.
Mesoderm	Notochord, lining of body cavity, muscles, bones, circulatory system
Ectoderm	Integument and associated glands, nervous system

TRANSPIRATION LABORATORY

This lab is another example of a controlled experiment to evaluate the effects of environmental variables on the rate of transpiration for a plant. The lab is easy to complete and requires simple equipment. The objectives are threefold: another opportunity to explore the scientific method through controlled experimentation, a chance to make direct observations on the phenomena immediately related to the physical properties of water, and an occasion to observe the environment's effects on an actual organism.

A suitable plant species is obtained and set up with a potometer, a fancy name for a tube that measures the amount of water taken up by a plant during transpiration. The water-filled tube is connected to the bottom of a plant so that the water leaving the tube and entering the plant due to transpiration can be measured. The potometer tube takes the place of soil as a source for water. The plant is then subjected to a number of treatment effects that might include increased light, darkness, heat, or air movement. Changes in transpiration, if any, are compared to the rate of transpiration of a control plant that is under stable conditions, usually room temperature and room lighting. The control plant serves as the benchmark to which all other treatment effects are compared.

Transpiration is affected by environmental conditions in a way very similar to human skin. Increasing the amount of heat around skin causes increased sweat and air movement, which increases evaporation on the skin surface; the result is increased water loss. The same is true for the leaf surface of a plant—increased heat causes an increase in water loss. Light or darkness in the absence of changes in temperature only affects a leaf by modifying the rate of photosynthesis. Increasing the light intensity should increase the rate of photosynthesis and the opening of stomata, thereby increasing transpiration and water loss. As light levels decrease in intensity, water loss decreases. Water loss at the leaf surface causes greater uptake of water through the potometer, which can be measured as movement of the meniscus through the tube.

PHYSIOLOGY OF THE CIRCULATORY SYSTEM LABORATORY

This lab allows you to detect the effects of different activities on the circulatory system as a measure of two variables: blood pressure and heart rate. Skill at using a sphygmomanometer is a lost art so, hopefully, there will be an adequate supply of electrical devices for recording blood pressure in the classroom. Heart rate is more easily monitored at either a brachial or carotid artery, but inexpensive pulse monitors are also widely used. Students are typically paired and asked to measure each other's blood pressure and heart rate to form a baseline or control. These values are often compared to standard values for height and weight. Changes in blood pressure and heart rate are then recorded for various activities that might include sitting, then standing, exercising, and/or submerging the face in cold water.

The human circulatory system is closed and closely monitored so that pressure and rate of flow is adjusted to meet the needs of the body. At rest, blood pressure and flow match the needs of a low level of activity. Fewer of the body's cells are at work, so the need for energy and gas exchange is low. Even the simple act of standing up from a horizontal position can cause a drastic change in pressure in the carotid artery of the neck due to the gravitational pull of blood to the legs and feet. Special receptors in the carotid artery detect the change in pressure and send a nervous signal to the heart to increase its rate. At the same time, peripheral vessels contract, forcing blood flow centrally to the brain and viscera and increasing blood pressure. If the response is too slow or cannot match the need for oxygen to the brain, the body forces itself to get to a horizontal position by blacking out or fainting. This is the cause of dizziness when a person gets up from a chair too quickly. If a person who stands up does nothing else, the body equilibrates blood pressure and heart rate to meet the need of standing. When measuring these changes, a quick increase in blood pressure should be evident followed by a decrease in blood pressure until a steady state is reached. The steady state while standing is likely to be higher than the steady state while lying down because the circulatory system has to work harder to pump blood up to the brain.

The physiological response to exercise is very similar to an increase in blood pressure due to standing up, depending on the duration of the exercise. At the initiation of exercise there should be a marked increase in blood pressure and heart rate in response to pressure changes in arterial flow. If a moderate exercise level is maintained, other means of keeping heart rate and blood pressure elevated take effect, such as pH of the blood and blood glucose levels. The measurement of the pH of the blood and blood glucose levels are beyond the scope of this laboratory. The change in heart rate in response to exposing the face to cold water is a relic of a possible aquatic ancestry. Seals can dive to great depths because their metabolic rate decreases when they submerge. Metabolic rate is controlled by nervous input from epidermal cells and causes a decrease in heart rate. Humans share a similar response that can be measured if a person's face is placed in a basin of cold water. Humans have the same type of nerve cells in the face and neck area as seals, so when a person's face is submerged in cold water, the person's heart rate should suddenly decrease.

After measuring how these variables are affected by circulatory system activity, you should gain an appreciation of how closely the body's responses are controlled. The quickness of variable responses should indicate to the student that they are nervous responses and not hormonal responses. Measurement of variable change is only one example of the many ways that the body maintains homeostasis through complex feedback mechanisms.

IF YOU LEARNED ONLY SIX THINGS IN THIS CHAPTER . . . .

1. The human body regulates its internal state via mechanisms including the hormone-receptor system and the nervous system.

2. Hormones are secreted by glands and travel via fluid throughout the body. They activate receptor cells that produce receptor proteins, creating a reaction.

3. In the nervous system, an electrochemical impulse travels from one neuron to another when an action potential occurs. Neurotransmitters cross the synaptic gap and bind to the receptors of the next neuron, propagating the action potential.

4. Plants produce energy through photosynthesis and lose water via transpiration. As water evaporates from the leaves, it pulls water up through channels in the xylem. The phloem carries nutrients throughout the plant. Plants can reproduce asexually via vegetative propagation. Sexual reproduction in plants takes place in the flower.

5. Animal body systems consist of specialized cells and tissues that perform a function. You should know the structure and function of the endocrine (hormone), respiratory, digestive, integumentary (skin), and skeletal systems in humans.

6. Sexual reproduction in animals consists of male and female gametes meeting and forming the zygote. Stages in embryonic development include cleavage, gastrulation, and organogenesis. The ectoderm forms the integument and nervous system, the mesoderm becomes the muscles, bones, and circulatory system, and the endoderm becomes the organs of the digestive tract.

APPLYING THE CONCEPTS: QUESTIONS AND EXPLANATIONS

1. Messages delivered by the nervous system are conveyed by series of neurons. A message is passed from one neuron to the next by

   (A) the release of Ca^+ ions by the presynaptic neuron that travel across the synaptic gap and initiate the uptake of Na^+ ions in the postsynaptic neuron
   (B) the release of neurotransmitters by the presynaptic neuron that travel across the synaptic gap and initiate an action potential in the postsynaptic neuron
   (C) the release of Na^+ ions by the presynaptic neuron that travel across the synaptic gap and initiate an action potential in the postsynaptic neuron
   (D) the mechanical deformation of the postsynaptic neuron by the presynaptic neuron
   (E) the depolarization of the presynaptic neuron, causing a slight depolarization in the postsynaptic neuron

2. Which of the following terms describes a plant's ability to grow toward or away from an environmental stimulus?

   (A) Taxis
   (B) Kinesthesis
   (C) Tropism
   (D) Vining
   (E) Tractability

3. All organisms have the ability to exchange gases needed for metabolism because they have

   (A) blood for circulation
   (B) muscles for air movement
   (C) hemoglobin molecules
   (D) a liquid medium for dissolving gases
   (E) an opening for ventilation

4. A scientist removes one of the four cells that have resulted from the first two divisions of an animal zygote during early embryonic development. The animal continues to develop, but is missing a considerable amount of organs and does not live. Which of the following statements can be said about the animal?

   (A) The animal was a protostome.
   (B) The animal was a deuterostome.
   (C) The animal had indeterminate cleavage.
   (D) The animal did not have a full complement of DNA in its remaining cells.
   (E) The animal did not have enough resources to develop fully.

5. Placing a ripe banana in a bag with an unripe avocado will quicken the ripening of the avocado. This is caused by the positive feedback mechanism initiated by which of the following plant compounds?

    (A) Auxins
    (B) Cytokinins
    (C) Ethylene
    (D) Gibberellins
    (E) Phytochrome

6. The creosote bush, *Larrea tridentata*, is found in circular stands of cloned individuals in the southwestern deserts of the United States. This is an example of propagation called

    (A) wind dispersal
    (B) hybrid fitness
    (C) polyploidy
    (D) alternation of generations
    (E) vegetative reproduction

7. Which of the following glands contributes to thermoregulation by adjusting the resting metabolic rate?

    (A) Pancreas
    (B) Thyroid
    (C) Anterior pituitary
    (D) Adrenal
    (E) Pineal

8. Nutrients that are produced as a result of photosynthesis in a leaf are delivered to the roots of a tree by the pull of

    (A) gravity
    (B) osmotic potential between the source and the sink
    (C) root pressure
    (D) transpiration of water through the stomata in the leaf
    (E) the partial pressure of CO_2 in the leaf

9. Deciduous trees have programmed death of their leaves in the fall and initiate this response with which hormone(s)?

    (A) Abscisic acid
    (B) Auxins
    (C) Cytokinins
    (D) Ethylene
    (E) Gibberellic acid

10. Carbon dioxide waste is transported from working cells in the body to the lungs in the form of

    (A) bicarbonate ions
    (B) oxygen
    (C) disassociated carbon and oxygen
    (D) glucose and water
    (E) hemoglobin

11. A doctor analyzes the urine of a patient and determines that the contents suggest the kidneys may not be functioning normally. Which of the following compounds present in urine indicates possible kidney dysfunction in a human?

    (A) Water
    (B) Urea
    (C) Salt
    (D) Sugar
    (E) Protein

12. Which of the following compounds is used by a plant to adjust to changing light levels as the seasons change?

    (A) Chlorophyll *a*
    (B) Chlorophyll *b*
    (C) Cytokinins
    (D) Gibberellic acid
    (E) Phytochrome

13. Guttation, the accumulation of water on the leaf surface of plants due to root pressure, would most likely occur under which of the following conditions?

    (A) Midday heat
    (B) Morning humidity
    (C) Afternoon wind
    (D) Overnight freeze
    (E) Midmorning sun

14. Which of the following terms describes the synchronized muscular contractions that force chyme through the digestive tract?

    (A) Peristalsis
    (B) Arrhythmia
    (C) Gastric reflux
    (D) Colitis
    (E) Pyloric emptying

Questions 15–19 refer to the following diagram of an embryo after early development.

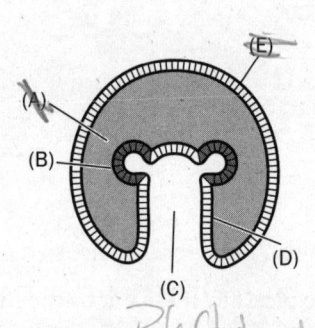

15. Archenteron
16. Ectoderm
17. Endoderm
18. Mesodermal pouch
19. Embryonic gut

Questions 20–24 refer to the following diagram of the structures of a flower.

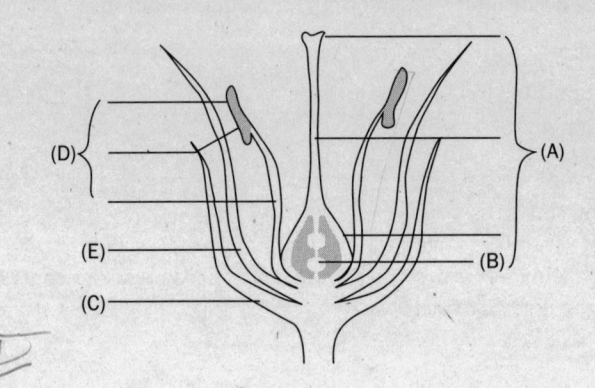

20. Petal
21. Sepal
22. Carpel
23. Stamen
24. Ovule

Questions 25–29 refer to the following list.

(A) Cell body
(B) Dendrite
(C) Axon
(D) Synapse
(E) Neurotransmitter

25. Propagates an action potential along the length of a neuron
26. Highly branched receptor of chemical and mechanical stimuli
27. Chemical messenger sent from one neuron to another
28. Contains the nucleus and organelles of the neuron
29. Gap junction between two adjacent neurons

Free-Response Question

Describe and **discuss** plant form in a terrestrial environment as it relates to the physiological functions listed below. Include in your discussion evolutionary comparisons with aquatic plants.

(a) Reproduction
(b) Dispersal
(c) Water balance
(d) Structural support

ANSWERS AND EXPLANATIONS

1. B

This question is asking about the transmission of a message in the nervous system between two neurons, so the message is passing across a synapse. This means that the signal is not electrical, but chemical and mediated by a neurotransmitter, not ions or electrical current. This rules out (A), (C), and (E). Be careful to note that Ca$^+$ ions play a part in the uptake of the neurotransmitters in the postsynaptic neuron, but it is the neurotransmitters that continue to relay the message. Some sensory neurons initiate action potentials by being mechanically deformed, but this is not a mechanism for transmission across neurons, so (D) is incorrect.

2. C

The term tropism comes from the Greek *trope* which means "to turn." Tropism is an automatic reaction in orientation or a positive or negative response to a stimulus.

3. D

This question should cause little difficulty as long as you don't rush to pick an answer. The question includes all organisms ranging from the smallest virus to the largest whale. Since not all organisms have blood, muscles, or hemoglobin, this rules out (A), (B), and (C). Some organisms respire directly across their skin so (E) is incorrect. Gas exchange requires that gasses be dissolved in a liquid medium, which requires at least a moist interface in organisms that aren't aquatic.

4. A

There are two answers in this list that may cause some confusion because they are duplicates. In addition, (E) is a possible condition presented to the animal, but there isn't enough information to determine if the animal has this condition. Choice (D) can't be correct because all of the cells have a full complement of DNA as a result of mitosis. The animal that is described above has determinate cleavage so (C) can't be correct and deuterostomes (B) have indeterminate cleavage, which is a duplicate of (C), which leaves (A).

5. C

This is a common trick used at home, but the reason why it works isn't usually understood. There are two good clues to the answer of this question. First, the question is asking for a compound that causes ripening. Second, the question is looking for a compound that creates a positive feedback loop. Most biological reactions are negative feedback mechanisms. The correct answer is ethylene (C). Ethylene is the plant hormone associated with ripening and is an excellent example of a positive feedback loop.

6. E

The key issue in this question is that the individuals are clones. Cloned organisms are genetic duplicates that arise from mitosis in some form of asexual reproduction. Creosote bushes spread by vegetative reproduction through a massive root system. Scientists believe that some of these clonal groups may be 9,000 years old.

7. B

This question asks precisely for a form of thermoregulation that involves metabolic rate. The body does regulate body temperature with the hypothalamus, but fortunately this isn't a possible response. Metabolic rate is closely tied to control exhibited by T_3 and T_4 produced by the thyroid, and people with hypo- or hyperthyroid conditions often have trouble with body weight and body temperature fluctuations.

8. B

You may automatically choose (A) because "gravity" is included in the common phrases "the pull of gravity" or "gravity's pull," and answer (A) may make some intuitive sense because the nutrients in question are moving down the plant; however, this is not the correct answer. Root pressure pushes water away from the roots instead of pulling it, so (C) can't be correct. Transpiration affects xylem flow (D) and the partial pressure of CO_2 is more important in animal respiration than in plants (E). The correct answer is (B). Phloem material is pulled from sources to sinks due to osmotic potential.

9. D

The one answer that may be misleading is abscisic acid (A). This plant hormone slows down growth, especially during periods of stress such as water loss. It also earned its name because scientists once thought it contributed to the abscission of leaves in the fall. The correct answer is (D), however. It is productive to think of ethylene as the ripening hormone and that ripening is a form of aging and death. The role of fruits is to die and fortify the seeds, after all.

10. A

Choice (B) might be tempting, since oxygen is converted to carbon dioxide in some processes, but this is not correct. If the amount of CO_2 in the blood is too high the pH drops too much, causing acidosis. The answer to this problem is to convert the CO_2 into bicarbonate until it reaches the lungs, where it is liberated again as gas.

11. E

Urine is a waste product of normal body function that contains large amounts of water, urea (a by-product of protein metabolism), and salts. This excludes (A), (B), and (C). The tricky answer choice is (D), sugar. Sugar in the urine is a concern and a possible indicator of diabetes, but it is not a sign of improper kidney function. The kidneys filter out large protein molecules in the nephron, so protein (E) in the urine would indicate a problem with the filtration process in the kidney.

12. E

Intuitively, one would think the compound must have some kind of sensitivity to light since it is the stimulus for the reaction. Photopigments [(A) and (B)] are certainly light sensitive, but the chlorophylls have not been discussed in connection with plant regulation other than photosynthesis. Because both cytokinins and gibberellic acid are plant hormones, you may be tempted to choose (C) or (D). The correct answer, however, is phytochrome. Phytochrome is not a plant hormone, instead it is the light sensor that triggers hormone response.

13. B
Guttation occurs under conditions that decrease transpiration. Transpiration is just like evaporation so is increased by heat and wind, things that make the air dry. Guttation would be most likely in conditions that would make the air wet and cool.

14. A
It takes more than gravity and the pressure of more food coming down the gullet to get the lumen moving through the digestive tract. The digestive organs are lined with smooth muscle that contracts under the control of the sympathetic nervous system in rhythmic contractions that form waves. This process is called peristalsis (A).

15. C
As with any diagram, start with the most obvious answers, then move to the answers that are less obvious. The most obvious answer in the diagram is likely ectoderm, since the ectoderm is the outside layer. The archenteron and the embryonic gut are synonymous terms.

16. E
The ectoderm is the outer layer of cells.

17. D
Endo- means inside, so "endoderm" is the inner layer of cells.

18. B
This structure even looks like a pouch and will go on to produce the mesoderm.

19. C
As mentioned in the answer to question 15, the archenteron and the embryonic gut are the same thing. You will just have to memorize the names of certain structures for the exam.

20. E
As with the previous Cluster Question, choose the most obvious answer first. The most obvious answer is likely petal, since it is a term that is familiar even outside of biological study. The petals are the external, showy parts of the flower.

21. C
One can think of a sepal as a green petal.

22. A
This is the collective name for the ovary, pistil, and stigma.

23. D
This is the collective term for the filament and anther.

24. B

This is the remaining option. You can remember its central location in the flower by thinking of its central role in angiosperm reproduction, as the location of the female gametophyte and the site where male and female gametophytes meet.

25. C

This structure forms the main body of the neuron.

26. B

The dendrites project from the cell body to receive chemical information from other neurons.

27. E

You can think of the neurotransmitter as the agent that "transmits" or carries chemical information between neurons, across the synaptic gap.

28. A

Even if you don't know the answer, you can make a guess based on the fact that the terms "nucleus" and "organelle" usually come up in the discussion of cells.

29. D

The synapse enables neurons to propagate action potentials in the presence of neurotransmitters and to remain at rest when neurotransmitter reuptake is complete.

Free-Response Answer Explanation

This question promotes to the student that there is a close association between form and function in biology, especially in the plant kingdom. Be prepared to assimilate concepts from other chapters with the processes presented in this chapter, in order to answer the question.

It is crucial to discuss structure as it relates to the given processes to receive many points. It is more important to discuss all of the processes in the question than to include more than one structure for any given process. For example, water balance in terrestrial plants is maintained by root uptake, the closing of stomata, and leaf structure, but not all of these parts of water uptake need to be discussed. It is more important to pick one part of water uptake and be specific about how it affects water balance, as well as how this plant structure has provided an adaptive advantage, allowing the colonization of a terrestrial environment.

Here is a possible response:

(a) Aquatic plants relied on water for the delivery of gametes and the dispersal of spores. Gametes and spores did not require structures for preventing desiccation in a terrestrial environment. Land plants produce seeds that possess a hard coat that originates from the integument of the ovule. At maturation, the seed has already lost the majority of its water and enters a stage of almost suspended activity until it germinates. The land plant has adapted the ability to produce a spore-like structure that can endure harsh, dry environments almost indefinitely until growing conditions are favorable by entering a dormant phase. When the proper growing conditions are available, the seed germinates, which is a continuation of growth of the embryo.

(b) Land plants meet the challenge of dispersing their offspring in many ways; one is the production of fruit. When a flower is pollinated, the walls of the ovary thicken into the pericarp of the fruit in coordination with seed development. If a flower is not pollinated, fruit set usually does not occur and the plant does not waste resources on a fruit without seeds. The fruit ripens and usually becomes attractive to an animal. Therefore, the fruit not only provides protection to the developing seeds, but aids in their dispersal. Some fruiting bodies resemble the sails of a boat, a parachute, or feathers so they catch the wind and "fly" away. Other fruiting bodies are sweet and succulent and attractive to animals. When an animal feeds on the fruit, it drops the seeds far from the parents.

(c) The most striking adaptations to land survival are evident in desert plants. For example the jojoba has an extremely waxy covering on the leaf surface that almost eliminates evaporative water loss. The waxy cuticle is composed of lipids that are hydrophobic and keep water from moving across the leaf surface. The jojoba also has leaves that grow with a vertical orientation to the sun. This reduces the amount of incident light they receive and reduces water loss due to increased transpiration at higher temperatures. Aquatic plants do not suffer from drought stress, so have thin gelatinous leaf tissue that is completely permeable to water. This leaf structure does not allow for a rigid plant that can grow on land. The aquatic plant would lie flat on the ground.

(d) The aquatic environment provided a buoyant medium upon which plants could rely for support. There are also factors that cause water movement so that plant structures have increased exposure to the water interface. On land, plants maximize their exposure to sunlight by producing leaves on branching and climbing stems, while the water-absorbent roots stay in the ground. This form of structural support necessary to maintain a rigid vertical orientation would not be possible without the cellular specialization of sclerenchyma cells with rigid cell walls composed of lignin. Sclerenchyma cells also have bundles of longitudinal fibers and a matrix of sclereids. Secondary growth of dicots by the lateral meristem allowed for the production of wood from xylem tissue, providing an additional form of support for shrubs and trees.

Chapter 11: **Ecology**

- It's Not a Vacuum Out There: How Organisms Interact with Each Other and Their Environment
- Animal Behavior Laboratory
- Dissolved Oxygen and Aquatic Primary Productivity Laboratory
- Applying the Concepts: Questions and Explanations

IT'S NOT A VACUUM OUT THERE: HOW ORGANISMS INTERACT WITH EACH OTHER AND THEIR ENVIRONMENT

The natural world is a harsh place. Organisms face the daily challenge of finding nutrients (often in the form of other organisms), safe havens, and mates. Organisms must not only live by and adjust to the abiotic (nonliving) factors presented by the environment, but also compete for resources with other organisms while they avoid becoming food themselves. Some scientists cite examples of cooperative exchanges among and between species, but others criticize ecological hypotheses that suggest any alternative explanation for these phenomena other than the need of an individual to survive and propagate its DNA. Fortunately, the College Board has concentrated their questions about this hypothesis-dense discipline on just a few concepts.

Primarily four concepts within the topic of ecology are found on the exam. The first concept covers the basics of population biology, including the abiotic and biotic factors that affect population size, growth, and decline. The second concept covers the diagnostic features of the different levels of environmental systems, or the "classification of the environment." The third concept deals with nutrient cycling and energy exchange, and the chapter finishes with a few examples of human impact on the global environment. Ecological terminology can be found both throughout the chapter and in the sample questions at the end of the chapter.

Population Dynamics

What do you get if you put one bacterium in a lot of space and give it an unlimited amount of resources and time to reproduce? You get a whole lot of bacteria, that's what! As a matter of fact you get an exponential rise in the number of bacteria according to the equation

$N_t = N_0 e^{rt}$ where

N_t is the number of bacteria at time t

N_0 is the number of bacteria at the beginning

r is the rate of population growth (a difference of the reproduction and death rates)

t is the number of chronological steps of reproduction (seconds, minutes, years, etc.) and

e is the constant 2.71828. . . .

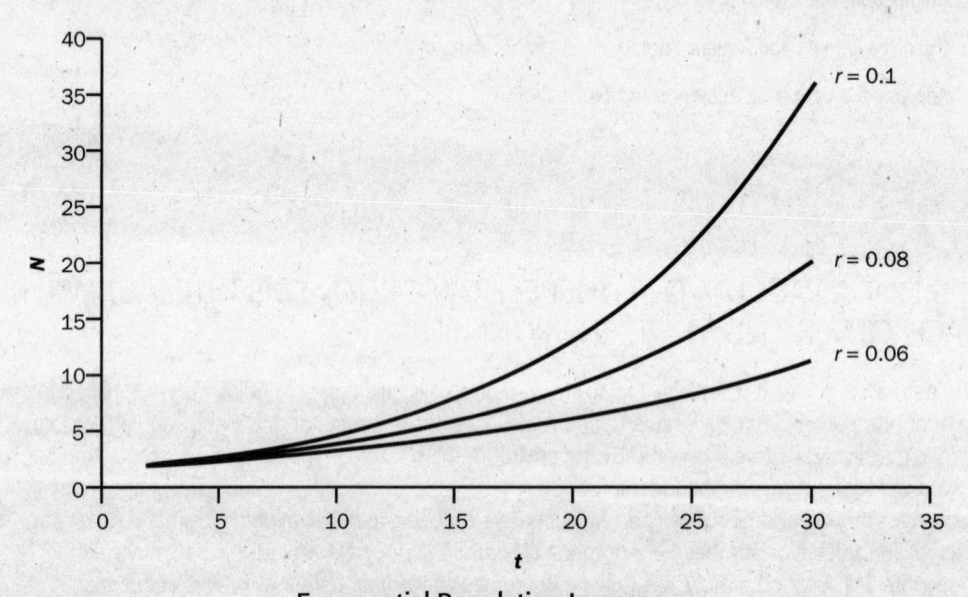

Exponential Population Increase

The three lines on the graph show the exponential increase in population size with no limits according to the equation $N_t = N_0 e^{rt}$. Each line represents a population increasing at the rate indicated, having started with two individuals.

This graph shows how rapidly populations can grow when the increase is exponential. Remember that t is a measure of the population's rate of producing another generation, which is very short for bacteria. Some bacteria can reproduce every 10 to 30 minutes, so 30 time steps would only take between 5 and 15 hours. What is really incredible is that a population of bacteria can grow to numbers in the millions in only a matter of hours.

The reality is that populations do not grow exponentially without limits. Most reach a size at which they plateau, called the **carrying capacity**. The rate of change in size of natural populations can be estimated by the equation $\frac{\Delta N}{\Delta t} = rN\left(\frac{K-N}{K}\right)$ where

$\frac{\Delta N}{\Delta t}$ is the change in the population size over the given time

N is the size of the population at the beginning of time t

r is the rate of population growth (again, a difference of the reproduction and death rates) and

K is the carrying capacity.

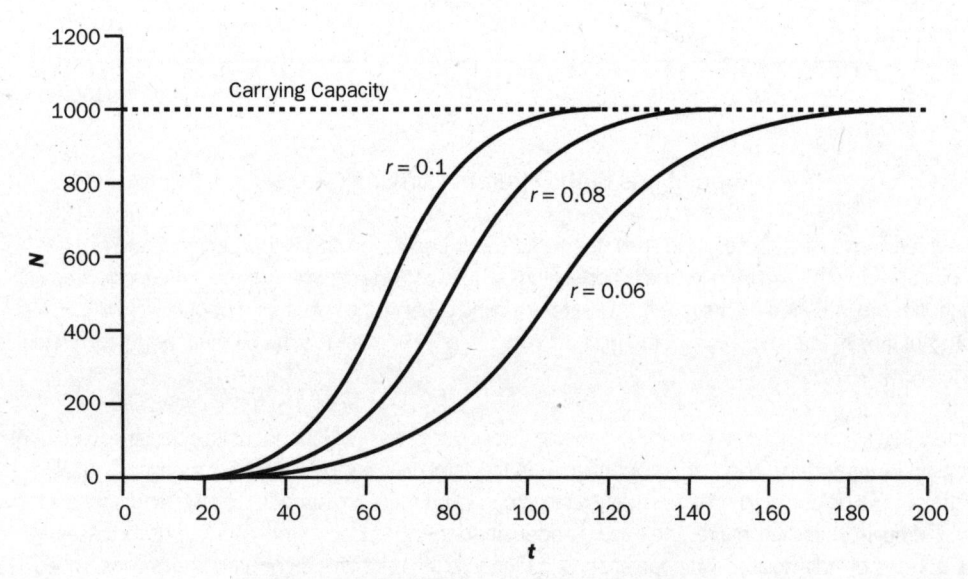

Population Increase to Carrying Capacity

These three lines show the increases in population size with the same rates of population growth as in the previous graph, but with a limited carrying capacity of 1,000 individuals. This graph shows that carrying capacity is the ultimate limit to population size, while the rate of population growth affects how quickly the population increases. A change in carrying capacity will change where the plateau occurs.

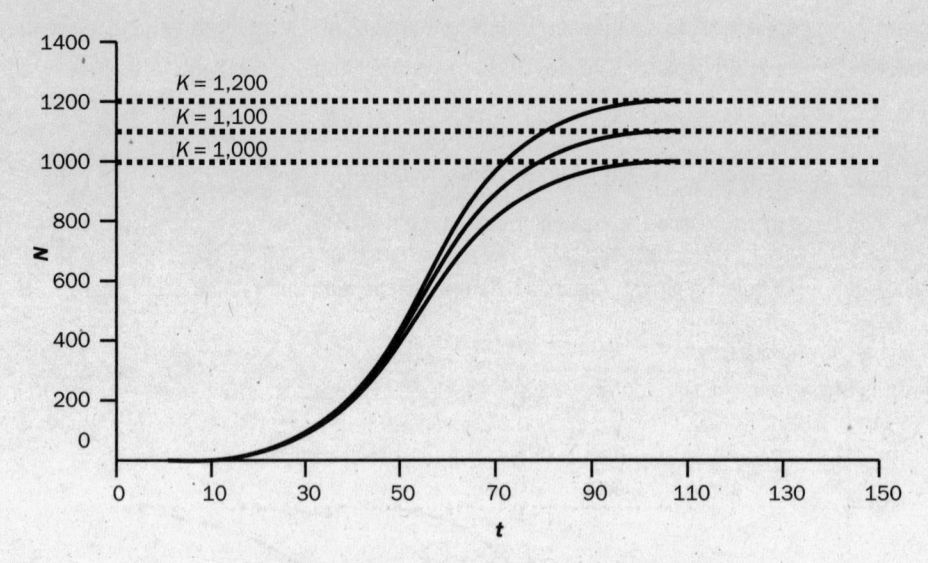

Populations With Different Carrying Capacities

The three lines in the above graph show the increases in population size with a constant rate of population growth ($r = 0.10$) for three populations whose carrying capacities are not the same. Carrying capacities are indicated on the graph. When the carrying capacity increases for a population with a constant growth rate, the shape of the **growth curve** hardly changes. It is the ultimate population size that is really affected.

Factors that affect population growth fall into two categories, density-independent and density-dependent. **Density-independent** factors affect populations in the same way regardless of how many individuals are in the population at that given time. These factors tend to be more catastrophic, abiotic factors such as flood, drought, hurricane, or fire. The impact of **density-dependent** factors on populations increases as the number of individuals in the population, and therefore the density in a given area, increases. These factors include variables such as competition among or between species, predation, or emigration.

Different **biotic and abiotic factors** in the environment have contributed toward adaptive strategies that allow organisms to maximize their reproduction. Species that monopolize rapidly changing environments, like disturbed habitats, produce many offspring quickly. These species are called ***r*-selected** or ***r*-strategists** because their strategy is to increase their *r*-value. *r*-strategists generally have short life spans, begin breeding early in life, and produce large numbers of offspring.

Species in more stable environments tend to produce fewer offspring and invest more resources into the success of each offspring. These species are called ***K*-selected** or ***K*-strategists**. *K*-strategists have longer life spans, begin breeding later in life, have longer generation times, and produce fewer offspring. They take better care of their young than do *r*-strategists; they have also evolved to be better at exploiting limited parts of their environments. There is a continuum between *r*-selected and *K*-selected species, as few communities are made up of strictly *r*-strategists or *K*-strategists.

Every species has a set of conditions that are optimal for its reproductive strategies; in any ecosystem, the most abundant species will be the one for which the environment most closely approximates this set of optimal conditions. If these conditions remain constant, all other things being equal, the distribution of organisms in an ecosystem should remain roughly the same. Without important changes to the environment or available resources, there is little reason to expect a population to boom or crash. However, these kinds of significant changes are inevitable over time, and they determine which species can thrive or dominate in an ecosystem. This change in the species makeup of an ecosystem is called ecological succession.

A Hierarchical Classification of the Environment

Just as every organism needs an identifier to allow effective discourse among the scientific community, the environment is also made up of named divisible units with diagnostic features. The largest division within the Earth is the **biome.** There is no consensus among the scientific community as to which regions constitute a biome. Deserts, grasslands, taigas, and oceans are all types of biomes. They are defined by large geographic regions with homogenous abiotic and biotic characteristics such as low rainfall, high altitude, dominated by short grasses, and so on. Biomes are divided into ecosystems, which defy standardizations of scale. An **ecosystem** can be an entire mountain range or a coral reef, but can also be an ephemeral pool of water in the middle of the desert. An ecosystem is a collection of communities and their abiotic environment, grouped around a geographic feature. The key is that ecosystems includes processes like nutrient flow. You will not likely be asked to recall any specific ecosystem on the exam, as you will with the biomes.

Biomes of the World

Biome	Climate	Flora and Fauna
Tropical Rainforest	Constant temperatures, high rainfall	Most diverse biome with many species from all domains
Deciduous Forest	Temperate, warm summers, cold winters with snow	Deciduous trees, herbs, and grasses, forest-dependent fauna, migratory birds
Taiga	Long, cold winters and short, cool summers	Conifers, short grasses, low diversity, mammals and birds, few reptiles and insects
Desert	Extremely low rainfall, temperature extremes	Shrubs, cacti, and succulents, reptiles, specialized adaptations to deal with extremes in temperature and drought
Grassland	Semi-arid to temperate, hot summers and cold winters	Long or short grasses, very low diversity
Tundra	Extremely short growing season, long cold winters	Lichens, mosses, sedges, dwarf shrubs, simplest biome with lowest diversity
Savanna	Tropical rainy season, arid dry season	Perennial grasses, shrubs, sparse trees, diverse in large mammals

There are three other terms with which you should be familiar. An ecosystem is composed of a habitat, populations, and communities. The **habitat** is the total of the various abiotic factors such as terrain, soil type, and water. **Populations** are all the organisms that occur in a specific habitat. **Communities** are all the populations of organisms that live in the habitat. The fauna, flora, and abiotic factors collectively make up an ecosystem.

A **niche** is a theoretical construct that is defined by a multidimensional space made up of all of the environmental factors, both abiotic and biotic, that define where and how a population of organisms lives. For example, if a population of adult pelagic fish feed at a depth of 10–20 m, that range of depths is part of the fish's niche. If a population of thermophiles thrives in liquid clay at 100°C, both the clay soil and the high temperature are part of its niche.

Environmental Dynamics

Environmental dynamics deals with the flow of energy and other nutrients through the environment. **Energy flow** is traditionally shown in the form of a food web. Arrows are drawn between different species of organisms, with the direction of the arrow indicating the direction of energy flow.

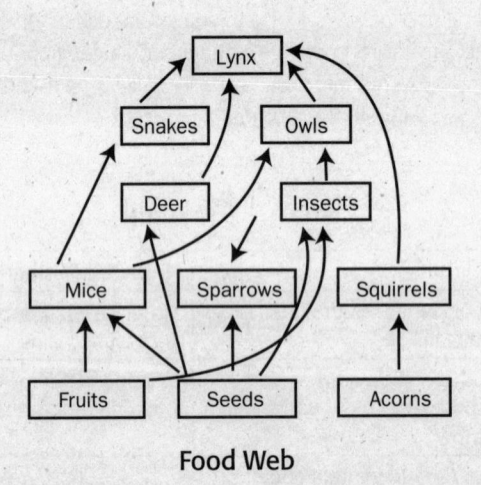

Food Web

In the illustration of a terrestrial food web, the arrows indicate the direction of energy flow between the organisms. The fruits, oaks, and grasses only have arrows pointing away from them, so they are primary producers. The lynx is the top predator (quaternary consumer) and only has arrows pointing toward it. There are no decomposers in this food web.

Food webs can quickly become quite complicated because there are so many different organisms interacting in any given ecosystem. Food webs are rarely entirely comprehensive for the same reason. As one kind of organism feeds another, nutrition moves through a food web from one **trophic level** to the next. Organisms included in a food web may be primary producers (**autotrophs**), which produce energy from sunlight, primary and secondary consumers, and decomposers. Because they make their own food, **primary producers** only have arrows pointing away from them in a food web. Organisms that get energy by consuming other living things (**heterotrophs**) have arrows pointing toward them, indicating the flow of energy from another trophic level. If **decomposers** are part of the food web, there will also be arrows leading from the various levels of heterotrophs to the decomposers. **Primary consumers** eat plants, while **secondary consumers**, **tertiary consumers**, and **quaternary consumers** feed on the levels below them, such as the primary consumers. **Herbivores** eat only plants (thus, they are always primary consumers). However, some secondary and tertiary consumers are **omnivores**, which eat both animals and plants. Omnivores in a food web are indicated by arrows pointing toward them from both primary producers and primary consumers.

Other simplified models illustrating how nutrients are cycled include those for water, nitrogen, carbon, and phosphorous. These systems are known as **biogeochemical cycles**.

In the **water cycle**, the sun's energy drives the movement of water through the environment. The sun causes surface water to evaporate and plants to transpire, releasing water vapor into the atmosphere. In the atmosphere, the water vapor cools and condenses into clouds and precipitation. Precipitation returns water to the Earth's surface, where it runs off into lakes and rivers and percolates into groundwater.

There is more nitrogen in the air than oxygen: nitrogen gas makes up over two thirds of Earth's air. However, most living things cannot get the nitrogen they need from nitrogen gas. Certain groups of bacteria in the soil, called **nitrogen-fixing bacteria** convert nitrogen gas into nitrates and nitrites; plants then take up the nitrites and nitrates from the soil and from fertilizers. Animals, including humans, get nitrogen by eating plants and other living or dead organic matter. When animals and plants die, groups of bacteria and fungi break down these organic remains and release nitrogen back into the soil. Animal waste contains nitrogen, as well. **Denitrifying bacteria** in the soil then change the nitrates and nitrites back into nitrogen gas, which returns to the air. Together, all of these steps make up the **nitrogen cycle**.

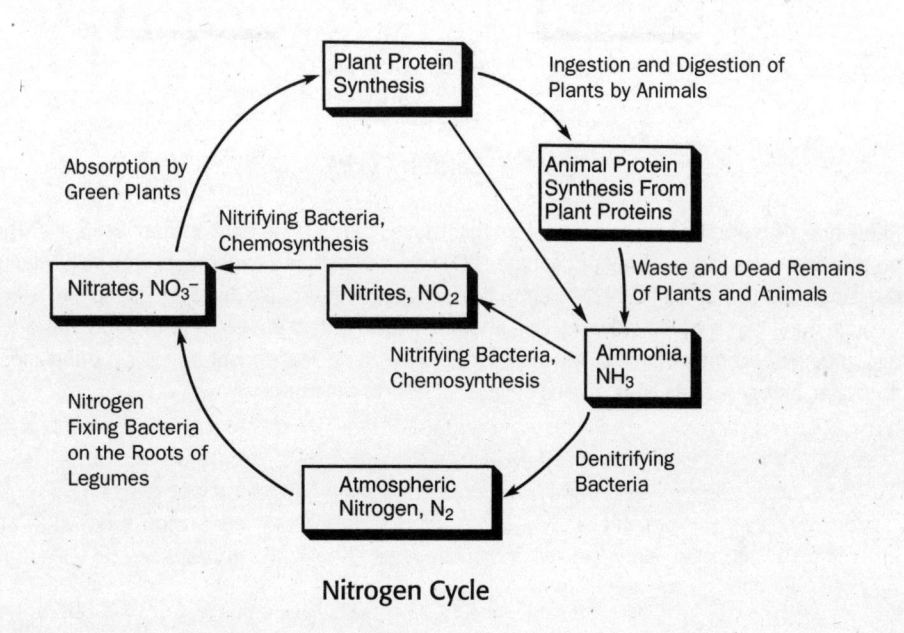

Nitrogen Cycle

Photosynthesis and respiration are the complementary reactions of the **carbon cycle**. In photosynthesis, carbon dioxide and water combine to form sugars and oxygen; the sun's energy is stored in the bonds among carbon atoms in the sugars. Respiration uses carbohydrates and oxygen to produce carbon dioxide, water, and energy; it releases the energy that is stored by photosynthesis. As animals eat the plants and one another, and breathe, carbon moves through the food web; the carbon that plants and animals contain is returned to the soil when they decompose. The carbon in the soil enters the cycle again, perhaps as part of a microorganism. A great deal of carbon is stored in the oceans and in minerals. Combustion is another way that carbon enters the atmosphere, in the form of carbon dioxide. The widespread and heavy use of fossil fuels has created concerns about how all of this combustion is affecting the environment.

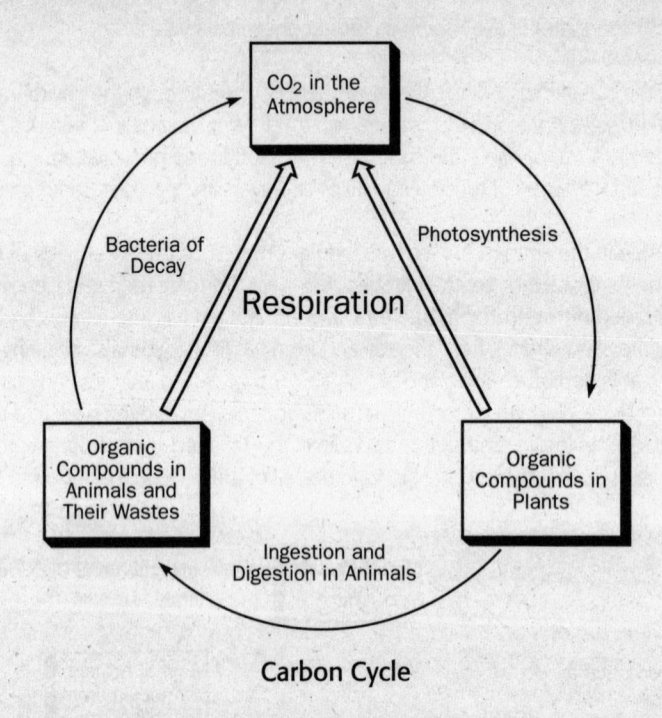

Carbon Cycle

Phosphorous is another important nutrient that moves through the environment in a cycle. The phosphorus cycle begins when phosphate (PO_4) from weathered rock moves into soils. Plants then take up phosphate and it becomes part of the living ecosystem. Like nitrogen, phosphate moves through living things as they feed on one another, and re-enters the ecosystem through the decomposers' action on their waste and their dead remains. The phosphorus cycle stands apart from the water, nitrogen, and carbon cycles because it has no gas phase.

Issues Affecting the Globe

Other than the major extinction events recorded every 100 million years or so in Earth's geological history, the only variables that seem to have global impact are related to human activity. You should be familiar with two of these variables.

The Earth is surrounded by a layer of ozone (O_3). The layer lies in the stratosphere, 10–17 km from the Earth's surface, and protects the surface from harmful UV rays emitted by the sun. Over the past 20 years, scientists have determined that the ozone layer is being depleted by compounds containing chlorine, fluorine, bromine, carbon, and hydrogen (halocarbons) that seem to be produced by human activity. One particular group of chemicals, CFCs (chlorofluorocarbons), has received a considerable amount of the blame for depletion of the ozone layer. The ozone layer may be depleted by as much as 60 percent over the Antarctic in the spring, causing the notorious "hole in the ozone layer." Scientists worry that without changes in human activity, the ozone layer will be depleted beyond recovery, preventing certain forms of life on Earth.

The other major global impact of human activity is the release of "greenhouse" gases (water vapor, carbon dioxide, methane, and nitrous oxide) into the atmosphere from the inefficient burning of fossil fuels (there are sources from manmade aerosols as well). As these gases become more dense in the Earth's atmosphere, it is assumed that they prevent heat from escaping the Earth's surface, causing "global warming." Scientists believe that the Earth's atmospheric temperature is steadily rising and that this will cause major shifts in seawater levels, local climates, and world weather patterns.

ANIMAL BEHAVIOR LABORATORY

The main purpose of this laboratory is to record direct observations of an organism's response to environmental stimuli. The laboratory includes a very simple procedure involving the use of isopods (sowbugs or pillbugs) and a makeshift apparatus that allows the animals to choose between two chambers based on the presence or absence of any number of environmental stimuli. These variables may include light versus dark, dry versus moist, acid versus base, or hot versus room temperature. Once the animals are turned loose in the choice apparatus, the number of animals in each chamber is counted at regular intervals for a period of time between 10 and 20 minutes. There is no difficult analysis involved unless the class performs several replicated trials and univariate statistics are calculated. The main purpose of the lab is to learn that, in fact, animals do react to environmental conditions.

In addition to the hands-on experience, you should learn a few ecological terms while completing this lab. The first two terms deal with how animals react to stimuli. If an organism reacts directly in response to the direction from which the stimulus originates, it is called **taxis**. When you put isopods into light and dark chambers, they react to the light by moving away and toward the dark. This is an example of taxis. The other term is kinesis. **Kinesis** is when there is reaction to a stimulus, but the reaction is random in regard to the stimulus.

Another type of response you should remember is an **agonistic response**, a response of aggression or submission between two organisms. When a domestic dog and a housecat interact, the cat may hunch up its back and its hair may stand on end in an aggressive response. This is an agonistic response. Likewise, if the dog rolls over on its back with its feet in the air and exposes its belly, it is being submissive. This is also an agonistic response.

DISSOLVED OXYGEN AND AQUATIC PRIMARY PRODUCTIVITY LABORATORY

There are two parts to this laboratory investigation, but both deal with environmental effects on the amount of oxygen that is available to organisms in water. The first part deals with the effects of an abiotic factor (temperature) on the concentration of dissolved oxygen in water. As you learned in chapter 3, life exists in a physical world dictated by physical laws. The amount of oxygen in water is affected by the water temperature, whether there are organisms living in the water or not. Simply put, as temperature increases, the amount of free oxygen in water decreases. It escapes to the atmosphere above the water column.

The second part of the lab focuses on a biotic component and the ability of an aquatic autotroph to maintain photosynthesis with reduced levels of light. The deeper in the water column an organism lives, the less available light there is. This limits the depths at which a photoautotroph can survive and contributes to the partitioning of a habitat based on the photosynthetic pigments present in the organisms. Both temperature and light levels affect aquatic ecosystems greatly, and this is a chance to demonstrate these effects.

You will be given a technique to detect the amount of dissolved oxygen in water. This might be a chemical analysis using titration, or you may have access to an electrochemical sensor. Either way, you will measure the amount of dissolved oxygen in water at various temperatures. You can then extrapolate the level of oxygen saturation of the water samples at each temperature using a tool called a nomogram.

The nomogram is a good example of some of the simple models that are used in ecology and physiology that require little technical proficiency. Many of these old models are still accurate, but they often are replaced by more expensive equipment with much greater accuracy and precision. The nomogram is simple to use: just line up a ruler at the temperature and dissolved oxygen concentration that was measured and read the oxygen saturation on the diagonal. The more important issues in this first part of the laboratory are why you calculate percent of oxygen saturation when you already have the dissolved oxygen, and why oxygen in the water is important.

The amount of oxygen in the water column is important as a measure of biological activity or potential biological activity. The more oxygen that is available in the ecosystem, the more non-autotrophic organisms the system will support. The reason you need percentage saturation is because if the amount of soluble oxygen in the aquatic ecosystem doesn't match what is expected for a given temperature and absolute measurement of oxygen, there must be some other factor contributing to the difference in the amount of available oxygen. The percentage saturation essentially provides a benchmark that can be used for comparison among bodies of water at any temperature.

The second part of the laboratory is perhaps the more interesting part. Here you will take samples of water from a natural system at different depths and measure the dissolved oxygen as you did above. You might also apply different light treatments to these water samples and measure the dissolved oxygen. Some classrooms will substitute a commercial algal colony rather than using naturally occurring autotrophs.

One of these treatments will be to place a sample of water in complete darkness for a period of time, eliminating photosynthesis. The assumption is that only respiration will affect any changes in oxygen concentration. The decrease in oxygen in the dark sample will be the normalizing quantity to accurately measure the amount of oxygen produced in the samples under light. After all, the same amount of respiration should be occurring in the samples under light as those in the dark (you used a similar control sample in the Cell Respiration Laboratory in chapter 5). You should discover that there is less oxygen production in the samples that have less light, which you probably assumed before the experiment was carried out.

IF YOU LEARNED ONLY SIX THINGS IN THIS CHAPTER . . . .

1. Populations with infinite resources increase exponentially in a J-shaped curve. Usually the environment has a carrying capacity, the maximum number of individuals it will support. Factors affecting population size can be density dependent (overcrowding) or density independent (a storm wipes out part of a population).

2. *r*-selected species have many offspring and provide almost no parental care (fish, insects). They succeed best in new habitats with many resources. *K*-selected species have few offspring and care for them for an extended period (elephants, humans). Those species do best in stable environments near carrying capacity.

3. A biome is a climatic zone with associated animals and vegetation (tundra, desert, etc.). An ecosystem comprises a community of living organisms and its habitat. Different populations of organisms make up the community. Each organism is adapted to a specific niche or role.

4. Food webs trace energy flow in a community. Different trophic levels consist of primary producers (plants), primary consumers (herbivores), secondary consumers (carnivores), and decomposers (bacteria).

5. Materials such as water, nitrogen, carbon, and phosphorus travel through the environment and are recycled in biogeochemical cycles, involving different life forms.

6. Human activity has affected the environment significantly, e.g., ozone depletion and global warming.

APPLYING THE CONCEPTS: QUESTIONS AND EXPLANATIONS

1. Mistletoe is a vascular plant that imbeds rootlike structures into the limbs of a tree, robbing it of moisture and nutrients. The mistletoe provides no resources to the tree on which it depends. The ecological relationship between the mistletoe and the tree is an example of

    (A) commensalism
    (B) herbivory
    (C) mimicry
    (D) parasitism
    (E) predation

2. Konrad Lorenz, a noted behavioral psychologist, was famous for conditioning a gaggle of geese at a particular age in their development to adopt his rubber boots as their mother. This kind of animal behavior results from

    (A) habituation
    (B) imprinting
    (C) instinct
    (D) nurturing
    (E) navigation

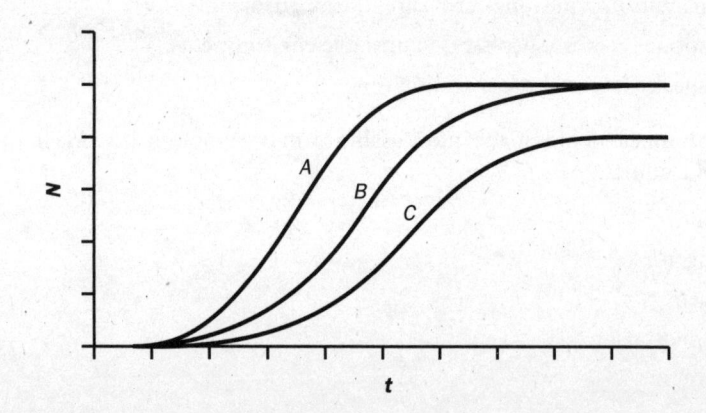

3. The diagram above shows the population growth curves for three different populations of an organism. Which of the following statements is true?

    (A) Population A has a higher carrying capacity than population B.
    (B) Population B has a higher growth rate than population A.
    (C) Population C has a higher growth rate than population A.
    (D) Population C started with fewer individuals than populations A or B.
    (E) Population B has a higher carrying capacity than population C.

4. Which of the following groups of organisms is the most likely primary producer in a tundra biome?

    (A) Conifer forests
    (B) Small shrubs
    (C) Fleshy succulents
    (D) Subterranean fungi
    (E) Green algae

5. According to scientists, the Earth's ozone layer is being depleted primarily by which of the following?

    (A) Greenhouse gases
    (B) Acid rain
    (C) UV rays from the sun
    (D) CFC (chlorofluorocarbon) molecules
    (E) O_2 oxidation

6. All of the following statements would be true of an *r*-selected species EXCEPT which?

    (A) The mother produces a large number of eggs that suffer heavy predation.
    (B) The mother reaches sexual maturity early in life.
    (C) The parents provide long-term care for the offspring.
    (D) The species is more successful in unstable environments.
    (E) The species has a short generation time.

7. This important element is made into a usable form by symbiotic bacteria in the root nodules of legumes.

    (A) Carbon
    (B) Nitrogen
    (C) Oxygen
    (D) Phosphorous
    (E) Potassium

8. Which would be the keystone species in the climax community of a savannah biome?

    (A) Grass
    (B) Moss
    (C) Pine
    (D) Lichen
    (E) Shrub

9. Which of the following statements is false?

    (A) Grass would be a primary producer within an ecological community.
    (B) Humans would be consumers within an ecological community.
    (C) Omnivores consume both plants and animals.
    (D) Herbivores consume animals.
    (E) Bacteria and fungi would be decomposers or saprophytes within an ecological community.

10. Gross primary productivity differs from net primary productivity in that

    (A) Gross primary productivity is the total chemical energy generated by producers while the net primary productivity subtracts out the loss of energy to respiration by plants.
    (B) Net primary productivity is the total chemical energy generated by plants while the gross primary productivity subtracts out the loss of energy to respiration by plants.
    (C) Gross primary productivity is the total chemical energy generated by producers while the net primary productivity subtracts out the loss of energy to respiration by animals.
    (D) Net primary productivity is the total chemical energy generated by plants while the gross primary productivity subtracts out the loss of energy to respiration by animals.
    (E) None of the above

11. If a toxin is concentrated in a food web and is present in primary producers

    (A) the toxin will not be found in organisms at the top of the food web
    (B) the toxin will be more concentrated in organisms at the top of the food web
    (C) the toxin will be less concentrated in organisms at the top of the food web
    (D) the toxin will be found in the same concentration in organisms at the top of the food web
    (E) None of the above

12. In a food chain which consists of grass → spiders → mice → snakes → hawks, the organism with the most biomass is the

    (A) grass
    (B) spiders
    (C) mice
    (D) snakes
    (E) hawks

Questions 13–16 refer to the following list.
    (A) Deciduous forest
    (B) Desert
    (C) Grassland
    (D) Savanna
    (E) Taiga

13. An area with extremely low rainfall, dominated by shrubs and succulents

14. A monocot-dominated area with cold winters and hot, arid summers

15. A subtropical expanse with the highest diversity of ungulates

16. Long, cold winters and short summers; low diversity and gymnosperms dominant

Questions 17–20. Use the diagram of the food web below to identify the placement of each of the following trophic levels. Each letter corresponds to a species in the food web.

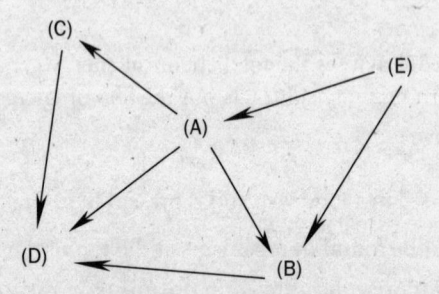

17. Primary producer

18. Omnivore

19. Decomposer

20. Top carnivore

Free-Response Question

A swab of bacteria is taken and placed in a petri dish containing the growth medium agar. Twenty bacteria were transferred into the petri dish by the swab.

(a) **Discuss** the rate of growth of bacteria if the bacteria divide every 15 minutes. **Calculate** how many bacteria will be in the petri dish after 15 minutes, if $r = 0.10$.

(b) **Explain** why the population of bacteria eventually will stop multiplying in the petri dish.

(c) **Calculate** the number of bacteria that will be present in the petri dish after 45 minutes if $r = 0.15$ and there is no limiting factor to consider.

ANSWERS AND EXPLANATIONS

1. D

This question includes some of the additional ecological terminology you might encounter. The way to approach problems that highlight interaction between organisms is to analyze the benefit or loss to each side of the interaction. In this case, the mistletoe is taking and not giving and the tree is giving and getting nothing. The mistletoe is a detriment to the tree, but it isn't ingesting the tree. Instead, the mistletoe is robbing the tree of its nutrients, sucking its blood like a mosquito. The mistletoe is a parasite (D).

2. B

Some animals have cognitive functions that fit within a developmental timeline. Humans, for example, learn languages most efficiently in the first four to five years of childhood. The brain during this period of time is "pliable" in terms of making an imprint on learning, and for many parts of the brain the process is not reversible. Habituation is when an organism loses sensitivity to certain stimuli, coming to ignore them. Instincts are behaviors that are not learned, and it is unlikely that geese innately will follow a pair of boots around. Nurturing is a generic term describing those behaviors that are learned in life, essentially the opposite of instinct. Navigation is the ability to orient to the environment. The correct answer is (B), imprinting.

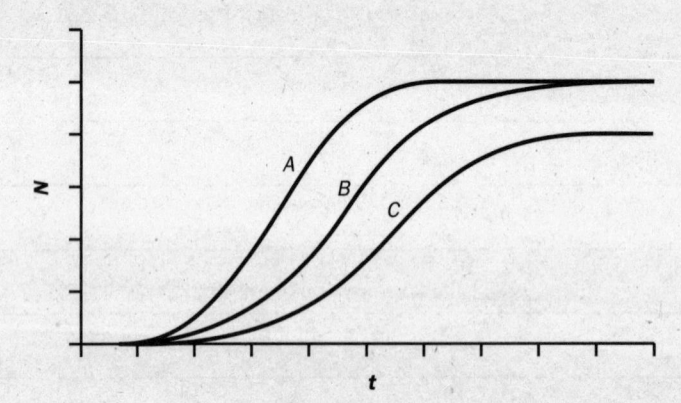

3. E

To answer questions of this type, you need to know how growth rate and carrying capacity affect the profiles of population growth curves. Logistic growth is typified by an ultimate plateau which is equal to the carrying capacity: the higher the plateau, the higher the carrying capacity. Populations A and B have the same carrying capacity, which is higher than it is for population C. This rules out (A). Growth rate affects the steepness of the exponential portion of the curve, the incline before the plateau. The quicker the population gets to its carrying capacity, the higher the growth rate. Population A has the highest growth rate, followed by B, then C. This rules out (B) and (C). All of the populations start with the same number of individuals at the far left of the curves, so (D) is wrong. By process of elimination and by referring back to the definition of carrying capacity, you can determine that the correct answer is (E).

4. B

You might initially confuse the tundra biome with the taiga biome because they are both cold regions with little biodiversity, but the important distinction is that taiga is dominated by gymnosperms. Watch out for (A), conifer forests, which are characteristic of taiga, not tundra. Succulents (C) are parts of arid zones with high sunlight, fungi (D) are not the primary producers of any biome, and green algae (E) are more likely in temperate zones and are usually not primary producers. The tundra is dominated by lichens, shrubs, and sedges, so (B) is correct.

5. D

All of the possible answers are connected with conservation issues, so you have to go beyond general familiarity. The ozone layer blocks UV rays, so you might be tempted to consider (C), but this is incorrect. Ultimately, barring being familiar with intricate atmospheric chemistry, you will simply need to remember that CFCs are harmful to the ozone layer (D).

6. C

While reviewing for the exam, it's often beneficial to pick a few examples of organisms that could be classified by particular terminology. This way you have an example to help spark your memory. When considering *r*-selected organisms you should focus on organisms that produce huge numbers of offspring and leave the offspring to fend for themselves, such as salmon. That way you can ask yourself, "Would salmon provide long-term care for their young?" The answer is no and the correct answer to this question is (C).

7. B

You might improve your memory of this phenomenon by considering farmers. Farmers rotate their crops to help limit the depletion of soil nutrients and often use beans in the rotation. The beans are legumes and contribute to nitrogen fixation. Nitrogen (B) is the correct answer.

8. A

Grasses are the dominant primary producers in grassland biomes including the savannah, making grass the keystone species.

9. D

Choice (D) is false because herbivores consume plants (not animals). Carnivores consume animals. The other choices are true: grass would be a primary producer within an ecological community (A); humans would be consumers within an ecological community (B); omnivores consume both plants and animals (C); and bacteria and fungi would be decomposers or saprophytes within an ecological community (E).

10. A

Gross primary productivity is the total chemical energy generated by producers while the net primary productivity subtracts out the loss of energy to respiration by plants. Choice (B) is incorrect because net primary productivity and gross primary productivity have been switched. Choices (C) and (D) are incorrect because primary productivity does not take into account losses to respiration by animals.

11. B

If a toxin is ingested by primary producers in a food web, the toxin will be more concentrated in organisms at the top of the food web. Primary consumers consume primary producers that contain the toxin; secondary consumers primary consume primary consumers that contain the toxin from consumed producers; thus, the concentration of toxin is greater the higher up on the food web an organism is.

12. A

Organisms at the top of the food chain have the least biomass, while organisms at the bottom have the greatest biomass. In this food chain, grass is at the bottom and has the greatest biomass.

13. B

The information in this Cluster Question is directly out of the review list of the world's biomes. Start by matching the obvious biomes like deserts or oceans, then work your way to the less obvious. This is the desert match.

14. C

If you know that grasses are monocots, you will easily find the correct answer.

15. D

Ungulates are hoofed, grazing animals. They are most diverse in the open tall savannas of Africa and Asia.

16. E

The long, cold winters clue is a giveaway, especially since tundra isn't one of the possible answers.

17. E

With a combination of knowledge of the trophic levels and a little bit of logic you should be able to figure out the answer to this type of question. Remember that primary producers are indicated by having only arrows that point away, either being eaten by herbivores or omnivores or providing biomass to decomposers. Once the primary producer is identified, you can identify an herbivore by those species that only ingest primary producers. Start with the autotrophs then work your way through the energy paths.

18. B

Omnivores will ingest primary producers as well as other species that are not primary producers.

19. D

You can distinguish a decomposer from a top carnivore in food webs containing decomposers because it will not have any arrows leading away from it. Be careful. Some food web diagrams do not include decomposers in the system. The College Board may include the position of decomposers in the food web as part of the question.

20. C

You can distinguish the top carnivore from other carnivores because the only energy flow away from a top carnivore is to a decomposer.

Free-Response Answer Explanation

Bacteria reproduce at an exponential rate, with the only limit on population growth being the amount of food available for bacteria to consume. Since the reproductive cycle of the bacteria in the question is so short, bacteria will continue to rapidly replicate until there is no food available to feed the growing number of bacteria.

Here is a possible response:

(a) Bacteria in the petri dish will grow at a steady exponential rate, provided there is food for them to consume. Since the rate of population growth is the difference of reproduction and death rates, the rate will continue to grow as long as the bacteria are able to continue to rapidly reproduce. Using the equation $N_t = N_0 e^{rt}$, where $N_0 = 20$ (the number of bacteria at the beginning, $r = 0.10$ (the rate of population growth), and $t = 1$ step of reproduction, the number of bacteria in the petri dish after 15 minutes will be $N_t = 20e^{(.10 \times 1)}$ or about 22 bacteria.

(b) The population of bacteria in the petri dish will eventually stop multiplying because there will not be enough food, or growth medium, to support an expanding population. Bacteria need food in order to produce energy and replicate. The number of bacteria that can grow in the petri dish indicates the "carrying capacity" of the area of growth. Once all the food (agar) in the dish is consumed and there is a high number of bacteria in the dish, bacteria will start dying off due to lack of food.

The rate of change in size of the bacterial population can be determined by the equation $\frac{\Delta N}{\Delta t} = rN\left(\frac{K-N}{K}\right)$ where $\frac{\Delta N}{\Delta t}$ is the change in the population size over the given time, N is the size of the population at the beginning of time t, r is the rate of population growth (again, a difference of the reproduction and death rates), and K is the carrying capacity. The carrying capacity is the limiting factor in the rate of change in size of the bacterial population.

(c) There are 20 bacteria in the petri dish to start with. Every 15 minutes is one time step, so 45 minutes equals three time steps (15 minutes + 15 minutes + 15 minutes). With a rate of population growth of 0.15, $N_t = 20e^{(.15 \times 3)}$ or about 31 bacteria.

| PART FOUR |

Practice Tests

HOW TO TAKE THE PRACTICE TESTS

The next section of this book consists of practice tests. Taking a practice AP exam gives you an idea of what it's like to answer these test questions for a longer period of time, one that approximates the real test. You'll find out which areas you're strong in, and where additional review may be required. Any mistakes you make now are ones you won't make on the actual exam, as long as you take the time to learn where you went wrong.

The two full-length practice tests in this book each include 100 multiple-choice questions and four free-response (essay) questions. You will have 80 minutes for the multiple-choice questions, a ten-minute reading period, and 90 minutes to answer the free-response questions. Before taking a practice test, find a quiet place where you can work uninterrupted for three hours. Time yourself according to the time limit at the beginning of each section. It's okay to take a short break between sections, but for the most accurate results you should approximate real test conditions as much as possible. Use the ten-minute reading period to plan your answers for the free-response questions, but don't begin writing your responses until the ten minutes are up.

As you take the practice tests, remember to pace yourself. Train yourself to be aware of the time you are spending on each question. Try to be aware of the general types of questions you encounter, as well as being alert to certain strategies or approaches that help you to handle the various question types more effectively.

After taking a practice exam, be sure to read the detailed answer explanations that follow. These will help you identify areas that could use additional review. Even when you've answered a question correctly, you can learn additional information by looking at the answer explanation.

Finally, it's important to approach the test with the right attitude. You're going to get a great score because you've reviewed the material and learned the strategies in this book.

Good luck!

Practice Test 1

Section I: Multiple-Choice Questions

Time: 80 Minutes
100 Questions

Directions: Choose the best answer choice for the questions below.

1. Achondroplasia is a form of dwarfism among humans. A male and female who both have achondroplasia and are under 3.5 ft tall produce an offspring of normal size. The alleles for the father, mother, and their child must be, respectively

   (A) *aa-aa-Aa*
   (B) *AA-aa-Aa*
   (C) *Aa-Aa-AA*
   (D) *Aa-Aa-Aa*
   (E) *Aa-Aa-aa*

2. Why is the muscle of the left ventricle of the heart thicker than the muscle of the right ventricle?

   (A) It only appears to be more muscular because the ventricular space is smaller.
   (B) It has to contain all of the blood collected from the rest of the body.
   (C) It has to pump blood against the pressure of the lungs.
   (D) It has to pump blood to the rest of body.
   (E) All organs are naturally asymmetrical.

3. Self-pollination in plants is an example of

   (A) asexual reproduction, because a single parent is involved
   (B) asexual reproduction, because offspring are genetically identical to the parent plant
   (C) asexual reproduction, because offspring are genetically unique
   (D) sexual reproduction, because offspring are produced via fusion of gametes
   (E) sexual reproduction, because offspring are genetically identical to the parent plant

4. How are bacteria used to produce human insulin?

   (A) They are grown on media rich in sugar, which stimulates insulin production in bacteria.
   (B) The DNA sequence that codes for human insulin production is inserted into the bacterial genome.
   (C) Human pancreas cells are grown in culture with bacteria and transformation occurs.
   (D) Specific bacteriophage viruses are used to produce the correct mutation in the bacterial genome.
   (E) Human plasmids introduced into the bacteria stimulate insulin production.

GO ON TO THE NEXT PAGE

Site	Plants Species	Plants Individuals	Amphibians Species	Amphibians Individuals	Reptiles Species	Reptiles Individuals	Mammals Species	Mammals Individuals	Total Species	Total Individuals
Bluewater Swamp	15	113	2	8	3	8	5	7	25	136
Papago Buttes	5	27	0	0	2	4	2	3	9	34
Beaver's Bend	8	121	2	2	0	0	3	18	13	141
Sherwood Forest	4	159	1	1	0	0	6	24	11	184
Tortilla Flats	4	63	0	0	3	24	1	5	8	92

5. Based on the table above, which site has the greatest species diversity?

   (A) Bluewater Swamp
   (B) Papago Buttes
   (C) Beaver's Bend
   (D) Sherwood Forest
   (E) Tortilla Flats

6. Which of the following is/are true about deoxyribonucleic acid (DNA)?

   I. DNA stores information.
   II. DNA codes for production of proteins.
   III. DNA is found in all living things.

   (A) I only
   (B) II only
   (C) III only
   (D) I and III only
   (E) I, II, and III

7. The three-chambered heart of amphibians was important in the evolution of terrestrial animals because

   (A) it facilitated pulmonary circulation
   (B) it facilitated single-loop circulation
   (C) it facilitated metamorphosis in tadpoles
   (D) it was more efficient than the four-chambered hearts of fish
   (E) it allowed diffusion of gases through the skin of frogs and newts

8. Allosteric regulation of enzyme activity involves

   (A) competitive binding at the enzyme active site
   (B) turning off genes that code for enzyme production
   (C) conformational change in the enzyme due to binding at the allosteric site
   (D) binding of allosteric inhibitors to the substrate
   (E) turning on genes that code for substrate production

9. What part of the vascular system of *Acer sacrum*, the sugar maple tree, would one tap to make maple syrup?

   (A) Tracheids
   (B) Vessel elements
   (C) Xylem
   (D) Phloem
   (E) Endodermis

10. Which of the following are classified in the phylum Arthropoda, animals with hard exoskeletons and jointed appendages?

    (A) Lobsters
    (B) Insects
    (C) Snails
    (D) A and B only
    (E) A, B, and C

11. Transpiration against the force of gravity is possible in trees 10[?] tall because

    (A) water is actively transported from roots to leaves
    (B) evaporation from stomata pulls water up through the tree
    (C) high pressure in the soil pushes the water up
    (D) gravity creates pressure in the xylem, squeezing water out of the stomata
    (E) photosynthesis in the leaves uses water and pulls it through the tree

12. Chimpanzees demonstrate insight learning when presented with a problem. Stacking boxes and climbing them to bat down a suspended banana with a stick demonstrates the ability to

    (A) combine separate experiences to solve novel problems
    (B) associate a new stimulus with a particular reward or punishment
    (C) form memories with no immediate consequence and use them later
    (D) learn to ignore stimuli that have no positive or negative associations or consequences
    (E) perform a particular act in order to receive a reward, which then acts as a reinforcer

13. What are the products of double fertilization in angiosperms?

    (A) An embryo and endosperm
    (B) Two seeds
    (C) A seed and a fruit
    (D) A fruit and a flower
    (E) Two pollen grains

14. Plants form close associations with mycorrhizae, fungi that colonize plant roots. The plant benefits because the fungus makes soil phosphorus available to the plant. The fungus benefits because the plant provides it with sugars. This is an example of

    (A) commensalism
    (B) competition
    (C) parasitism
    (D) mutualism
    (E) predation

15. Internal fertilization probably evolved to allow vertebrates to

    (A) develop two different sexes
    (B) conserve energy as they produce larger eggs
    (C) identify their own species
    (D) reproduce faster than invertebrates
    (E) rely less on water as they invaded the land

16. In the very early evolution of animals, bilateral symmetry led to the formation of left- and right-sidedness, dorsal and ventral surfaces, and anterior and posterior ends. This trend was followed by

    (A) appendages
    (B) cephalization
    (C) gills
    (D) pseudopodia
    (E) an oral opening

17. Why does acid rain affect evergreen trees (e.g., conifers) more adversely than it does deciduous trees (e.g., oaks)?

    (A) Evergreen leaves have thinner cuticles and are more sensitive to acid.
    (B) Acid rain occurs more frequently in habitats dominated by evergreens.
    (C) Deciduous trees are adapted to very acidic conditions.
    (D) Loss of leaves each year reduces long-term effects in deciduous trees.
    (E) Evergreens are unable to filter rain as efficiently as deciduous trees.

18. If the pH of a solution has decreased from 6 to 2, how has the H^+ ion concentration changed?

    (A) It has decreased by a factor of 4.
    (B) It has increased by a factor of 4.
    (C) It has decreased by a factor of 10,000.
    (D) It has increased by a factor of 10,000.
    (E) It has decreased by a factor of 40,000.

19. What kind of organic macromolecules are formed when amino acids undergo dehydration synthesis and the resulting polypeptide chains twist and fold into three-dimensional structures?

    (A) Proteins
    (B) Carbohydrates
    (C) Lipids
    (D) Nucleic acids
    (E) Sugars

20. A cell from an unknown organism is examined under a microscope. It has a membrane-bound nucleus, a large central vacuole, chloroplasts, and a cell wall. What type of cell is it?

    (A) Prokaryotic
    (B) Bacterial
    (C) Fungal
    (D) Plant
    (E) Animal

21. An enzyme speeds up chemical reactions by

    (A) giving up electrons in reduction reactions
    (B) gaining electrons in oxidation reactions
    (C) acting as a reactant to form new products
    (D) acting as a catalyst by loosening chemical bonds
    (E) acting as a substrate and being used up in the reaction

22. Which kingdom is the oldest?

    (A) Monera
    (B) Protoctista
    (C) Fungi
    (D) Plantae
    (E) Animalia

GO ON TO THE NEXT PAGE

23. A pea plant that is heterozygous for alleles that control pea color will express the dominant phenotype and produce yellow peas. If the heterozygous plant is cross pollinated with one that produces the recessive phenotype (green peas) and 800 offspring are produced, how many would you expect to have green peas?

    (A) 100
    (B) 200
    (C) 400
    (D) 600
    (E) 800

24. A man who has a sex-linked recessive disorder carries the gene for the condition on his X chromosome. If he marries a woman who does not have the gene on either of her X chromosomes, what are the chances that their sons will have the disease?

    (A) 100 percent
    (B) 75 percent
    (C) 50 percent
    (D) 25 percent
    (E) 0 percent

25. Which of the following relationships is NOT an example of symbiosis?

    (A) A tick sucking blood from a dog
    (B) A clownfish living in the tentacles of a sea anemone for protection
    (C) Ants farming aphids to feed on their honeydew
    (D) Cows eating grass and leaving manure behind as fertilizer
    (E) Mistletoe parasitizing an oak tree

26. During the light reactions of photosynthesis, water is split to donate electrons to a transport chain and create reducing power, and the protons thereby released fuel a membrane-bound proton pump in a process called photophosphorylation. What are the direct products of the light reactions of photosynthesis?

    (A) Glucose and oxygen
    (B) CO_2 and water
    (C) ATP and NADPH
    (D) ADP and P
    (E) Glucose and water

27. DNA replication is described as being semi-conservative, which means that

    (A) one strand is identical to the parent molecule and one is unique
    (B) one strand of DNA in a chromosome comes from the male gamete and the other from the female gamete
    (C) half of the genetic code is conserved when the molecule replicates
    (D) a replicated DNA molecule is composed of one old strand and one new strand
    (E) during the course of evolution, only about half of the genetic code in DNA is conserved

28. Water has extraordinary properties, not the least of which is that it is lighter in its solid form than in its liquid form. This allows ice to float, so that the greatest volume of water

    (A) freezes from the top down
    (B) freezes from the bottom up
    (C) remains unfrozen
    (D) freezes more quickly
    (E) freezes more slowly

GO ON TO THE NEXT PAGE

29. The sequence of nucleotides in DNA codes for production of proteins, and that code is carried to the ribosome as mRNA. Which of the following protein structures represents the longest strand of nucleotide bases?

    (A) The gene for hemoglobin, a protein which consists of 20 amino acids
    (B) The normal chloride channel protein gene associated with cystic fibrosis, which consists of 63 nucleotide bases
    (C) The mutated sickle cell form of the hemoglobin gene, which has a single substituted base.
    (D) The mutated cystic fibrosis mRNA strand, which has a base insertion
    (E) The mRNA for hemoglobin, which consists of 21 codons

30. According to endosymbiont theory, mitochondria and chloroplasts evolved from prokaryotes engulfed by early eukaryotic cells. Which of the following statements supports this theory?

    I. Mitochondria and chloroplasts have unique DNA not found in the cell nucleus.
    II. Mitochondria and chloroplasts are produced via division and not via synthesis by the cell.
    III. The inner mitochondrial and chloroplast membranes are similar to those of prokaryotes.

    (A) I only
    (B) II only
    (C) III only
    (D) I and II only
    (E) I, II, and III

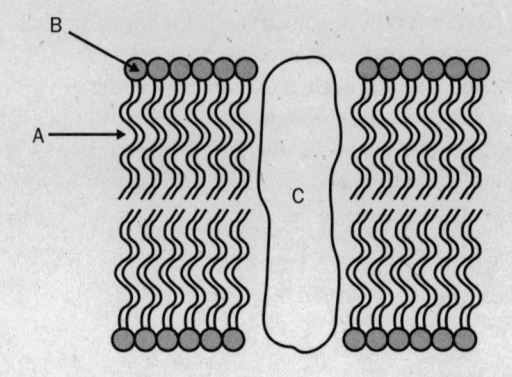

31. Which of the following statements is true?

    (A) The region of the membrane labeled A is nonpolar.
    (B) The region of the membrane labeled B is hydrophobic.
    (C) The structure labeled C is a complex carbohydrate.
    (D) Polar molecules such as water diffuse directly through the membrane bilayer.
    (E) The structure labeled C is a phospholipid.

32. Human white blood cells can "crawl" to damaged tissues to contact and phagocytize bacteria at the site of the damage. Which of the following processes facilitates this type of movement?

    (A) Addition of phospholipids to the cell plasma membrane
    (B) Rapid formation and deformation of actin filaments in the cytoskeleton
    (C) Beating of cilia against vessel and tissue surfaces
    (D) Whip-like motion of the cell's flagellum
    (E) Cellular expansion following uptake of water into the central vacuole

33. In humans, short-term energy supplies are stored as glycogen and long-term energy supplies are stored as fat. Which of the following statements explain(s) this phenomenon?

    I. Carbohydrates are more readily soluble in water than lipids.
    II. Lipids are more easily broken down by mitochondria than carbohydrates.
    III. Glycogen is more readily converted to glucose than is fat.

    (A) I only
    (B) II only
    (C) III only
    (D) I and III
    (E) I, II, and III

34. Which of the following statements is NOT TRUE about DNA and RNA?

    (A) DNA is double stranded and RNA is single stranded.
    (B) Adenine is a nucleotide common to both DNA and RNA.
    (C) Guanine is a nucleotide common to both DNA and RNA.
    (D) Both DNA and RNA can be found in the cell nucleus.
    (E) Both DNA and RNA can be found in the cell cytoplasm.

35. Cellulose, a complex carbohydrate found in plant cell walls, is the most abundant polymer on earth. Which of the following materials would you expect to contain cellulose?

    I. Cotton
    II. Wood
    III. Cork

    (A) I only
    (B) II only
    (C) III only
    (D) I and II
    (E) I, II, and III

36. During complete aerobic cellular respiration, each molecule of glucose broken down in the mitochondria can yield 36 molecules of ATP. What conditions might lead to a decrease in the amount of ATP produced in a given system?

    (A) An increase in the amount of glucose added to the system
    (B) A decrease in the amount of light the system is exposed to
    (C) A decrease in the amount of oxygen available in the system
    (D) A decrease in the amount of carbon dioxide available in the system
    (E) An increase in the amount of ADP in the system

37. Which of the following are most closely related to humans?

    (A) Lobsters and crabs
    (B) Clams and mussels
    (C) Bees and wasps
    (D) Sea stars and sea uchins
    (E) Jellies and corals

38. While exploring a tropical ecosystem, a student encountered an unidentified plant that had xylem and phloem, a free-living haploid generation, and spore-producing structures called "sori" on the underside of compound leaves. What type of plant did the student find?

    (A) A bryophyte
    (B) A fern
    (C) A gymnosperm
    (D) An angiosperm
    (E) A nonvascular plant

GO ON TO THE NEXT PAGE

39. Junipers are conifers that produce seeds with fleshy coats that resemble blueberries. Therefore it is NOT TRUE that junipers

    (A) produce fruits
    (B) reproduce sexually
    (C) are gymnosperms
    (D) are naked-seeded plants
    (E) bear seeds in cones

40. As a science project, you are assigned to observe the embryo of an unknown animal as it develops. During the course of your observations, you note that the cells undergo radial cleavage and that the coelom forms from splitting of the mesoderm. What type of animal might the embryo be?

    (A) A sponge
    (B) A mollusk
    (C) An insect
    (D) An earthworm
    (E) An echinoderm

41. If an animal's jaw contains teeth with broad, rigid surfaces, you can conclude that the animal was probably

    (A) an herbivore
    (B) a producer
    (C) a top consumer
    (D) a carnivore
    (E) a secondary consumer

42. Amphibians require standing water for successful reproduction, while reptiles do not. Which of the following factors contribute(s) to this phenomenon?

    I. External fertilization in amphibians
    II. Amniotic eggs of reptiles
    III. Metamorphosis in reptiles

    (A) I only
    (B) II only
    (C) III only
    (D) I and II only
    (E) I, II, and III

43. Birds are thought to be very closely related to crocodiles. What is a characteristic shared by birds and crocodiles that is NOT shared by birds and turtles?

    (A) A four-chambered heart
    (B) Amniotic eggs
    (C) Homeothermy (warm-bloodedness)
    (D) Deuterostome development
    (E) Hollow bones

44. Grasslike plants such as wheat and rice are monocots, while deciduous trees and most woody shrubs are dicots. Based on this information, which of the following statements is FALSE?

    (A) The leaves of grasses have parallel venation.
    (B) Deciduous trees are flowering plants.
    (C) The seeds of woody shrubs have a single cotyledon.
    (D) Wheat's flower parts occur in multiples of three.
    (E) Rice has diffusely scattered vascular bundles.

45. An unknown animal is observed to have bilateral symmetry. What other characteristic would you expect this animal to have?

    (A) Multicellularity
    (B) Diploblastic development
    (C) A notochord
    (D) Deuterostome development
    (E) All of the above

46. When glucose is broken down into carbon dioxide, water releases about 686 kilocalories of energy, most of which is lost as heat. If 7.5 kilocalories are conserved per ATP molecule, the efficiency of cellular respiration is about

    (A) 87 percent
    (B) 1 percent
    (C) 56 percent
    (D) 39 percent
    (E) 3 percent

47. Tapeworms are endoparasites, which means that they

    (A) live attached to the outside of their hosts
    (B) cannot live outside the body of a host
    (C) live inside the body of a host
    (D) must reproduce within the body of a host
    (E) cannot live inside the body of a host

48. A defective protein has been produced due to an error in gene transcription. The molecule that contained the error was most likely

    (A) DNA
    (B) rDNA
    (C) rRNA
    (D) mRNA
    (E) tRNA

49. Which of the following statements is/are true about the wings of birds, bats, and insects?

    I. They are homologous structures.
    II. They are analogous structures.
    III. They are examples of divergent evolution.

    (A) I only
    (B) II only
    (C) III only
    (D) I and III only
    (E) II and III only

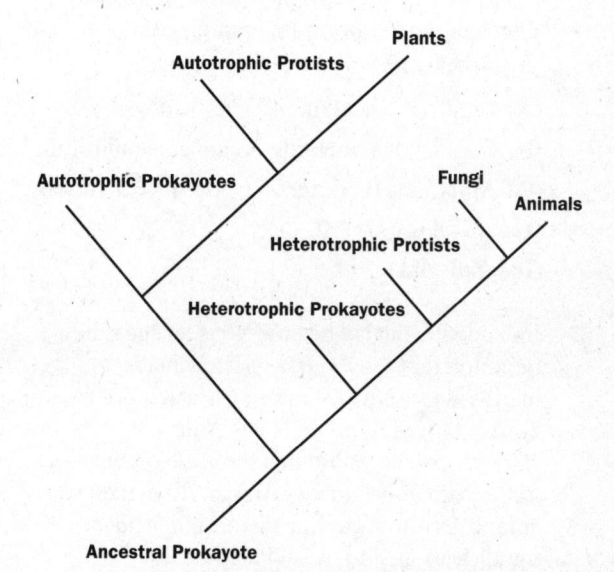

50. The phylogenic analysis shown above suggests that

    (A) animals are descended from fungi
    (B) fungi are more similar to animals than to plants
    (C) protists are a monophyletic group
    (D) the first organisms were eukaryotes
    (E) prokaryotes are a monophyletic group

GO ON TO THE NEXT PAGE

51. Which of the following is an example of post-zygotic reproductive isolation?

    (A) Species in the same area occupying incompatible habitats
    (B) Species separated geographically by a physical barrier such as an ocean
    (C) Prevention of gamete fusion
    (D) Incompatible mating rituals
    (E) Production of sterile hybrid offspring

52. Evolution may be defined as changes over time in the allele frequency of a population or species. Which of the following are examples of evolutionary processes?

    (A) Artificial selection in domestic dogs
    (B) Populations in Hardy-Weinberg equilibrium
    (C) Mutations that decrease reproductive fitness
    (D) A and C only
    (E) A, B, and C

53. Individuals who are heterozygous for the gene mutation that causes sickle cell anemia have increased resistance to malaria. Malaria occurs in central Africa but not in North America; researchers have found that the allele occurs much more frequently in native Africans than it does among African Americans. The high incidence of the allele in Africa is most likely due to

    (A) nonrandom mating
    (B) genetic drift
    (C) stabilizing selection
    (D) migration
    (E) macroevolution

54. Archaebacteria are unique prokaryotes that are thought to be more closely related to eukaryotes than they are to other bacteria. Which characteristic listed below supports this idea?

    (A) Many archaebacteria are adapted to extreme environments, such as deep-sea thermal vents.
    (B) The cell walls of archaebacteria lack peptidoglycans.
    (C) The start codon in archaebacteria is the same as that of eukaryotes.
    (D) Archaebacteria lack a membrane-bound nucleus and organelles.
    (E) Archaebacteria inhabit the guts of cattle and produce methane gas.

55. Organisms that reproduce sexually exhibit zygotic, gametic, or sporic meiosis. One way to determine the type of life cycle an organism has is by

    (A) observing embryonic development
    (B) comparing the diploid and haploid forms of the organism
    (C) determining when in the life cycle fertilization occurs
    (D) determining if gametes are multicellular or unicellular
    (E) counting the number of times meiosis occurs

56. Which of the following animals is most closely related to an octopus?

    (A) Whale
    (B) Trout
    (C) Snake
    (D) Spider
    (E) Clam

57. Which shows the correct order of hierarchy from simple to complex?

    (A) Molecule → tissue → cell → organism
    (B) Cell → tissue → organ → organism
    (C) Organism → species → population → ecosystem
    (D) Molecule → cell → organism → tissue
    (E) Species → organism → community → population

58. Rabbits are herbivores. What trophic level would a rabbit occupy in an ecosystem?

    (A) Primary producer
    (B) Primary consumer
    (C) Secondary consumer
    (D) Top consumer
    (E) Decomposer

59. Organisms transform energy when they ingest food, breaking it down into nutrients used to build tissues and make repairs. Each energy transfer increases the universe's level of

    (A) order
    (B) stability
    (C) disorder
    (D) energy
    (E) structure

60. What is the primary danger of a population relying on an agricultural monoculture for its dietary staple?

    (A) People get tired of eating the same thing.
    (B) Children develop aversion to foods that appear too often in their diet.
    (C) Food allergies develop after repeated exposure.
    (D) Genetic similarity makes an entire crop vulnerable to a single pest or pathogen.
    (E) Genetic variability within a single crop makes production unreliable.

Directions: For the questions below, there are five lettered choices followed by a phrase or sentence for each question. Choose the letter that best corresponds to the phrase or sentence for each question and fill in that answer on your answer grid. Each choice can be used once, two or more times, or not at all for the following questions.

Questions 61–65

   (A) Mosses
   (B) Ferns
   (C) Gymnosperms
   (D) Monocots
   (E) Dicots

61. Plants that produce seeds but not fruit
62. Plants with flower parts in multiples of four or five
63. Seedless vascular plants
64. Plants lacking vascular tissue
65. Plants with a dominant haploid generation

Questions 66–68

   (A) Bacterium
   (B) Protist
   (C) Fungus
   (D) Plant
   (E) Animal

66. Multicellular heterotroph with cell walls containing chitin
67. Multicellular terrestrial autotroph with cell wall containing cellulose
68. Multicellular heterotrophic organism lacking cell walls

GO ON TO THE NEXT PAGE

Questions 69–72

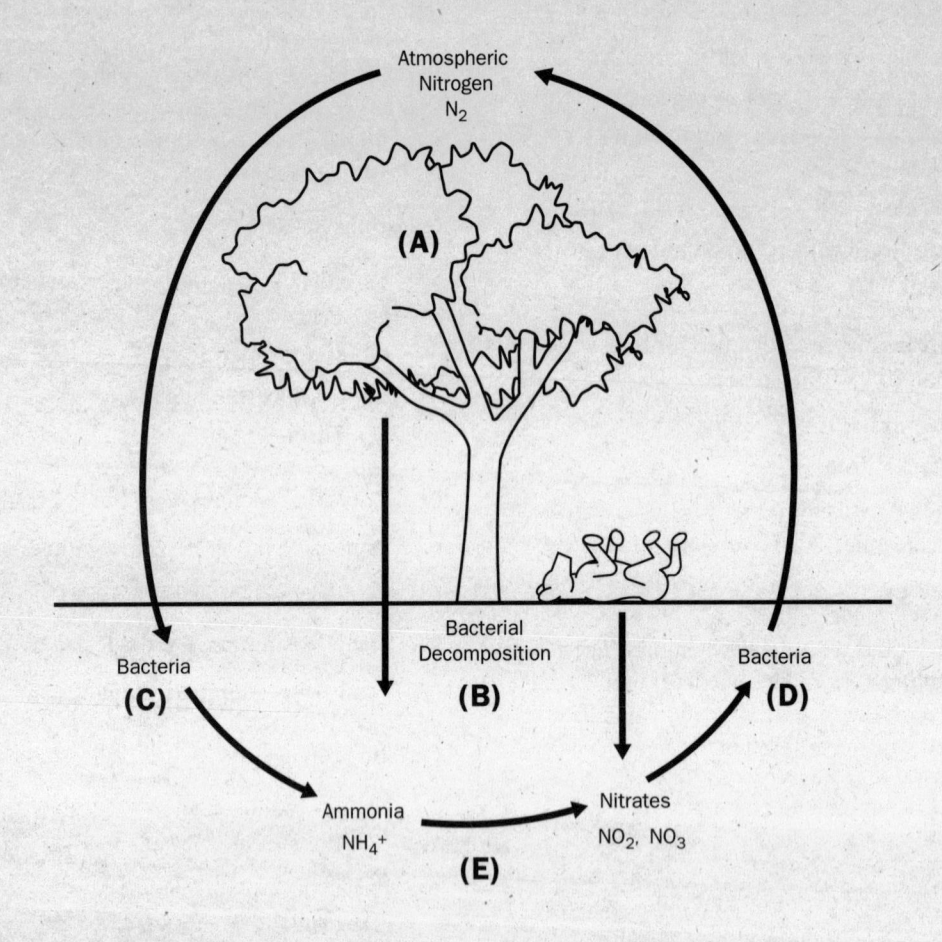

69. Nitrogen fixation by soil bacteria

70. Incorporation of nitrogen in rubisco

71. Denitrification of nitrates via reduction by soil bacteria

72. Conversion from organic to inorganic nitrogen

Questions 73–75

(A) DNA polymerase
(B) RNA polymerase
(C) Ligase
(D) Helicase
(E) Restriction enzymes

73. Enzyme that unzips the DNA molecule prior to replication

74. Records the genetic code by catalyzing the process of transcription

75. Catalyzes addition of nucleotides within replication forks

Directions: Each group of questions below concerns an experiment or laboratory situation or data. In each case, first study the description of the situation or data. Then choose the best answer to each question that follows and fill in the corresponding oval on the answer sheet.

Questions 76–80

Your biology teacher has created a computer game that simulates the function of a living cell. The object for the player is to act as the command center for the cell and to send the appropriate information for protein production and organelle function. For each situation below, select the appropriate cellular structure or functional response needed from the command center.

76. More ATP is needed.

(A) Increase protein synthesis.
(B) Increase mitochondrial activity.
(C) Produce more cytoskeletal elements.
(D) Initiate cell division.
(E) Release more peroxisomes.

77. The cell is in the S stage of interphase.

(A) Replicate DNA.
(B) Align chromosomes at the cell equator.
(C) Rest.
(D) Pull sister chromatids to opposite poles of the cell.
(E) Form a cleavage furrow.

78. The cell has been dropped into a hypertonic solution.

(A) Take up water through aquaporins.
(B) Close all membrane-bound proteins.
(C) Pump out solutes.
(D) Lose water through aquaporins.
(E) Explode.

79. Amino acids are needed at ribosomes for protein production.

(A) Send rDNA.
(B) Send mRNA.
(C) Send rRNA.
(D) Send tRNA.
(E) Send mDNA.

80. The cell has engulfed particles that need to be disposed of.

(A) Initiate chloroplast activity.
(B) Stimulate the ER to synthesize lipids.
(C) Increase lysosome activity.
(D) Shut down enzyme activity.
(E) Increase cellular respiration.

GO ON TO THE NEXT PAGE

Questions 81–84

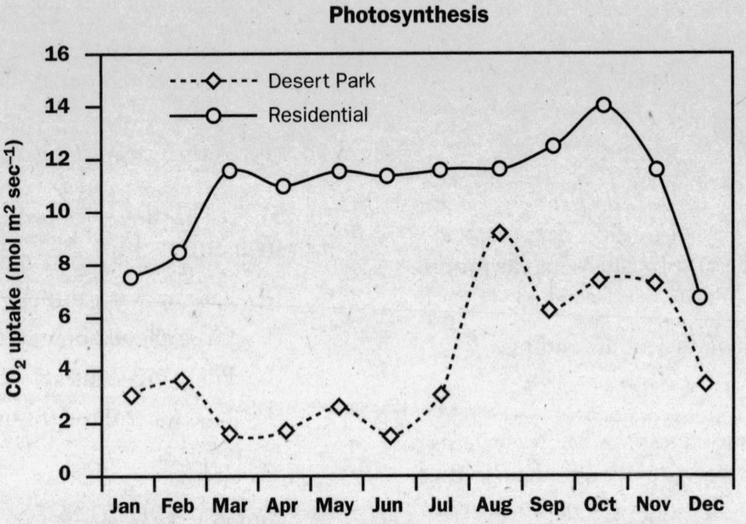
Photosynthesis
Desert Park
Residential
CO_2 uptake (mol m² sec⁻¹)
16
14
12
10
8
6
4
2
0
Jan Feb Mar Apr May Jun Jul Aug Sep Oct Nov Dec

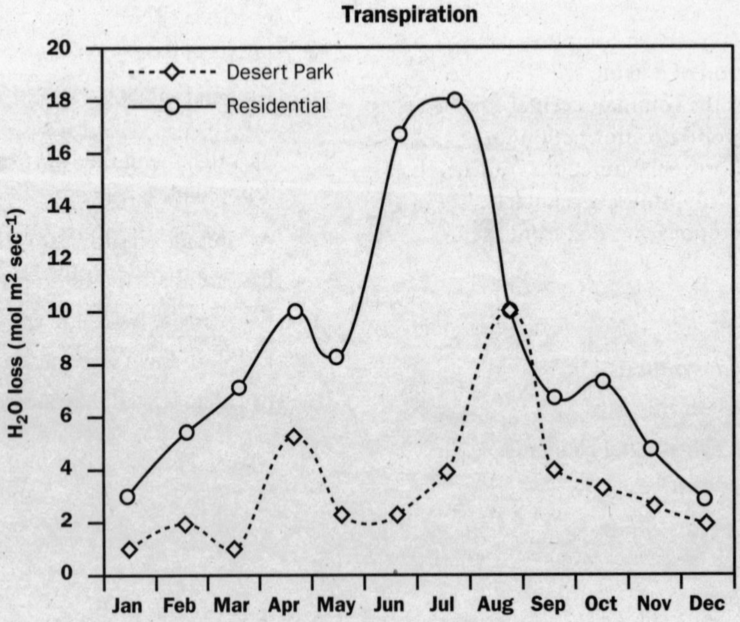
Transpiration
Desert Park
Residential
H_2O loss (mol m² sec⁻¹)
20
18
16
14
12
10
8
6
4
2
0
Jan Feb Mar Apr May Jun Jul Aug Sep Oct Nov Dec

GO ON TO THE NEXT PAGE

Scientists conducting research on the ecology of cities made a comparison of rates of net photosynthesis and transpiration of plants in irrigated suburban residential and unirrigated desert park sites in Phoenix, Arizona for one year. Phoenix is a very hot and arid city with a pattern of infrequent, light rains during the spring and periodic heavy monsoon rains during late summer and fall. Total rainfall is typically less than 7 inches per year. A variety of tree and shrub species at three desert parks and three residential sites were chosen for study. Dawn-to-dusk measurements of photosynthesis and transpiration were averaged to a monthly mean for each site type. Results for the year are shown on the opposite page.

81. Based on the results, what conclusion might be drawn about the photosynthesis of plants in the two types of sites?

    (A) In irrigated plants, photosynthesis is highest during summer months.
    (B) Plants in desert sites have lower rates of photosynthesis because stomata close to conserve water.
    (C) Productivity is greatly reduced by irrigation in residential sites.
    (D) Water availability limits photosynthesis in irrigated residential plants during the summer.
    (E) High rates of transpiration inhibit photosynthesis in irrigated plants during the summer.

82. Transpiration rates in unirrigated desert plants follow a pattern consistent with the seasonal periods of rainfall that occur in Phoenix, but rates of photosynthesis do not. Why might this be the case?

    (A) Hot weather in late summer increases photorespiration in desert plants and net photosynthesis is reduced.
    (B) Rates of photosynthesis would not be expected to be related to transpiration rate.
    (C) Growth respiration during spring reduces net photosynthesis despite water availability.
    (D) Increased rates of photosynthesis are expected to lag several months behind increased transpiration rate.
    (E) The stomata of desert plants do not open fully during spring, thereby reducing photosynthesis.

83. Water-use efficiency (WUE) is calculated as photosynthesis divided by transpiration. What conclusion about plant WUE might be drawn from these data?

    (A) WUE is much higher in desert than in residential plants.
    (B) WUE is much higher in residential than in desert plants.
    (C) WUE is higher in summer than in winter for plants at both site types.
    (D) WUE is higher in winter than in summer for plants at both site types.
    (E) It is impossible to make any conclusions about WUE based on these data.

84. What might be a source of error in the experimental design in terms of comparison of the two site types?

    (A) The same species of plants were not used for both site types.
    (B) The measurements were not made on the same days for both site types.
    (C) It is not valid to compare irrigated and unirrigated plants.
    (D) A and B only
    (E) B and C only

Questions 85–89

The pedigree below shows the occurrence of a genetic disorder through several generations and several lineages of a family. The squares represent males and the circles represent females; the semicircles represent females heterozygous for, but unaffected by, the defective gene. None of the males are heterozygous for the condition. The symbols G1, G2, and G3 represent the first, second, and third generations subsequent to the initial mating pair. The symbols L1, L2, and L3 represent the three lineages of descent from the initial generation.

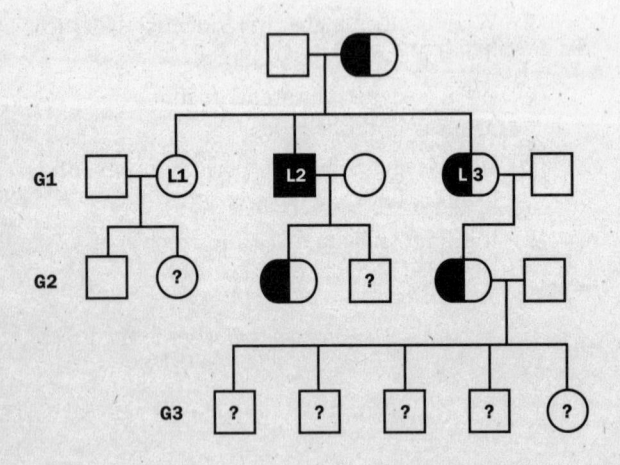

85. What is the probability that the female offspring in the second generation of the first lineage carries the defective gene?

    (A) 100 percent
    (B) 50 percent
    (C) 25 percent
    (D) 0 percent
    (E) Impossible to determine from the data given

86. What is the probability that the male in the second generation of the second lineage has the genetic disorder?

    (A) 100 percent
    (B) 50 percent
    (C) 25 percent
    (D) 0 percent
    (E) Impossible to determine from the data given

87. What type of disorder is most likely shown by this pedigree?

    (A) A sex-linked dominant disorder
    (B) A sex-linked recessive disorder
    (C) A non-sex-linked dominant disorder
    (D) A non-sex-linked recessive disorder
    (E) A disorder associated with multiple genes

88. Of the five children in the third generation of the third lineage, how many of the children would be expected to be affected by the disorder?

    (A) Only the girl
    (B) All of the boys
    (C) Half of the boys
    (D) One of the boys
    (E) None of the children

89. Which of the hereditary diseases listed below might this pedigree represent?

    (A) Tay-Sachs disease
    (B) Huntington's disease
    (C) hemophilia
    (D) sickle cell anemia
    (E) achondroplasia

GO ON TO THE NEXT PAGE

Questions 90–93

		Patient Value	Normal Value
Hematology	Hemoglobin	15	13–18 gm/dl
	White Blood Cell Count % Neutrophils % Lymphocytes	36,000 97 3	5000–10,000 µl/mm^3 48–73% 18–48%
	Platelet Count	175,000	150,000–350,000/ml
Blood Chemistry	Glucose	444	70–110 mg/dl
	Blood Urea	87	7–18 mg/dl
	Ethanol	0.1	0–0.1 mg/dl
	Blood pH	7.0	7.35–7.45

A man is found unconscious on the sidewalk and a passerby rushes him to a nearby hospital emergency room. The patient has no identification and no medical history is available. On initial examination, the ER physician finds that the man's temperature is elevated and that his blood pressure is slightly decreased. His EKG is normal and there are no obvious signs of trauma on his body. A laboratory work-up is ordered and the above test results are obtained.

90. Which of the following is LEAST likely to be the man's problem?

    (A) Metabolic acidosis
    (B) Sepsis
    (C) Kidney failure
    (D) Anemia
    (E) Hyperglycemia

91. Which of the following diagnoses is/are supported by the lab results?

    (A) Diabetes
    (B) Chronic alcoholism
    (C) Metabolic acidosis
    (D) A and B only
    (E) A and C only

92. Based on the lab results, the physician wants to order more tests to rule out infection. Which of the laboratory results has led her to this decision?

    (A) The red blood cell count
    (B) The white blood cell count
    (C) The blood glucose level
    (D) The blood urea level
    (E) The blood pH

93. Based on results of a urinalysis, the physician decides to order bacterial blood and urine cultures on the patient. She also wishes to get some idea of what the causative organism is before culture results are available. What would be a good additional test for her to order?

    (A) A gram stain of urine
    (B) An acid fast stain of the blood
    (C) A screen for protozoan parasites in the urine
    (D) A spinal tap
    (E) A serological analysis of the blood

Questions 94–96

In a long-term project studying the interactions of several species of animals on an isolated island, scientists counted the number of individuals of each species visiting a site on the island over the course of several days every summer for 100 years. The results from that study are shown below.

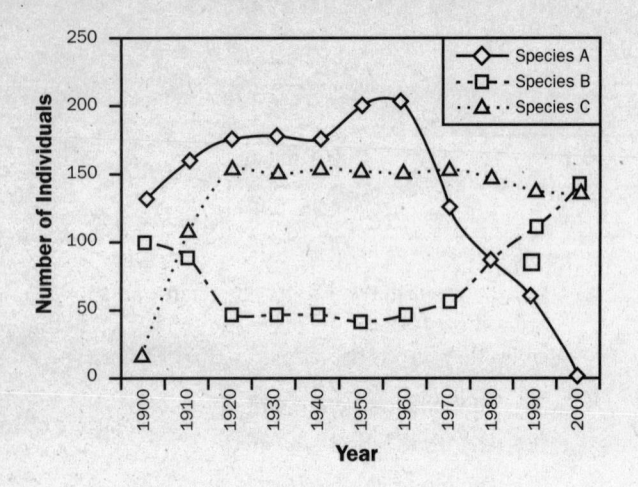

94. Which of the following statements is supported by the figure?

    (A) Species A and species C probably compete for limited food supplies.
    (B) Species B might be a predator of species A.
    (C) Species C will probably increase in number during the 21st century.
    (D) Species C has reached the carrying capacity of the system.
    (E) Species A has reached the carrying capacity of the system.

95. The population of species C remained fairly constant until the 1980s, when numbers began to gradually decline. Based on the figure, which of the following is NOT a possible explanation for the decline of species C?

    (A) Species C has a mutualistic relationship with species A.
    (B) Species C is a predator of species B.
    (C) Species C is a predator of species A.
    (D) Species C is prey for species B.
    (E) Species C was influenced by another organism not included in the study.

96. During the years 2000–2004, a new species (species D) began to appear at the site. The new species is known to have a competitive relationship with both species A and B. What changes in the population dynamics might be expected over time?

    (A) A gradual reappearance of Species A
    (B) A gradual increase in the population of Species B
    (C) A gradual increase in the population of Species C
    (D) A gradual decrease in the population of species B
    (E) No change would be expected

Questions 97–100

The diagram below shows energy transformations within a cell. Each form of energy is represented by the symbols E I–E IV. Two cellular organelles are represented by the letters A and B. Answer the following questions about the various processes depicted in the diagram and about the cell in which they are occurring.

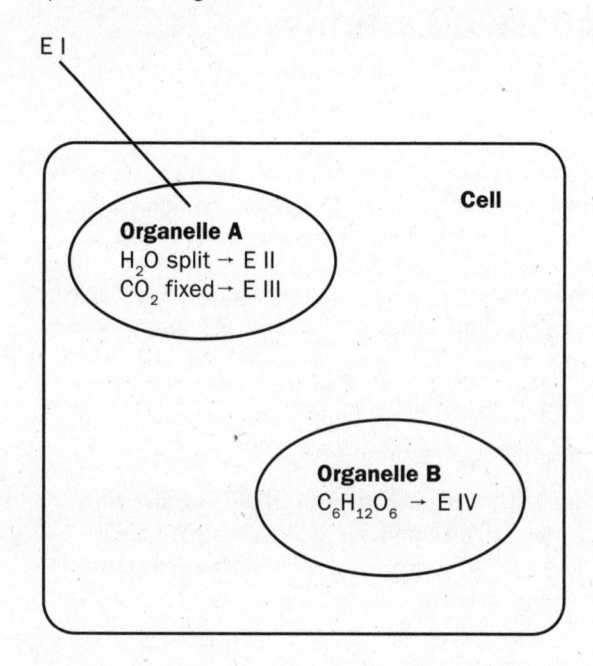

```
E I
         ┌─────────────────────────────────┐
         │                            Cell │
         │   ╱───────────╲                 │
         │  │ Organelle A │                │
         │  │ H₂O split → E II             │
         │  │ CO₂ fixed → E III            │
         │   ╲───────────╱                 │
         │                                 │
         │              ╱───────────╲      │
         │             │ Organelle B │     │
         │             │ C₆H₁₂O₆ → E IV    │
         │              ╲───────────╱      │
         └─────────────────────────────────┘
```

97. What form of energy is represented by E II?

    (A) Radiant energy in the form of photons
    (B) Chemical energy being stored in the bonds of glucose
    (C) Chemical energy in the form of ATP
    (D) Chemical energy released by glycolysis
    (E) Reducing power in the form of ADP

98. If the transformation depicted in organelle B requires oxygen, what form of energy is represented by E IV?

    (A) Radiant energy in the form of photons
    (B) Chemical energy being stored as glycogen
    (C) Chemical energy in the form of ATP
    (D) Chemical energy released by glycolysis
    (E) Reducing power in the form of ADP

99. What cellular organelles are represented as A and B, respectively?

    (A) The nucleus and the ribosome
    (B) The mitochondrion and the chloroplast
    (C) The mitochondrion and the ribosome
    (D) The chloroplast and the Golgi complex
    (E) The chloroplast and the mitochondrion

100. Which kind of organism could the cell shown belong to?

    (A) A photosynthetic bacteria
    (B) A photosynthetic protist
    (C) A plant
    (D) B and C only
    (E) A, B, and C

IF YOU FINISH BEFORE TIME IS CALLED, YOU MAY CHECK YOUR WORK ON THIS SECTION ONLY. DO NOT TURN TO ANY OTHER SECTION IN THE TEST.   **STOP**

Ten-Minute Reading Period

Take the next ten minutes to glance over the four questions that comprise Section II of this test. You can take notes in the margins, but you may not begin answering any of the questions in any way whatsoever.

When the ten-minute period is over, you may begin answering the four free-response questions.

Section II: Free-Response Questions

Time: 90 Minutes
4 Questions

Directions: Answer each part of the following free-response questions with complete sentences. Answers should be in essay form. Make sure to provide a detailed response to each part of each question. Diagrams and other figures may be included to demonstrate knowledge of a topic, but they should not be the only answer you provide.

1. In a particular species of guppy, tails can either be long or short and either feathered or straight. A cross between a true-breeding long, feather-tailed male guppy and a true-breeding short, straight-tailed female guppy produces progeny that all have short, straight tails.

    (a) **Express** the genotypes of the adults in the cross and the genotypes of all the progeny.

    (b) **Determine** the phenotypic frequencies of the progeny in an F2 generation.

    (c) A mating between a short, feather-tailed female and a short, straight-tailed male from the F1 generation produces 383 short, straight-tailed guppies, 367 short, feather-tailed guppies, 118 long, straight-tailed guppies, and 132 long, feather-tailed guppies. **Analyze** the data using a chi-squared fit test to **determine** if the frequencies of phenotypes are what would be expected from this cross.

The formula for chi-squared is: $\chi^2 = \sum \frac{(Observed - Expected)^2}{Expected}$

Probability (p)	Degrees of Freedom (d.f.)				
	1	2	3	4	5
0.05	3.84	5.99	7.82	9.49	11.1

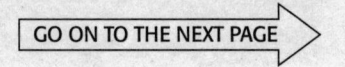

GO ON TO THE NEXT PAGE

2. All living organisms contain genetic information that provides several functions inherent to the individual organism and the perpetuation of its species.

   (a) **Describe** how genetic material contributes to the regulation of physiological function and development.

   (b) **Discuss** how the nature of genetic material both perpetuates the identity of an individual and provides for high biodiversity.

3. The human body manifests physiological changes in every organ system throughout the process of eating, from the time a person first anticipates food to the elimination of waste.

   (a) **Design** a controlled experiment where you test the hypothesis that the body's reaction to eating begins before food enters the body.

   (b) **Describe** physiological changes that occur in three different organ systems when food is ingested and how these systems coordinate a response in the human body.

4. Human impact on the Earth's environment is poignantly demonstrated in the Mississippi River drainage into the Gulf of Mexico. Every spring, high levels of nitrogen and phosphorous wash into the river from the Midwestern United States farming region and into the northern Gulf. These nutrients supply a huge algal bloom at the water surface creating a dangerously destructive hypoxia that all but destroys the marine ecosystem.

   (a) **Discuss** the cycling of oxygen in the natural marine ecosystem and how this cycle is upset during an algal bloom.

   (b) Pick two organisms from two different trophic levels in the marine ecosystem and **explain** the impact of the algal bloom on their ecology.

   (c) It has been suggested that the hypoxic condition in the Gulf might be contributing to an increased number of shark attacks along the Texas coast. **Explain** the thinking behind this suggestion.

Practice Test I: Answer Key

1. E	26. C	51. E	76. B
2. D	27. D	52. D	77. A
3. D	28. C	53. C	78. D
4. B	29. D	54. C	79. D
5. A	30. E	55. B	80. C
6. E	31. A	56. E	81. B
7. A	32. B	57. B	82. C
8. C	33. D	58. B	83. D
9. D	34. E	59. C	84. D
10. D	35. E	60. D	85. D
11. B	36. C	61. C	86. D
12. A	37. D	62. E	87. B
13. A	38. B	63. B	88. C
14. D	39. A	64. A	89. C
15. E	40. E	65. A	90. D
16. B	41. A	66. C	91. E
17. D	42. D	67. D	92. B
18. D	43. A	68. E	93. A
19. A	44. C	69. C	94. D
20. D	45. A	70. A	95. B
21. D	46. D	71. D	96. D
22. A	47. C	72. B	97. C
23. C	48. D	73. D	98. C
24. E	49. B	74. B	99. E
25. D	50. B	75. A	100. D

ANSWERS AND EXPLANATIONS

Section I: Multiple-Choice Questions

1. E

The allele that causes achondroplasia is dominant, which can be inferred from the fact that both parents are affected and the offspring is not. To produce unaffected offspring, both parents must be heterozygous *Aa*, and their normal-sized offspring must be *aa*.

2. D

The left ventricle has to pump blood to the most distal parts of the body. The right atrium is the collecting site of blood from the body (B), and the right ventricle pumps blood to the lungs (C), necessarily under far less pressure. Choices (A) and (E) are untrue.

3. D

By definition, asexual reproduction is production of offspring genetically identical to the single parent, primarily via cell division. Sexual reproduction involves fusion of unique haploid gametes to produce genetically unique diploid offspring. Self-pollination is a form of sexual reproduction. A single parent plant produces both male and female gametes, but each is the product of meiosis and genetically unique. Fusion of two gametes produces a unique diploid zygote, which will develop into a unique adult.

4. B

The only way a bacterium can be induced to produce a human protein such as insulin is to have the gene that codes for production of that protein inserted into the bacterial genome. Choice (A) is clearly wrong; bacteria are cultured on sugar-rich media routinely and it does not confer human abilities upon them. Both (C) and (D) suggest technically feasible possibilities, except that the chances of such a beneficial transfer of genetic information occurring in either case would be virtually zero. Choice (E) is incorrect because human cells do not have plasmids; the plasmids used to insert human genes are of bacterial origin.

5. A

Species diversity is a measure of the number of different kinds of organisms present at a site. Papago Buttes (B) does not have very many species or many individuals. Beaver's Bend (C) has a lot of different species, but the overwhelming majority of animals there are mammals. Although Sherwood Forest (D) has the most plants and animals, there are few species. The same is true of Tortilla Flats (E).

6. E

DNA stores the information necessary for building cells and carrying out their functions, and it allows that information to be passed on to subsequent generations. DNA codes for production of proteins including enzymes, which regulate all of the other metabolic activities of the cell, such as synthesis of organic molecules. Traditionally, scientists have held that the presence of DNA is one of the defining characteristics of life; anything that lacks DNA is not alive. This is one reason why the status of viruses as living organisms is in doubt: some only have RNA instead of DNA.

7. A

Amphibians and many reptiles have a three-chambered heart that allows circulation to the lungs where blood is oxygenated, a vital process in the evolution of land animals. Choice (B) is incorrect because single-loop circulation is associated with fish. Choice (C) is incorrect, because metamorphosis is genetically regulated and not related to the cardiovascular system. Fish do not have four-chambered hearts (D). Although frogs and newts can exchange gases through the skin, this process bypasses the lungs and so is unrelated to the three-chambered heart (E).

8. C

Allosteric regulators may be inhibitors or activators, but by definition they always change the shape of the protein due to binding at an allosteric site, and not by binding at the enzyme active site (A). Allosteric inhibitors bind to the enzyme allosteric site and not the substrate (D). They change the shape of the enzyme so that the substrate cannot be bound at the enzyme active site. Activators change the shape of the enzyme to allow binding of the substrate. Allosteric regulation is unrelated to genetic regulation, ruling out (B) and (E).

9. D

Tracheids (A) and vessel elements (B) are cellular components of xylem (C), which transports water and minerals up from the soil through the plant body. Phloem (D) is the vascular tissue responsible for sugar transport, which is tapped to collect maple syrup. Endodermis (E) lies within roots and regulates uptake of minerals.

10. D

Lobsters and insects are arthropods. Snails are mollusks. Their shells are not exoskeletons, and they lack jointed appendages.

11. B

Transpiration is a passive process, one that does not require energy expenditure by the plant, ruling out (A). Although positive pressure in the roots can cause some water loss, it is not the typical case and does not occur in very tall trees, so (C) is incorrect. Gravity (D) could only exert pressure downward, and the amount of water actually used in photosynthesis (E) is tiny relative to the amount lost via transpiration. Because water is cohesive and adhesive, evaporation literally pulls water through the tree, much like sucking up liquid through a straw.

12. A

Insight learning is the ability to exhibit a useful or productive behavior in a situation with which the animal has no experience. Choice (B) is classical conditioning, choice (C) is observational learning, choice (D) is habituation, and choice (E) is operant conditioning.

13. A

Angiosperms, or flowering plants, are considered a monophyletic group based on the characteristic of double fertilization. One sperm fertilizes the egg of a mature ovule, and the other one fertilizes the polar nuclei. The ovule develops into an embryo, and the polar nuclei develop as endosperm, food reserves for the embryo. Although many seeds may be produced (B) and a fruit surrounds the seed or seeds (C), each seed with its embryo is a product of a double fertilization event. Flowers (D) are where fertilization occurs, and pollen grains (E) produce the sperm.

14. D

Commensalism (A) is a form of symbiosis in which one organism benefits and the other is neither benefited nor harmed. Competition (B) occurs when two organisms both use the same limited resources, such as food or sunlight. Parasitism (C) is another form of symbiosis, one in which one organism benefits while the other is harmed. Predation (E) involves one organism consuming another. Mutualism (D) is the form of symbiosis in which both organisms involved benefit.

15. E

The two primary forms of fertilization are external and internal. External fertilization requires that eggs be deposited in safe environments, protected from heat and desiccation, so that eggs are invariably deposited in water or areas that are reliably damp. Internal fertilization frees animals from these constraints.

16. B

Cephalization was the evolutionary development in animals that created an anterior end in which sensory equipment (e.g., for taste) was localized. Appendages (A) and gills (C) came after the head. Single cells (such as amoebae and macrophages) move by way of pseudopodia (D), and oral openings (E) are also found in animals with radial symmetry.

17. D

Acid rain is harder on evergreen trees because their leaves have to endure many years of cumulative effects. Evergreens typically have leaves with thicker cuticles than deciduous trees, so (A) is incorrect. Choice (B) is incorrect because acid rain is not a localized phenomenon. Although (C) might be a reasonable guess, one must consider that acid rain is composed of sulfuric acid, and very few living things can tolerate regular sulfuric acid showers. Although plant canopies do some "filtering" of precipitation (E), it would not do much good in the case of acid rain, because most of the damage is done to the leaves and not via uptake of acid by the plant roots.

18. D

The pH scale is negative and logarithmic. Therefore, for each unit that pH decreases, H^+ ion concentration increases by a factor of 10. A fourfold decrease in pH translates to a ten-thousand fold (10^4) increase in H^+ ion concentration.

19. A

Amino acids are the monomers of proteins. All organic macromolecules are formed via dehydration synthesis of monomers. For carbohydrates (B), they are monosaccharides and for lipids (C), fatty acids and glycerol. Nucleotides are the monomers of DNA and RNA (D). Only proteins are made of amino acids, and polypeptides fold and twist into distinct three-dimensional shapes.

20. D

Choices (A) and (B) are incorrect because the organism has a membrane-bound nucleus, making it a eukaryote. Both fungi (C) and animal (E) cells lack chloroplasts. Only a plant cell would have all of the structures mentioned.

21. D

Enzymes catalyze chemical reactions but are not used up (E) or changed by the reactions they catalyze. Either oxidation (A) or reduction (B) would change the enzyme, as would its being chemically altered into a new product (C).

22. A

Kingdom Monera includes prokaryotes such as bacteria. Scientists believe these prokaryotic organisms evolved prior to eukaryotic organisms (including all the other answer choices).

23. C

The Punnet square for the described cross would show *Yy* crossed with *yy*. The offspring produced would be 50 percent *Yy* and 50 percent *yy*, so half would show the recessive phenotype.

24. E

This disorder is sex-linked and X-linked, passed on the X chromosome. The father can pass the gene on to his daughters, but not to his sons. Since boys have one X and one Y chromosome and mothers lack Y chromosomes, a boy always gets his Y chromosome from his father. Sons would therefore never get the X-linked gene. In this instance, no daughters would have the disorder either, since they would all get one "healthy" X chromosome from their mother. Because it is a recessive disorder, the condition would not affect the daughters.

25. D

Symbiosis is simply defined as organisms living in close association with each other. Choices (A) and (E) are examples of parasitism, (B) of commensalism, and (C) of mutualism, all forms of symbiosis. Choice (D) is an example of predation, one organism feeding on another, which is not a form of symbiosis.

26. C

Although oxygen is a by-product of the light reactions of photosynthesis, glucose is not produced until the light-independent reactions occur, so (A) is incorrect. Choices (B) and (E) are incorrect because CO_2 and water (respectively) are not products of photosynthesis at all, they are reactants. During the light reactions, ATP is produced by a membrane-bound proton pump via photophosphorylation (ADP + P $\rightarrow$ ATP). The electrons released by water ultimately are accepted by $NADP^+$ to produce NADPH. Choice (D) is a reaction that occurs during the light-independent reactions to provide energy for glucose production.

27. D

DNA replication produces a double-stranded molecule identical to its parent because after unzipping the double helix, a complementary strand is produced for each strand. This results in a new molecule composed of one old strand and one new complementary strand. All other answer choices have implications for a replicated DNA molecule different from the original.

28. C
Ice acts as an insulating cap on a body of water, allowing the bulk of water to remain unfrozen, which protects water-dwelling organisms. If ice sank and bodies of water froze from the bottom up, then they would become solid, unable to support much life. Choices (D) and (E) are variations on the theme of overall freezing.

29. D
Both the non-mutated hemoglobin and cystic fibrosis gene sequences consist of 20 amino acids plus the stop codon. Since each amino acid has a three-nucleotide codon, both are composed of 63 bases, so (A) and (B) would be the same length. Choice (E) is also the same length; the stop codon is incorporated in the mRNA. Choice (C) is incorrect because the mutation associated with sickle cell is a substitution; one nucleotide replaces another, so the length of the molecule is the same. The cystic fibrosis mutation involves an insertion—the addition of a nucleotide—making it the longest molecule.

30. E
All of the characteristics described lead biologists to believe that mitochondria and chloroplasts both represent incorporation of prokaryotes into early eukaryotic cells.

31. A
The region labeled A is composed of long fatty acid tails of the phospholipids that make up the cell plasma membrane. The lipid end of the molecule is nonpolar and so hydrophobic, which prevents polar molecules such as water from crossing the membrane, so (D) is incorrect. Region B is the polar, hydrophilic region of the molecule. The structure labeled C is a membrane-bound protein, making (C) and (E) incorrect.

32. B
Most cell movement is associated with the cytoskeleton. Changes in the plasma membrane (A) are associated with endocytosis and exocytosis. Cilia (C) and flagella (D) are not associated with the amoeboid movement of human white blood cells, and movement via cellular expansion (E) is most often associated with plants.

33. D
Carbohydrates serve as quick, short-term energy reserves and are soluble in water relative to insoluble fats. Glycogen is a type of complex carbohydrate that is converted to glucose as needed. The mitochondria act primarily on carbohydrates; using fat as an energy source is less efficient.

34. E
Double-stranded DNA is confined to the nucleus, while single-stranded RNA (A) can be found in either location [(D) and (E)]. Adenine (B) and guanine (C) are monomers of both molecules; thymine is unique to DNA and uracil is unique to RNA.

35. E
All of the materials listed are of plant origin and composed of plant cells. Cotton bolls are a kind of fruit, wood is secondary xylem, and cork is the bark of an oak tree.

36. C
Complete cellular respiration is an aerobic (oxygen-requiring) process. In the absence of oxygen, fermentation occurs and the amount of ATP yielded decreases. Adding glucose to the system (A) would not decrease the amount of ATP produced; nor would it increase the amount produced per glucose molecule. Light (B) and carbon dioxide (D) are important to photosynthesis, not cellular respiration. ADP (E) is converted to ATP during cellular respiration, so increasing the amount available could not decrease ATP production.

37. D
Sea stars and sea urchins are the only deuterostomes given in the answer choices. Despite the radial symmetry of the adults, their larvae are bilaterally symmetrical (making them members of Bilateria). All of the others in the list are protostomes, except for jellies and corals, which diverged much earlier on and exhibit radial symmetry in addition to being members of Radiata.

38. B
Bryophytes (A) are nonvascular plants (E) and so lack xylem and phloem. Gymnosperms (C) and angiosperms (D) are both seed plants and the haploid forms are not free living. Ferns are the seedless vascular plants that bear spores in sori.

39. A
Junipers are conifers and therefore gymnosperms (C), or naked-seeded plants (D). This tells us that they do not produce fruits; only angiosperms produce fruits. Junipers reproduce sexually (B) and the berrylike structures on junipers are a form of strobilus, or cone (E).

40. E
Echinoderms have deuterostome development, characterized by radial cleavage and formation of a true coelom from the mesoderm. Sponges (A) are acoelomates, while mollusks (B), insects (C), and earthworms (D) are coelomates with protostome development.

41. A
Herbivores tend to have broad teeth, which are good for grinding up plant material; they are primary consumers feeding on producers. Producers (B) are plants, while top consumers (C), carnivores (D), and secondary consumers (E) tend to have pointed teeth for tearing meat.

42. D
Amphibians must have water to successfully reproduce because eggs are fertilized outside the body in water. Reptiles have internal fertilization and can lay eggs away from water because the embryo is enclosed in a watertight structure. Metamorphosis is associated with amphibians and not with reptiles.

43. A
Both birds and crocodiles have four-chambered hearts, while most other reptiles, such as turtles, have three-chambered hearts. Birds, crocodiles, and turtles all lay amniotic eggs (B) and have deuterostome development (D). Of the three, only birds are warm-blooded (C) with hollow bones (E).

44. C
Monocots such as grasses (A), wheat (D), and rice (E) tend to have a single cotyledon, parallel leaf venation, flower parts in threes or multiples of three, and diffuse vascular bundles. Dicots such as woody shrubs (C) have two cotyledons. Both monocots and dicots (such as deciduous trees) can be flowering plants.

45. A
By definition, all animals are multicellular. Diploblastic development (B) is associated with Cnidarians, which have radial symmetry. Both a notochord (C) and deuterostome development (D) are associated with only two phyla, the echinoderms and chordates. There are many phyla that have bilateral symmetry but are not necessarily echinoderms or chordates (such as arthropods and mollusks).

46. D
The missing number in this calculation is 36, the number of ATP molecules produced by glycolysis. That means 36×7.5 kilocalories are conserved. Divide by 686×1 (for 100 percent) to get 39 percent.

47. C
Ectoparasites such as ticks live attached to the outside of a host body (A). Endoparasites can live outside the body of a host (B) and often reproduce (D) outside the host's body; however, they live much of their lives and feed attached to the inside of the host's body, ruling out (E) and making (C) the best choice.

48. D
An error in DNA would be considered a mutation and not a transcription error, ruling out (A). rDNA is recombinant DNA, associated with genetic engineering, so (B) is incorrect. Errors in rRNA or tRNA would lead to translational errors, ruling out (C) and (E). Transcription is the job of mRNA.

49. B

Homologous structures are those that are derived from the same structure, but don't necessarily have the same function, such as the foreleg of a horse and arm of a human. Homologous structures are the result of divergent evolution. Structures that have a similar function but which are not derived similarly are analogous, such as the wings of birds and insects.

50. B

Cladograms are intended to show the phylogenic relationships among organisms, not to imply ancestral descent, so (A) is wrong. This cladogram implies that protists are polyphyletic, ruling out (C), and that the first organisms were prokaryotes, which are also polyphyletic, ruling out (D) and (E).

51. E

Post-zygotic reproductive isolation is that which precludes successful reproduction at some point of development in a diploid offspring, after the fertilization event. Incompatible mating rituals (D), geographic separation [(A) and (B)], and prevention of gamete formation (C) are all pre-zygotic factors. Only sterile hybrids (such as mules) represent post-zygotic examples of reproductive isolation.

52. D

Artificial selection (A) is a process that can act on gene mutations (C), both factors contributing to evolution. Populations in Hardy-Weinberg equilibrium (B) are the antithesis of evolutionary processes; they represent stable allele frequencies and little change over time.

53. C

Stabilizing selection occurs when some factor causes an allele to stay in the gene pool despite being disadvantageous when homozygous. Choices (A), (B) and (D) all reflect occurrences that could affect African Americans and Africans. Macroevolution (E) is change on a grand scale, such as speciation, an event not likely to occur in human populations anytime soon.

54. C

All of the answer choices state something true about archaebacteria, but only (C) relates them to eukaryotes.

55. B

Life cycles are characterized by the timing of meiosis and the characteristics of the diploid and haploid generations. Embryonic development (A) does not reveal either of these characteristics, nor does fertilization (C). All gametes are unicellular, ruling out (D), and meiosis always only occurs once, ruling out (E).

56. E

An octopus is classified in the phylum Mollusca, as are bivalves such as clams. Whales (A), trout (B), and snakes (C) are all vertebrates, a subgrouping of the phylum Chordata. Spiders (D) are arthropods.

57. B

Choice (A) is incorrect because tissues are composed of cells. In (C), species should follow population, a subset of any given species. Choice (D) shows tissues being made up of organisms, and the opposite is true. Choice (E) should show community as the most complex entity in the series.

58. B

Plants (A) are ecosystem primary producers, animals are not. Primary consumers are herbivores (B); they eat the primary producers. Secondary and top consumers [(C) and (D)] eat other consumers, such as rabbits. Decomposers (E) such as bacteria and fungi break down the organic materials in dead organisms.

59. C

The second law of thermodynamics says that every change in energy or energy transfer contributes to the entropy of the universe. Entropy is disorder. Choice (D) is wrong because it violates the first law of thermodynamics: energy can be neither created nor destroyed.

60. D

Potato blight caused the Irish famine in the 1800s, destroying the population's dietary staple. All potato plants were susceptible to the blight because they were so genetically uniform. Genetic uniformity is the hallmark of monocultures, assuring that they grow, bloom, and produce at the same time. For that reason, (E) is incorrect. Choice (A) may be true but is irrelevant. Choices (B) and (C) are incorrect.

61. C

Gymnosperms are seed plants [mosses (A) and ferns (B) are not], but only angiosperms [monocots (D) and dicots (E)] produce fruit.

62. E

Dicots tend to have flower parts that occur in fours and fives, while the flower parts of monocots (D) generally occur in threes. Mosses (A), ferns (B), and gymnosperms (C) are not flowering plants.

63. B

Mosses (A) are nonvascular. Ferns are vascular (i.e., they have xylem and phloem) like all higher plants, but reproduce via spores and not seeds.

64. A

Mosses lack xylem and phloem, and so are nonvascular plants.

65. A

Mosses have dominant haploid forms and only transiently produce diploid spore-forming structures. Ferns may have free-living haploid forms, but the diploid form is larger and grows from the haploid form. Seed plants have a dominant diploid generation and the haploid form is not free living.

66. C

Although there are unicellular fungi, most are multicellular, all are heterotrophic, and all have cell walls containing chitin. Bacteria (A) have cell walls of peptidoglycan, and protists (B) lack chitin. Plant (D) cell walls contain cellulose, and animal cells lack walls (E).

67. D

All plants are multicellular, and all have cellulose cell walls. Some algae (protists) share these characteristics, but none are terrestrial.

68. E

All animal cells lack a cell wall, and all are heterotrophic. Some protists share these characteristics, but those that do are all unicellular.

69. C

Nitrogen fixation involves conversion of atmospheric N_2 into ammonia by specialized bacteria.

70. A

Rubisco (ribulose bisphosphate carboxylase/oxygenase) is the most abundant protein on earth; it is the enzyme responsible for fixation of carbon via photosynthesis and is found in the chloroplasts of leaves.

71. D

Soil bacteria break nitrates down into N_2 gas, releasing it back into the atmosphere.

72. B

Bacteria also convert organic forms of nitrogen tied up in plant and animal biomass into inorganic forms such as ammonia and nitrates. Choice (E) refers to the process of nitrification, which is also carried out by bacteria.

73. D

Helicase unzips the DNA double helix at replication forks to enable DNA replication. Enzymes are often named for the reactions they catalyze; this is true for helicase, which works on the helix.

74. B

During transcription, RNA nucleotides complementary to the DNA strand are joined to form a single-stranded mRNA for transport to the ribosomes in the cytoplasm. RNA polymerase is the enzyme that catalyzes polymerization of the mRNA strand.

75. A

Complementary nucleotides are laid down within replication forks, a reaction catalyzed by DNA polymerase. Another aptly named enzyme, it creates polymers of DNA from monomers.

76. B

When cells need ATP, it is time for the organelles responsible for cellular respiration—the mitochondria—to go to work. ATP is produced when glucose is broken down and the energy stored in its covalent bonds is released.

77. A

Most of the cell cycle is spent in interphase, which is anything but a resting stage. One of the most important events to occur during interphase, which consists of the G1, S, and G2 stages, is DNA replication.

78. D

In a hypertonic solution, a cell will lose water through aquaporins via osmosis. This is because the solute concentration is higher outside the cell relative to inside the cell. Cells cannot simply close membrane protein channels to prevent this loss. They can alter their internal tonicity, a process which would involve moving solutes (not water) into the cell.

79. D

Protein production occurs at ribosomes when the genetic code recorded on mRNA is translated. In order to physically build proteins, amino acid monomers must be shuttled to the ribosome. This is accomplished by transfer RNA (tRNA), which has an anticodon complementary to the mRNA strand. The anticodon ensures that the correct amino acid will be transported to the ribosome.

80. C

Lysosomes are the garbage collectors and recyclers of the cell. They contain enzymes that help break down particles engulfed by the cells so that the molecules they contain can be recycled.

81. B

The figure shows markedly decreased photosynthesis throughout the year in plants in the unirrigated desert sites relative to irrigated plants at residential sites. Choice (A) is incorrect because the figure shows the highest rates of photosynthesis at residential sites during early spring and late fall, with somewhat lower rates during the summer. Choice (C) is incorrect because one would expect higher productivity in plants with higher rates of photosynthesis. Choice (D) is incorrect because the transpiration figure suggests that irrigated plants are not limited by water availability; Choice (E) is incorrect because high rates of transpiration imply wide-open stomata, which would promote and not inhibit photosynthesis.

82. C

Although hot weather (A) might be expected to increase photorespiration and reduce net photosynthesis, the figure shows the highest rates of photosynthesis for unirrigated plants during late summer and early fall, coinciding with the seasonal rains. Choice (B) is incorrect because high rates of transpiration imply open stomata, which promotes photosynthesis. Choices (D) and (E) are incorrect; environmental factors that influence transpiration simultaneously influence photosynthesis and season has nothing to do with opening of stomata. Rapid growth during spring is likely to increase respiration, effectively decreasing net photosynthesis (net photosynthesis = gross photosynthesis—respiration).

83. D

Although water-use efficiency (WUE) is not specifically shown in the graph, it can be inferred from the patterns of photosynthesis and transpiration. In general, the low rates of transpiration during the winter make WUE highest during that season for plants at both site types, and lowest during the hot summer. WUE is similar for plants in both site types most of the year.

84. D

Both (A) and (B) are possible sources of error, since it is desirable to hold all factors constant except for those being tested. Different species might show different responses that are unrelated to irrigation and site type. Also, differing weather conditions on a particular day when measurements were made might influence a plant's physiological response. Although the experiment is aimed at comparing plants in desert and residential sites, irrigation is the factor most expected to influence the outcome of the study, so it is, in effect, the main factor being tested.

85. D

Since the female parent in the second generation of the first lineage (labeled L1) does not carry the defective allele, none of her offspring will be affected.

86. D

While the male parent is homozygous for the disorder, the female parent does not carry the allele for the disorder, so none of their offspring will be affected.

87. B

The answer is actually in the introduction to this group of questions and not necessarily in the pedigree itself. None of the males are heterozygous because males carry only one X chromosome. The females who are heterozygous do not have the condition because they have two X chromosomes, and the dominant form blocks expression of the defective allele, so it must be a sex-linked recessive disorder.

88. C

In a sex-linked recessive disorder, about half of the children would carry the allele, but only the males would be affected by the condition.

89. C

Of the conditions listed, only hemophilia is a sex-linked recessive disorder. All of the other conditions are the result of genes linked to somatic chromosomes.

90. D

The lab results show that the patient has a normal value for hemoglobin, which suggests that anemia is not his problem. All of the other results point to diabetes (high glucose is hyperglycemia), which is often associated with kidney failure (high blood urea), ketoacidosis (low blood pH), and chronic infections (sepsis).

91. E

Although the results do indicate the patient has probably ingested alcoholic beverages, the value is not out of normal range and offers no indication of chronic abuse. Both (A) and (C) are consistent with the lab values reported.

92. B

White blood cells are one of the first lines of defense in infection. Neutrophils are most often associated with bacterial infections, while lymphocytes are associated with immune response (production of antibodies). None of the other lab results would be associated with infection.

93. A

Since the urinalysis has prompted the orders for bacterial cultures, it is likely that microscopic examination and chemical testing of the urine have indicated the presence of bacteria. To determine the type of bacteria present, a gram stain would be the most useful test. Acid fast stains (B) are used to detect unusual organisms such as the infective agent of tuberculosis. The other tests suggested are not appropriate.

94. D

Choice (A) is incorrect because competition between species A and C would not lead to increases in both populations over the course of 60 years. Species B could not be a predator of A (B), because increases in A should have promoted similar increases in B, until the numbers of species A fell enough to promote a similar decline in species B. There are no data to suggest that species C should increase (C), in fact, the opposite trend is shown. When carrying capacity is reached, the characteristic leveling off shown by the curve for species C is seen, not the erratic pattern demonstrated by species A (E).

95. B

If species C were a predator of species B, that population would be expected to grow in response to an increase in prey (species B) availability, rather than showing the gradual decline indicated by the figure. All of the other explanations are feasible; loss of mutualistic species A might negatively impact species C, albeit slowly. Loss of prey (species A) might have a similar effect, as an increase in the number of predators (species B) might. Choice (E) is also a valid possibility; just because data on other species are not presented, it does not mean they do not exist.

96. D

Species A is effectively out of the picture in 2000, so (A) is not a good answer. Species B would not be expected to increase (B), because of increased competition with the new species. Based on the data given, effects on species C could not be predicted, but some change in the community dynamics would be expected.

97. C

The reaction occurring in organelle A is photosynthesis, as indicated by the splitting of water and fixation of CO_2. The first reaction in this organelle represents the light reactions, in which the sun's energy is converted to chemical energy in the form of ATP (E II).

98. C

The reaction occurring in organelle B is cellular respiration, as indicated by the release of energy stored in the bonds of glucose. Since the question specifies that oxygen is required, the process depicted must show aerobic respiration rather than glycolysis. The form of energy released in aerobic respiration is chemical energy in the form of ATP (E IV).

99. E

The organelle labeled A is transforming energy via the process of photosynthesis, a process that occurs in the chloroplasts. The organelle labeled B is transforming energy via cellular respiration, which takes place in the mitochondria.

100. D

Although all of the organisms listed are capable of photosynthesis, bacteria lack true organelles such as chloroplasts and mitochondria. Both photosynthetic protists and plants can carry out photosynthesis and have true organelles.

Section II: Free-Response Questions

1. The first two parts of this question should give you no problem. Determine a terminology for the alleles that makes sense to you and start constructing a Punnett square. The sexes of the progeny are never mentioned so there is no concern for sex linkage. You should include in your answer all you know about the genes, the alleles, and the traits. For example, is the gene autosomal or sex-linked? Is it a two-allele or three-allele system? Part (c) will be the most difficult. There is not only the chance that you will forget how to do a chi-squared analysis or construct a null hypothesis, but the cross is not straightforward. Figuring out the expected frequencies could take a while. Come back to this one if you are running short on time. Notice that the F1 male must be heterozygous for each allele. The female must be homozygous for feathered tail, and heterozygous for the short-tailed allele because at least some of the progeny have long tails.

    (a) Key points to include: genotype, phenotype, correct presentation of alleles

Here is a possible response:

The cross demonstrates that there are two autosomal dominant genes involved with the two traits in question that are represented by two alleles for each gene. The trait of tail length can be represented by the alleles L^S for the dominant short-tailed allele and L^L for the recessive long-tailed allele. The gene for tail shape can be represented by S^S for the dominant straight-tailed allele and S^F for the recessive feather-tailed allele. The true-breeding female would have the genotype $L^S L^S S^S S^S$ and the true-breeding male would have the genotype $L^L L^L S^F S^F$. The progeny would all be heterozygous for both genes so their genotypes would be $L^S L^L S^S S^F$.

    (b) Key points to include: an accurate Punnett square, dihybrid cross, Mendel's Laws, 9:3:3:1 ratio

Here is a possible response:

The genotypes and phenotypes of the F2 generation can be easily figured by use of a dihybrid cross of a male and a female of the F1 generation using a Punnett square as follows.

	$L^S S^S$	$L^S S^F$	$L^L S^S$	$L^L S^F$
$L^S S^S$	$L^S L^S S^S S^S$	$L^S L^S S^S S^F$	$L^S L^L S^S S^S$	$L^S L^L S^S S^F$
$L^S S^F$	$L^S L^S S^S S^F$	$L^S L^S S^F S^F$	$L^S L^L S^S S^F$	$L^S L^L S^F S^F$
$L^L S^S$	$L^S L^L S^S S^S$	$L^S L^L S^S S^F$	$L^L L^L S^S S^S$	$L^L L^L S^S S^F$
$L^L S^F$	$L^S L^L S^S S^F$	$L^S L^L S^F S^F$	$L^L L^L S^S S^F$	$L^L L^L S^F S^F$

The frequencies of the progeny exhibit the classic 9:3:3:1 ratios that typify Mendel's two laws of segregation and independent assortment.

    (c) Key points to include: Punnett square, correct genotypes of offspring, degrees of freedom, null hypothesis, correct chi-square calculations

Here is a possible response:

This cross can also be demonstrated effectively using a Punnett square. The male from the F1 generation must be heterozygous for each allele because it is a progeny of parents with true-breeding lines. The female must be homozygous for the feather-tailed allele because it expresses that recessive trait. Because there are several progeny that express the long-tailed trait, the female cannot be homozygous dominant for the short-tailed allele. The cross would look like this:

	$L^S S^F$	$L^S S^F$	$L^L S^F$	$L^L S^F$
$L^S S^S$	$L^S L^S S^S S^F$	$L^S L^S S^S S^F$	$L^S L^L S^S S^F$	$L^S L^L S^S S^F$
$L^S S^F$	$L^S L^S S^F S^F$	$L^S L^S S^F S^F$	$L^S L^L S^F S^F$	$L^S L^L S^F S^F$
$L^L S^S$	$L^S L^L S^S S^F$	$L^S L^L S^S S^F$	$L^L L^L S^S S^F$	$L^L L^L S^S S^F$
$L^L S^F$	$L^S L^L S^F S^F$	$L^S L^L S^F S^F$	$L^L L^L S^F S^F$	$L^L L^L S^F S^F$

The progeny would have phenotypes according to the following frequencies: three short, straight-tailed; three short, feather-tailed; one long, straight-tailed; and one long, feather-tailed. There are a total of 1,000 progeny as a result of this cross so the null hypothesis would be that we would expect to see 375 short, straight-tailed; 375 short, feather-tailed; 125 long, straight-tailed; and 125 long, feather-tailed. The alternative hypothesis is that the phenotypic frequencies are not apportioned according to the frequencies in the null hypothesis. Degrees of freedom are calculated by taking the number of possible categories, in this case four, and subtracting one. We will test the calculated chi-squared value against a critical value taken from a chi-squared distribution with three degrees of freedom and an acceptable probability of 0.05.

We construct the chi-squared analysis as follows.

$$\chi^2 = \frac{(383-375)^2}{375} + \frac{(367-375)^2}{375} + \frac{(118-125)^2}{125} + \frac{(132-125)^2}{125} = 1.125$$

Because the calculated chi-squared statistic does not exceed the critical value for the chi-squared distribution with three degrees of freedom for a probability of 0.05, we accept the null hypothesis that the frequencies we observed were the same as what we expected.

2. This question covers the topic of molecular genetics because of its role in regulation, physiology, development, and such. Why doesn't the question just ask about DNA? RNA can be used to perform most of the same functions as DNA in many organisms and systems. The question could be reworded as "How is DNA involved in the expression of proteins that control physiological functions? How is gene regulation or gene expression controlled? What is special about the structure of DNA and RNA that makes it ideal for replication? How does the genetic code, despite its simplicity and ability to replicate, still account for all of the species that have ever existed?" Focus on key points of each concept, add as much detail as you can remember, and form a cohesive set of statements.

   (a) Key points to include: DNA, mRNA, tRNA, protein synthesis, operon, inducer

Here is a possible response:

Genetic material is stored in organisms as nucleic acids linked together to form either DNA or RNA. DNA and RNA are links of base pairs, which include adenine, cytosine, guanine, thymine (in DNA), and uracil (in RNA). These links of base pairs, or nucleic acids, form chains of code that act as a template for the expression of all of the proteins that carry out biological functions. In bacteria, the code is present and transcribed in the cytoplasm, while in eukaryotic cells DNA is present in the nucleus. In eukaryotic cells the genetic code must be delivered from DNA to the cytoplasm through an intermediary, mRNA. Genes are present on the DNA strands with a promoter and operator region to which RNA polymerase attaches and travels down the strand, creating a mirrored template from the original DNA strand. The structural genes are downstream from the promoter so the mRNA lengthens as it codes for more and more genes included in the same operon, until it reaches a terminator region. The mRNA is released from the DNA strand and undergoes several modifications as introns are spliced out before the mRNA travels to the cytoplasm.

In the cytoplasm, the mRNA acts as a template for ribosomal units to translate protein using amino acids delivered by tRNA. Each three-base-pair set is a codon for a specific amino acid, although there are many more codons possible than amino acids, making the genetic code redundant. The ribosome travels down the mRNA template strand, elongating the protein as peptide bonds are formed between each amino acid. The finished polypeptide strand then undergoes some additional modification, such as removal of terminal amino acids, before the strand is a functional protein.

The expression of these proteins can be controlled at several levels such as during transcription, movement into the cytoplasm, translation, mRNA modification, and protein modification. Every cell in an organism contains the entire genome for that organism. Several local factors control gene expression, such as substrates in the cellular matrix that act as inducers to the expression of certain operons. Other genes are synchronized with particular events, almost as though an internal clock were present. The gene that expresses the proteins for an eye cell is still present in an animal's foot, but these genes are not expressed because of complex genetic regulation of a simple DNA code.

(b) Key points to include: genes, base pairs, replication, proteins, diversity

Here is a possible response:

DNA is composed of only four different base pairs and there is only one different base pair, uracil instead of thymine, in RNA. This has created a very simple system for duplication. When a cell replicates its DNA during mitosis, it simply unzips the double helix and adds nucleic acids along its length until a duplicate set of complementary strands is created. The base pairs bond with hydrogen bonds, with cytosine bonding to guanine and adenine bonding to thymine. DNA molecules are very long, but because the code only involves four different base pairs, it can easily be duplicated. The DNA strands are actually copied in small segments and linked together to form an entire strand.

Although there are only four different base pairs, the longer the strand gets, the more combinations of base pairs can be created. These long series of base pairs code for longer and more complex proteins that can differ in minor or very significant ways. For example, sickle cell anemia is caused by only one base pair difference in the genetic code for hemoglobin, but causes an incredible difference in the function of the protein. The high number of potential combinations from only four base pairs and the impact of small changes in the sequence of base pairs contributes to the diversity of proteins in an organism. This diversity of proteins contributes to the diversity of organisms that exist because physiological function is essentially protein driven. In essence, long and diverse code leads to diverse proteins, which leads to diverse organisms.

3. The first part of this question allows you some creativity in the response as long as the basics of experimental design are followed. The most important thing to include is the development of a testable hypothesis. Include a control, an independent variable, and a dependent variable to demonstrate a familiarity with the experimental process. The most likely response variables to measure in this experiment are salivation and peristalsis, but gastric secretion is also a candidate. If an unsuitable response variable is chosen, a considerable number of points may still be available for good design. It is not necessary for you to be familiar with any particular type of device or that any devices used in the design be potentially viable. It is most important that a particular variable be identified and measured. In order to get the most points, you will be expected to know the body's reactions to food stimulus as well as how to test a hypothesis. Part (b) should be answered with specific information about body systems. The focus should be integration so that coordination can be included in the response, but the response still should include specifics about each system. The question does not ask for all the responses of three different systems, so the response should not be comprehensive for any one system. There is likely to be one specific response for three different systems.

(a) Key points to include: testable hypothesis based on a specific example of the body's response to food, control, dependent variable, independent variable

Here is a possible response:

As Pavlov showed with his dogs and the trained response to a sound, there is a salivation reaction to food stimulus before food comes in contact with the mouth. When food is ingested, the salivary glands produce saliva that includes the enzyme amylase, which begins the process of digesting starch. The salivary response can also be initiated by the smell or sight of food or simply thinking about food. The hypothesis for this experiment is that only the smell of food will cause an increased volume of saliva to be produced in the mouth. The independent variable will be the smell of food and the dependent variable will be the amount of saliva produced. To measure the saliva, a device could be attached to a subject's mouth that would collect saliva, such as a suction tube used by a dentist. Subjects would be randomly assigned to a control group that would receive no smell stimulus and an experimental group that would receive the smell stimulus. The amount of saliva produced by the control and experimental group could be compared directly or with replicates, using a t-test.

(b) Key points to include: identify three systems, provide specific response for each, homeostasis, nutrient absorption

Here is a possible response:

As the body tries to maintain homeostasis, almost no organ system responds independently. When food is ingested, it causes acute changes in the digestive system, but also changes in the nervous and circulatory systems. The digestive system responds to the presence of nutrients in the lumen in different ways within different organs, but the majority of digestion takes place in the duodenum of the small intestine. In the duodenum, hydrolytic enzymes secreted by the pancreas after the release of chyme through the pyloric sphincter of the stomach do the majority of the work in breaking down macronutrients. The majority of absorption also takes place in the duodenum, as micronutrients are made available to the bloodstream.

The circulatory system closely follows digestive activity controlled by nervous impulses to the smooth muscle tissue that lines arteries and capillaries. These nervous impulses are coordinated by the hindbrain and take place without conscious thought. The constriction of vessels in the appendages and skin shunts blood flow away from the periphery and into the core near the organs of absorption, such as the duodenum. At the same time that nutrients are broken down in the small intestine, blood flow is increased around the small intestine. This increases the volume and speed of nutrient delivery to the rest of the body.

In addition to the change in blood flow, the action of the digestive system imparts a change in the endocrine system at the pancreas. Glucose absorbed by the small intestine and delivered by the increased blood flow causes an elevated blood glucose volume that is detected by the B-cells in the pancreas. These cells produce and secrete insulin in response to the glucose in the bloodstream. Insulin is delivered by the same increased blood flow throughout the body, where it is detected by receptor cells in the liver and muscles. The response is for the muscle and liver cells to absorb glucose and convert it to glycogen for storage.

4. This question is unlike many of those previously presented in its specificity, but it still requires an intricate level of synthesis. You are expected to direct your response to a particular system, but are required to pull on knowledge from several different areas to create a comprehensive answer. You should begin with the primary producers, giving specific taxa at every level if possible, and trace the path of oxygen through the ecosystem. A diagram may not be a bad idea here, but don't let the diagram stand alone. While composing this portion, you should be able to pinpoint trophic levels that would be affected greatly by hypoxia. In part (b) you should include aspects of the organisms' ecology outside of the oxygen cycle. If you are using an omnivore for your example, consider factors such as what organisms it eats, where it lives, and how it finds a mate. Part (c) expands the negative impact on the oxygen cycle into the food web since sharks are top-predator carnivores. Remember, sharks still need oxygen to breathe, but they also need nutrients to grow and for metabolism.

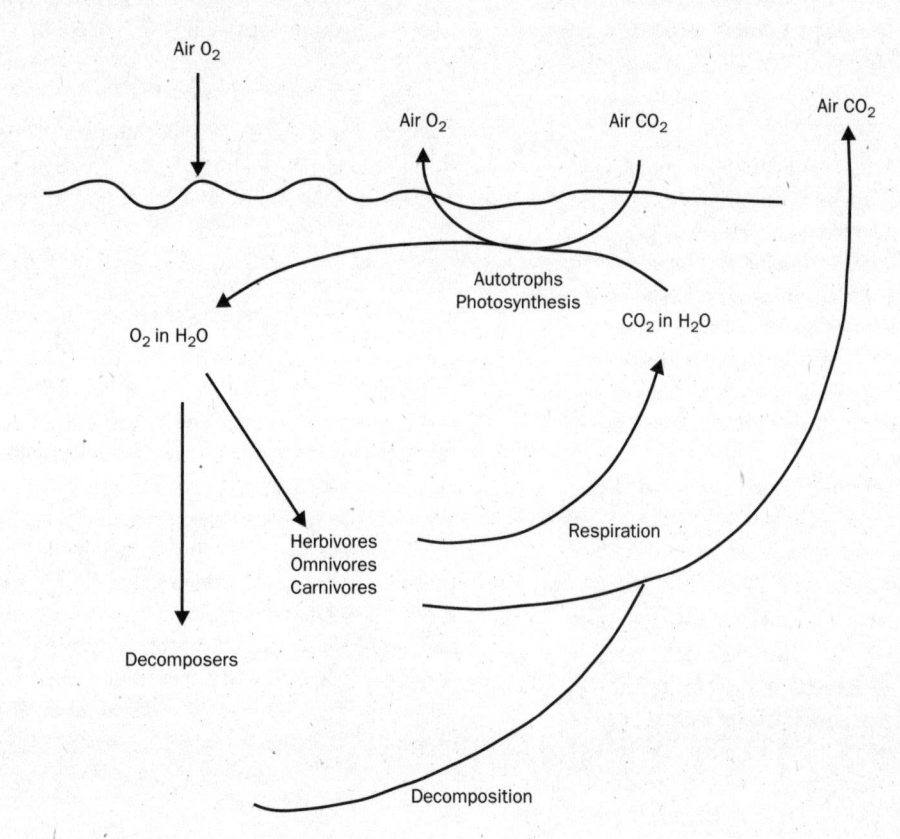

Diagram: Oxygen/carbon cycle in a marine system showing Air O₂ entering water, O₂ in H₂O feeding Herbivores/Omnivores/Carnivores and Decomposers; Autotrophs Photosynthesis producing O₂ and consuming CO₂ in H₂O; Respiration and Decomposition returning CO₂ to air.

(a) Key points to include: primary producers, hypoxia, dissolved oxygen, photosynthesis, respiration, trophic levels

Here is a possible response:

Except for trace amounts of free oxygen that originate from the air-water interface, free oxygen exists as a result of photosynthesis by autotrophic organisms using energy from the sun to fix carbon from atmospheric CO_2 while releasing H_2O and O_2. All other forms of life needing oxygen use this reservoir. In the marine system, the primary producers are various forms of seaweed alga and pelagic photosynthetic plankton like diatoms. The oxygen these organisms generate is released into water where some escapes to the atmosphere depending on the atmospheric pressure. Dissolved oxygen in the water table is utilized for metabolic respiration by every other member of the ecosystem such as carnivorous and herbivorous fish, herbivorous echinoderms, and detritivorous crustaceans. Dissolved oxygen is also necessary for certain chemical reactions involving the breakdown of organic tissue during decomposition. The processes of respiration and decomposition recycle the oxygen into CO_2, which enters the water table and the atmosphere where it is again utilized by the autotrophs.

This cycle is upset in the Gulf of Mexico because the nitrogen- and phosphorous-rich fresh water entering the marine system supports a surface-level algal bloom that blocks sunlight from reaching the natural primary producers on the marine floor and in the water table. The biomass of the alga increases beyond the system's ability to produce oxygen so that the rate of O_2 production can't meet the need for respiration and decomposition. As a result, the organisms needing oxygen and sunlight die until the algal bloom disappears. If enough of the organisms from the natural system can survive, the system begins to recover until the next spring.

(b) Key points to include: two example organisms, primary producer, consumers, respiration, energy flow in community

Here is a possible response:

The algal bloom affects the autotrophs most significantly by blocking sunlight. This prevents the organisms from producing oxygen themselves, but also prevents them from producing energy in the form of storage carbohydrates. All organisms in the ecosystem are connected somehow and autotrophs are no different. As organisms in the ecosystem die, the resources they provide disappear. Autotrophs rely on corals for substrate anchoring points. They rely on predators to decrease herbivory. As oxygen levels decrease in the water due to lower photosynthetic rates, slow-growing corals die and predators leave the area causing an increase in herbivory.

Herbivores like sea urchins are affected as the number of grazing spots decreases due to the death and decreased productivity of their algal food source. Herbivores also depend on their algal forest for protection from ocean currents and predators. As oxygen levels decrease in the water, herbivores with limited mobility, like sea urchins, suffer increased death rates as they can't move to more productive areas.

(c) Key points to include: top predator, trophic levels, oxygen deprivation, decreased food supply

Here is a possible response:

Two factors may be contributing to a relocation of a top predator like a shark. First, the oxygen content of the water may be unsuitable for a shark population. Normal function is not possible when the oxygen supply is limited. Sharks that are mobile can migrate to an area that isn't affected or is less affected by the algal bloom. The second factor is the food supply. As primary productivity decreases, it affects all levels of the ecosystem. Herbivore, omnivore, and primary carnivore species are affected so either die or migrate to more suitable habitat. All of these trophic levels are part of the shark's nutrient reservoir. A shark would have to increase its hunting range to encounter more food items. A shark might migrate with its prey items if they move to a more suitable habitat. Finally, a shark might need to switch to a different food source if its naturally occurring sources of nutrients disappear. Unfortunately, human bathers may become an alternate food source for larger sharks that are normally not considered man-eaters.

Practice Test 2 Answer Grid

1. Ⓐ Ⓑ Ⓒ Ⓓ Ⓔ
2. Ⓐ Ⓑ Ⓒ Ⓓ Ⓔ
3. Ⓐ Ⓑ Ⓒ Ⓓ Ⓔ
4. Ⓐ Ⓑ Ⓒ Ⓓ Ⓔ
5. Ⓐ Ⓑ Ⓒ Ⓓ Ⓔ
6. Ⓐ Ⓑ Ⓒ Ⓓ Ⓔ
7. Ⓐ Ⓑ Ⓒ Ⓓ Ⓔ
8. Ⓐ Ⓑ Ⓒ Ⓓ Ⓔ
9. Ⓐ Ⓑ Ⓒ Ⓓ Ⓔ
10. Ⓐ Ⓑ Ⓒ Ⓓ Ⓔ
11. Ⓐ Ⓑ Ⓒ Ⓓ Ⓔ
12. Ⓐ Ⓑ Ⓒ Ⓓ Ⓔ
13. Ⓐ Ⓑ Ⓒ Ⓓ Ⓔ
14. Ⓐ Ⓑ Ⓒ Ⓓ Ⓔ
15. Ⓐ Ⓑ Ⓒ Ⓓ Ⓔ
16. Ⓐ Ⓑ Ⓒ Ⓓ Ⓔ
17. Ⓐ Ⓑ Ⓒ Ⓓ Ⓔ
18. Ⓐ Ⓑ Ⓒ Ⓓ Ⓔ
19. Ⓐ Ⓑ Ⓒ Ⓓ Ⓔ
20. Ⓐ Ⓑ Ⓒ Ⓓ Ⓔ
21. Ⓐ Ⓑ Ⓒ Ⓓ Ⓔ
22. Ⓐ Ⓑ Ⓒ Ⓓ Ⓔ
23. Ⓐ Ⓑ Ⓒ Ⓓ Ⓔ
24. Ⓐ Ⓑ Ⓒ Ⓓ Ⓔ
25. Ⓐ Ⓑ Ⓒ Ⓓ Ⓔ
26. Ⓐ Ⓑ Ⓒ Ⓓ Ⓔ
27. Ⓐ Ⓑ Ⓒ Ⓓ Ⓔ
28. Ⓐ Ⓑ Ⓒ Ⓓ Ⓔ
29. Ⓐ Ⓑ Ⓒ Ⓓ Ⓔ
30. Ⓐ Ⓑ Ⓒ Ⓓ Ⓔ
31. Ⓐ Ⓑ Ⓒ Ⓓ Ⓔ
32. Ⓐ Ⓑ Ⓒ Ⓓ Ⓔ
33. Ⓐ Ⓑ Ⓒ Ⓓ Ⓔ
34. Ⓐ Ⓑ Ⓒ Ⓓ Ⓔ
35. Ⓐ Ⓑ Ⓒ Ⓓ Ⓔ
36. Ⓐ Ⓑ Ⓒ Ⓓ Ⓔ
37. Ⓐ Ⓑ Ⓒ Ⓓ Ⓔ
38. Ⓐ Ⓑ Ⓒ Ⓓ Ⓔ
39. Ⓐ Ⓑ Ⓒ Ⓓ Ⓔ
40. Ⓐ Ⓑ Ⓒ Ⓓ Ⓔ
41. Ⓐ Ⓑ Ⓒ Ⓓ Ⓔ
42. Ⓐ Ⓑ Ⓒ Ⓓ Ⓔ
43. Ⓐ Ⓑ Ⓒ Ⓓ Ⓔ
44. Ⓐ Ⓑ Ⓒ Ⓓ Ⓔ
45. Ⓐ Ⓑ Ⓒ Ⓓ Ⓔ
46. Ⓐ Ⓑ Ⓒ Ⓓ Ⓔ
47. Ⓐ Ⓑ Ⓒ Ⓓ Ⓔ
48. Ⓐ Ⓑ Ⓒ Ⓓ Ⓔ
49. Ⓐ Ⓑ Ⓒ Ⓓ Ⓔ
50. Ⓐ Ⓑ Ⓒ Ⓓ Ⓔ
51. Ⓐ Ⓑ Ⓒ Ⓓ Ⓔ
52. Ⓐ Ⓑ Ⓒ Ⓓ Ⓔ
53. Ⓐ Ⓑ Ⓒ Ⓓ Ⓔ
54. Ⓐ Ⓑ Ⓒ Ⓓ Ⓔ
55. Ⓐ Ⓑ Ⓒ Ⓓ Ⓔ
56. Ⓐ Ⓑ Ⓒ Ⓓ Ⓔ
57. Ⓐ Ⓑ Ⓒ Ⓓ Ⓔ
58. Ⓐ Ⓑ Ⓒ Ⓓ Ⓔ
59. Ⓐ Ⓑ Ⓒ Ⓓ Ⓔ
60. Ⓐ Ⓑ Ⓒ Ⓓ Ⓔ
61. Ⓐ Ⓑ Ⓒ Ⓓ Ⓔ
62. Ⓐ Ⓑ Ⓒ Ⓓ Ⓔ
63. Ⓐ Ⓑ Ⓒ Ⓓ Ⓔ
64. Ⓐ Ⓑ Ⓒ Ⓓ Ⓔ
65. Ⓐ Ⓑ Ⓒ Ⓓ Ⓔ
66. Ⓐ Ⓑ Ⓒ Ⓓ Ⓔ
67. Ⓐ Ⓑ Ⓒ Ⓓ Ⓔ
68. Ⓐ Ⓑ Ⓒ Ⓓ Ⓔ
69. Ⓐ Ⓑ Ⓒ Ⓓ Ⓔ
70. Ⓐ Ⓑ Ⓒ Ⓓ Ⓔ
71. Ⓐ Ⓑ Ⓒ Ⓓ Ⓔ
72. Ⓐ Ⓑ Ⓒ Ⓓ Ⓔ
73. Ⓐ Ⓑ Ⓒ Ⓓ Ⓔ
74. Ⓐ Ⓑ Ⓒ Ⓓ Ⓔ
75. Ⓐ Ⓑ Ⓒ Ⓓ Ⓔ
76. Ⓐ Ⓑ Ⓒ Ⓓ Ⓔ
77. Ⓐ Ⓑ Ⓒ Ⓓ Ⓔ
78. Ⓐ Ⓑ Ⓒ Ⓓ Ⓔ
79. Ⓐ Ⓑ Ⓒ Ⓓ Ⓔ
80. Ⓐ Ⓑ Ⓒ Ⓓ Ⓔ
81. Ⓐ Ⓑ Ⓒ Ⓓ Ⓔ
82. Ⓐ Ⓑ Ⓒ Ⓓ Ⓔ
83. Ⓐ Ⓑ Ⓒ Ⓓ Ⓔ
84. Ⓐ Ⓑ Ⓒ Ⓓ Ⓔ
85. Ⓐ Ⓑ Ⓒ Ⓓ Ⓔ
86. Ⓐ Ⓑ Ⓒ Ⓓ Ⓔ
87. Ⓐ Ⓑ Ⓒ Ⓓ Ⓔ
88. Ⓐ Ⓑ Ⓒ Ⓓ Ⓔ
89. Ⓐ Ⓑ Ⓒ Ⓓ Ⓔ
90. Ⓐ Ⓑ Ⓒ Ⓓ Ⓔ
91. Ⓐ Ⓑ Ⓒ Ⓓ Ⓔ
92. Ⓐ Ⓑ Ⓒ Ⓓ Ⓔ
93. Ⓐ Ⓑ Ⓒ Ⓓ Ⓔ
94. Ⓐ Ⓑ Ⓒ Ⓓ Ⓔ
95. Ⓐ Ⓑ Ⓒ Ⓓ Ⓔ
96. Ⓐ Ⓑ Ⓒ Ⓓ Ⓔ
97. Ⓐ Ⓑ Ⓒ Ⓓ Ⓔ
98. Ⓐ Ⓑ Ⓒ Ⓓ Ⓔ
99. Ⓐ Ⓑ Ⓒ Ⓓ Ⓔ
100. Ⓐ Ⓑ Ⓒ Ⓓ Ⓔ

Practice Test 2

Section I: Multiple-Choice Questions

Time: 80 Minutes
100 Questions

Directions: Choose the best answer choice for the questions below.

1. Jaws and teeth provide information about how an animal lives and what it eats because

   (A) jaws evolved from gill arches, so the structure of an animal's jaws and teeth show whether it is ancient or more recently derived
   (B) carnivores' teeth have broad, hard surfaces used for grinding, distinguishing meat eaters from plant eaters
   (C) diet influences the way an animal's skull and dentition evolve, providing clues to its position in the food chain
   (D) herbivores have pointed front teeth adapted for tearing that distinguish their skulls and jaws from those of carnivores
   (E) animals that do not have teeth cannot break down and digest solid food

Year	N	ΔN
1	50000	20000
2	70000	28000
3	98000	39200
4	137200	54880
5	192080	76832
6	268912	107564.8
7	376477	150590.7
8	527066	210827
9	737895	295157.8
10	1033052	413220.9

2. If these population data were plotted as a population growth curve, you would see

   (A) a logistic population growth curve
   (B) a reduction of population size as fecundity decreases
   (C) a positive relationship between generation time and intrinsic rate of growth
   (D) an exponential growth curve
   (E) population cycling over time

GO ON TO THE NEXT PAGE

3. Independent assortment of homologous chromosomes, crossing over, and random fertilization generate which of the following?

   (A) Haploid gametes
   (B) Diploid gametes
   (C) Genetic variability in sexually reproducing organisms
   (D) Genetic variability in asexually reproducing organisms
   (E) Random mutations

4. Which of the following is a function of enzymes?

   (A) They lower activation energy.
   (B) They are catalysts.
   (C) They do not change ΔG for a reaction.
   (D) A and B only
   (E) A, B, and C

5. Genetically engineered crop plants

   (A) are still very rare
   (B) are usually more difficult to create than genetically engineered animals
   (C) may hybridize with wild relatives, allowing novel genes to "escape" from crops
   (D) require periodic inoculation with recombinant plasmids in order to retain bioengineered genes
   (E) tend to rot more quickly after several generations

6. Three different lizard species live in sympatry. Each species specializes in feeding on a different size of prey. This would be an example of

   (A) the competitive exclusion principle
   (B) resource partitioning
   (C) symbiosis
   (D) interference competition
   (E) sympatric symmetry

7. Free and bound ribosomes

   (A) are structurally identical
   (B) are identified by where they are found
   (C) tend to produce proteins destined for different locations
   (D) B and C only
   (E) A, B, and C

8. What assumption underlies the methods used to construct a genetic map?

   (A) Recombination frequencies are directly proportional to the distance between genes on a chromosome.
   (B) Recombination frequencies are inversely proportional to distance between genes on a chromosome.
   (C) Linked genes never cross over.
   (D) Recessive genes are less common than dominant genes.
   (E) Recessive genes are more common than dominant genes.

9. Elephants utilitze which of the following?

   (A) Enternal fertilization
   (B) *r*-selected strategy
   (C) Amnotic eggs
   (D) *K*-selected strategy
   (E) Exponential population growth

10. What is the difference between apical and lateral meristems?

    (A) Apical meristems allow for plant lengthening, and lateral meristems allow for thickening.
    (B) Apical meristems allow for primary growth, and lateral meristems allow for secondary growth.
    (C) Apical meristems are located at the tips of roots and shoots, and lateral meristems are located along the length of roots and shoots.
    (D) All of the above
    (E) None of the above

11. Which of the following reduces weight in birds, making flight more efficient?

    (A) A gizzard instead of stomach and intestines
    (B) Reduced pectoral muscles
    (C) Air sacs instead of lungs
    (D) Only two chambers in the heart
    (E) Bones with a honey-combed internal structure

12. Indigenous rainforest people have used toxins from the brightly colored poison arrow frogs for hunting large animals. These types of tree frogs show

    (A) Müllerian mimicry
    (B) Batesian mimicry
    (C) aposematic coloration
    (D) cryptic coloration
    (E) predatory coloration

13. Which of the following is true of Plasmodesmata?

    (A) They allow water to pass from cell to cell.
    (B) They allow cytosol to pass from cell to cell.
    (C) They allow cells to be organized into tissues.
    (D) They are found only in plant cells.
    (E) All of the above

14. Formic acid, which is a carboxylic acid, is represented by

    (A) H_3C-CH_2-OH (ethanol structure shown)
    (B) $(H_3C)_2C=O$ (acetone structure shown)
    (C) $HC(O)-CH_2-CH_3$ (structure shown)
    (D) $H-C(=O)-OH$ (structure shown)
    (E) H_3C-SH

15. What component of a plant's photosystem loses electrons to the primary electron acceptor?

    (A) The thylakoid membrane
    (B) Antenna pigment molecules
    (C) The photon
    (D) The absorption spectrum
    (E) Chlorophyll *a* at the reaction center

16. Which of the following does NOT occur in meiosis II?

    (A) Synapsis
    (B) Cytokinesis
    (C) Separation of sister chromatids
    (D) Formation of the spindle apparatus
    (E) Formation of daughter cells

17. What do nicotine, morphine, cloves, and cinnamon have in common?

    (A) They emit chemical signals that prevent competing plants from germinating.
    (B) They provide mechanical defense against herbivores.
    (C) They function as chemical defenses against herbivores.
    (D) They reduce the plant's need for nutrients.
    (E) They increase the plant's drought tolerance.

18. A plant cell's central vacuole may store the following substances EXCEPT for which?

    (A) Cytosol
    (B) Proteins
    (C) Pigments
    (D) Inorganic ions
    (E) Herbivore toxins

19. Competitive inhibitors of enzymes can be reversed by

    (A) increasing the pH above the enzyme's optimal range
    (B) increasing the concentration of substrate
    (C) adding noncompetitive inhibitors
    (D) lowering the temperature below the enzyme's optimal range
    (E) removing the cofactors

20. If a reaction results in $\Delta G = -625$ kcal/mol, one can conclude that

    (A) free energy was absorbed
    (B) the system is closed
    (C) the system is at equilibrium
    (D) the reaction is exergonic
    (E) the reaction is endergonic

21. Which of the following best corresponds to the four levels of protein structure, moving from primary to quaternary structure?

    (A) Polypeptide folds, aggregation of polypeptide subunits, side-chain bonding, sequence of amino acids
    (B) Polypeptide folds, side-chain bonding, sequence of amino acids, aggregation of polypeptide subunits
    (C) Sequence of amino acids, side-chain bonding, aggregation of polypeptide subunits, polypeptide folds
    (D) Sequence of amino acids, polypeptide folds, side-chain bonding, aggregation of polypeptide subunits
    (E) Sequence of amino acids, aggregation of polypeptide subunits, polypeptide folds, side-chain bonding

GO ON TO THE NEXT PAGE

22. Clumps of overlapping mammalian cells would indicate a problem with cell cycle regulation because

    (A) normal cells use cyclin-dependent kinases for regulation
    (B) normal cell division is stimulated by active anaphase-promoting complex
    (C) normal cells show density-dependent inhibition and anchorage dependence
    (D) normal cell division is stimulated by the presence of growth factor
    (E) normal cells spend most of the time in interphase.

23. When ATP is used for transport, mechanical, or chemical work,

    (A) it is the adenosine that performs the work
    (B) it is the phosphorylates that perform the work
    (C) it is the precursor, ADP, that performs the work
    (D) it is the energy released when ATP leaves the mitochondria that performs the work
    (E) None of the above

24. Cell A and Cell B differ in the following ways: Cell A contains a cell wall. Cell B contains centrioles. Based only on this information, what can be concluded about these cells?

    (A) Cell A prokaryotic and Cell B is eukaryotic.
    (B) Cell A is eukaryotic and Cell B is prokaryotic.
    (C) Cell A is from a plant or prokaryote and Cell B is from an animal.
    (D) Cell B is from a plant or prokaryote and Cell A is from an animal.
    (E) Both cells are eukaryotic.

25. Deep-diving air breathers (e.g., seals, whales, penguins) have numerous adaptations that allow them to dive to great depths and remain underwater for long periods of time. Which of the following would be LEAST likely to be found in a deep diver?

    (A) Decreased O_2 storage in the lungs and increased O_2 storage in the blood
    (B) A large spleen
    (C) A large volume of blood per body mass ratio
    (D) Efficient use of buoyancy to aid locomotion
    (E) Low concentrations of myoglobin in the muscles

26. Rubisco is the most abundant protein in chloroplasts. What does it do?

    (A) It is an enzyme that catalyzes carbon fixation in phase 1 of the Calvin cycle.
    (B) It is the reducing agent that produces the sugar product in phase 2 of the Calvin cycle.
    (C) It regenerates RuBP in phase 3 of the Calvin cycle.
    (D) It is essential in the Krebs cycle for regeneration of ATP.
    (E) It is used in light reactions to produce NADPH.

27. What is the function of the electron transport chain in cellular respiration?

    (A) To break a large free-energy drop into smaller energy-releasing steps
    (B) To make ATP
    (C) To store electrons for use in the Krebs cycle
    (D) To convert pyruvate to acetyl CoA
    (E) To return Krebs-cycle enzymes to the mitochondrial matrix

28. A way in which seeds contained in mature plant ovaries are dispersed is

    (A) by passing through animal digestive systems
    (B) by insect pollination
    (C) by flagella or cilia
    (D) by phototropism
    (E) All of the above

29. Most biomass pyramids show a rapid decrease in biomass as trophic level increases. In aquatic systems, however, this pattern may be reversed so that one observes a larger standing crop of consumers compared to producers. What explains this pattern?

    (A) Aquatic producers tend to have larger body sizes than terrestrial producers.
    (B) Water is an easier medium to live in and aquatic organisms require less food.
    (C) Biomass in aquatic systems cannot be measured accurately.
    (D) Phytoplankton is rapidly consumed, but it has a high turnover rate.
    (E) Zooplankton reproduce quickly, but have poor survival success.

30. Sweat cools the human body because

    (A) water has a high heat of vaporization
    (B) water has a low heat of vaporization
    (C) water dissolves salts
    (D) water adheres to heat receptors
    (E) water reduces dehydration

31. Which property of water allows ice to float?

    (A) Relatively low surface tension
    (B) Relatively high heat of vaporization
    (C) Attraction to hydrophilic materials
    (D) Specific heat
    (E) Hydrogen bonding

32. Which sequence correctly depicts the order of events in the cell cycle? Assume that these events are continuous (i.e., after the last phase in each list, the cell returns to the first phase listed and again proceeds through the order of events).

    (A) G1 → Cytokinesis → G2 → Mitosis → S
    (B) G1 → G2 → S → Mitosis → Cytokinesis
    (C) S → G2 → Cytokinesis → Mitosis → G1
    (D) G2 → S → Mitosis → Cytokinesis → G1
    (E) S → G2 → Mitosis → Cytokinesis → G1

33. The main difference between the lytic and lysogenic cycles of phage reproduction is that

    (A) the lysogenic cycle alters the bacteria while the lytic cycle does not
    (B) the lytic cycle alters the bacteria while the lysogenic cycle does not
    (C) the lysogenic cycle kills the host cell while the lytic cycle does not
    (D) the lytic cycle kills the host cell while the lysogenic cycle does not
    (E) None of the above

34. During photosynthesis

    (A) light reactions produce sugar, while the Calvin cycle produces O_2
    (B) light reactions produce NADPH and ATP, while the Calvin cycle produces sugar
    (C) light reactions photophosphorylate ADP, while the Calvin cycle produces ATP
    (D) the Calvin cycle produces both sugar and O_2
    (E) light reactions produce both sugar and O_2

35. Primary succession will occur after

    (A) the climax community is established
    (B) a mudslide covers and kills the existing community of plants
    (C) *K*-selected species outcompete *r*-selected species
    (D) inhibition by colonizing species
    (E) the formation of a new volcanic island

36. All of the following are processes that can alter chromosome structure EXCEPT for which?

    (A) inversion
    (B) reduction
    (C) deletion
    (D) duplication
    (E) translocation

37. Why do larger organisms tend to have MORE cells (and not LARGER cells) than smaller organisms?

    (A) Very large cells would require more cytosol, which is difficult to produce.
    (B) The endoplasmic reticulum would not be able to retain its extensive folded structure within a very large cell.
    (C) There would not be enough DNA to fill the nucleus of a very large cell.
    (D) The metabolic requirements of very large cells could not be met with the available surface area of the plasma membrane.
    (E) Not enough evolutionary time has passed—larger organisms will evolve to have fewer, larger cells in the future.

38. The oldest fossils observed are prokaryotes, then fish, amphibians, reptiles, and finally mammals and birds (youngest fossils). This pattern

    (A) is an example of how comparative embryology supports Darwinian evolution
    (B) supports the idea of macroevolution
    (C) supports the idea of microevolution
    (D) is an example of how Lamarckian evolution has modified species over time
    (E) is inconsistent with other evidence supporting evolution

39. The idea that "ontogeny recapitulates phylogeny"

    (A) arose from the observation that vertebrate embryos look very similar
    (B) explains the evolution of birds from dinosaurs
    (C) depends on the presence of cytochrome *c*
    (D) supports the biogeographical distribution of marsupials
    (E) None of the above

40. A potential disadvantage of asexual reproduction when compared to sexual reproduction is that

    (A) asexual reproduction requires more resources than sexual reproduction
    (B) asexual reproduction requires that an organism remain immobile
    (C) it tends to be easier to produce numerous offspring quickly with asexual reproduction
    (D) asexual reproduction tends to limit the amount of possible genetic variation
    (E) asexual reproduction promotes inbreeding

41. Which sequence best describes what happens when a retrovirus infects a host cell?

    (A) RNA-DNA hybrid → reverse transcriptase → viral proteins → provirus
    (B) reverse transcriptase → RNA-DNA hybrid → provirus → viral proteins
    (C) provirus → viral proteins → RNA-DNA hybrid → reverse transcriptase
    (D) viral proteins → RNA-DNA hybrid → reverse transcriptase → provirus
    (E) reverse transcriptase → provirus → viral proteins → RNA-DNA hybrid

42. Which of the following does not have the potential to regulate gene expression in eukaryotes?

    (A) Alternative splicing during RNA processing
    (B) Repressor activity on operons
    (C) Chromatin structure
    (D) Translation initiation factors
    (E) Activators and enhancers

43. Members of a species provide parental care and rear few offspring during a given reproductive cycle. Which type of survivorship curve would be most consistent with these characteristics?

    (A) Type I
    (B) Type II
    (C) Type III
    (D) Cohort specific
    (E) Uniform

44. The effects of a hurricane on an island population of butterflies would be

    (A) an example of both a density-dependent and a density-independent factor because the population cycles over time
    (B) an example of a density-dependent factor because the storm will be more severe when the population is near carrying capacity
    (C) an example of a density-dependent factor because populations at higher density will be better able to recover from the storm
    (D) an example of a density-independent factor because population growth rate will increase following mortality from the storm
    (E) an example of a density-independent factor because the population size has nothing to do with the occurrence or intensity of the storm

45. What source of energy does ATP synthase use to generate ATP from ADP and inorganic phosphate in cellular respiration?

    (A) Light energy
    (B) Fermentation
    (C) Chemiosmosis of H^+
    (D) Enzymes
    (E) Glycolysis

46. Cellular respiration produces much more ATP per glucose molecule than fermentation because

    (A) respiration uses glycolysis to oxidize glucose
    (B) oxygen is necessary to release energy stored in pyruvate
    (C) fermentation uses NAD^+ as the oxidizing agent in glycolysis
    (D) respiration produces 2 ATP via glycolysis
    (E) All of the above

47. How many possible combinations of chromosomes are there during metaphase I if $2n = 6$?

    (A) 3
    (B) 6
    (C) 8
    (D) 10
    (E) 12

48. The logistic equation

    I. is a J-shaped curve that represents how population size changes with time
    II. incorporates "carrying capacity," which represents limits on the number of individuals the environment can support
    III. fits data from most real populations because its assumptions are easily met

    (A) None are correct.
    (B) Only I is correct.
    (C) Only II is correct.
    (D) Only III is correct.
    (E) All are correct.

49. Which organelle modifies products from the endoplasmic reticulum (ER) and then transports these products to other parts of the cell?

    (A) The Golgi apparatus
    (B) The lysosomes
    (C) The mitochondria
    (D) The ribosomes
    (E) The vacuoles

50. NAD^+ is _____ to NADH by _____, which releases _____ into the surrounding solution.

    (A) Oxidized; dehydrogenase; an electron
    (B) Oxidized; hydrogenase; a proton
    (C) Reduced; dehydrogenase; a proton
    (D) Reduced; dehydrogenase; an electron
    (E) Reduced; hydrogenase; a proton

51. Why do many herbivores have symbiotic microorganisms in their alimentary canals?

    (A) To break down chitin that the herbivore ingests with plant material
    (B) To regulate the herbivore's fecal production
    (C) To balance pH in the herbivore's intestine
    (D) To provide food (i.e., cud) that the herbivore can use at a later time
    (E) To make energy and nutrients in cellulose available to the herbivore

52. Root crops (e.g., beets and turnips) are harvested before they flower because

    (A) the root part of the plant extends deeper into the ground during flowering
    (B) flowering causes the root part of the plant to become soft and begin decaying
    (C) the root part of the plant stores extra water during flowering
    (D) flowering uses food energy stored in the root part of the plant
    (E) the root part of the plant synthesizes toxins during flowering

53. In their respective natural habitats, due to osmosis, saltwater fish continuously _____ water and _____ salts, and freshwater fish _____ water and _____ salts.

    (A) gain, lose; lose, gain
    (B) gain, lose; gain, gain
    (C) gain, gain; gain, lose
    (D) lose, gain; gain, lose
    (E) lose, lose; gain, gain

54. What ion channel(s) is/are required for an action potential to travel along the axon of a neuron?

    (A) Na^+, K^+, Ca^{2+}
    (B) K^+
    (C) Na^+
    (D) Na^+, K^+
    (E) Na^+, K^+, Cl^-

55. In an alternation of generations life cycle, the sporophyte refers to the _____ individual and the gametophyte refers to the _____ individual. _____ have an alternation of generations life cycle.

    (A) diploid; haploid; Some algae
    (B) haploid; diploid; Some algae
    (C) fruiting; germinating; Some plants
    (D) germinating; fruiting; Some plants
    (E) germinating; fruiting; Some slime molds

56. The tradeoff between water loss and CO_2 uptake has resulted in the evolution of what adaptation in many plants that experience a hot, dry environment?

    (A) C4 and CAM photosynthesis
    (B) C3 and CAM photosynthesis
    (C) C3 and C4 photosynthesis
    (D) Photorespiration that generates ATP
    (E) Stomata that are impermeable to water

57. Which situation illustrates a reaction norm?

    (A) Adopted and biological children in a single family make the same grammatical errors.
    (B) Skin color depends on multiple inherited genes.
    (C) Soil acidity affects whether hydrangea flowers of the same genotype will be violet-blue or pink.
    (D) Boldness in mice is correlated with tendency to explore new environments.
    (E) Plants in deserts in North America and in Africa have waxy, reduced leaves.

58. If an egg were to be fertilized by multiple sperm cells, one would hypothesize that

    (A) there was an abnormality in the acrosomal or cortical reaction
    (B) the egg's metabolism was not activated by the increase in Ca^{2+}
    (C) the nucleus of the first sperm did not fuse with the egg's nucleus rapidly enough
    (D) the egg or sperm contained aberrant chromosome numbers
    (E) the acrosomal process of the first sperm was incompatible with the egg's receptor molecules

59. How do natural killer (NK) cells and macrophages contribute to the body's defense?

   (A) NK cells attack infected cell membranes. Macrophages ingest invading organisms.
   (B) NK cells ingest invading organisms. Macrophages reduce inflammation.
   (C) NK cells release histamine. Macrophages ingest invading organisms.
   (D) NK cells produce T cells. Macrophages secrete antibodies.
   (E) NK cells produce B cells. Macrophages ingest invading organisms.

Directions: For the questions below, there are five lettered choices followed by a phrase or sentence for each question. Choose the letter that best corresponds to the phrase or sentence for each question and fill in that answer on your answer grid. Each choice can be used once, two or more times, or not at all for the following questions.

Questions 60–64

   (A) Protobionts
   (B) Panspermia
   (C) Homologous
   (D) Analogous
   (E) Adaptive radiation

60. Whale and bat forelimbs
61. Aggregates of abiotically produced molecules that may have preceded living cells
62. A single species evolving into many daughter species
63. Hypothesis that organic molecules were originally brought to Earth by meteorites and comets
64. Bird wings and insect wings

Questions 65–68

   (A) Fungi
   (B) Gymnosperms
   (C) Angiosperms
   (D) Animalia
   (E) Bryophytes

65. Pollinators and dispersers for many flowering plants
66. Have "naked seeds" (e.g., *Ginko*)
67. Decomposers that gain nutrients through absorption
68. Most ancient of the listed plant groups

Questions 69–71

   (A) Primase
   (B) Histone acetylation
   (C) Codons
   (D) Okazaki fragments
   (E) Leading strand

69. Affect(s) availability of genes for transcription
70. Joins RNA nucleotides during DNA replication
71. Base triplets along an mRNA molecule

Questions 72–74

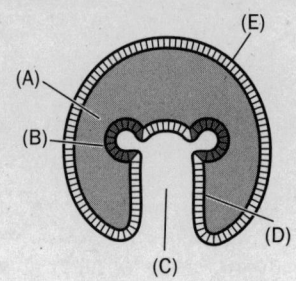

72. Germ layer from which the liver originates
73. Germ layer from which the nervous system originates
74. Germ layer from which the integument originates

Questions 75–79

(A) Gravitropism
(B) Positive feedback
(C) Self-incompatibility
(D) Photoperiodism
(E) Indeterminate growth

75. Meristems allow for this.
76. Ethylene production exhibits this.
77. A plant that requires nine hours of continuous darkness to flower exhibits this.
78. Roots growing into the soil and shoots growing toward the sun exemplify this.
79. Pollen grains unable to grow pollen tubes on stigmas with matching alleles exemplify this.

Directions: Each group of questions below concerns an experimental or laboratory situation or data. In each case, first study the description of the situation or data. Then choose the best answer to each question following it and fill in the corresponding oval on the answer grid.

Questions 80–82

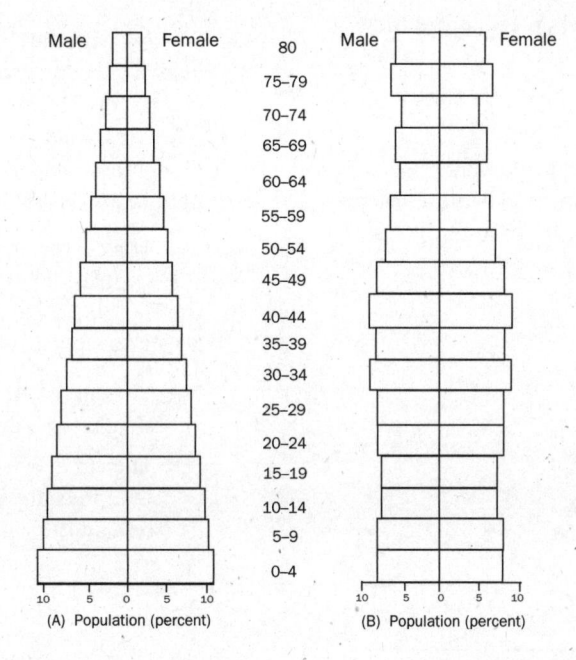

80. Which most accurately describes age-structures A and B?

    I. A is the population with a higher fertility rate
    II. B is the population with a higher fertility rate
    III. A represents a structure commonly observed for countries in North America and Western Europe
    IV. B represents a structure commonly observed for countries in North America and Western Europe

    (A) II and III
    (B) I and IV
    (C) II and IV
    (D) I and III
    (E) None are correct

81. Which social issue would you most expect to see in a country with an age-structure like A?

    (A) high unemployment for working-age people
    (B) delayed retirement
    (C) increased child mortality rate
    (D) decreased work-related injury rate
    (E) lack of access to education

82. Would a population structured like A or B be more likely to contribute to the global problems of overpopulation and overconsumption of resources?

    (A) A would contribute to overpopulation and B to overconsumption.
    (B) A would contribute to overconsumption and B to overpopulation.
    (C) A and B would contribute equally to these problems.
    (D) Neither population type would contribute to these problems.
    (E) It is impossible to determine what populations A and B would contribute to because scientists know little about patterns of resource availability or population growth rates.

Questions 83–85

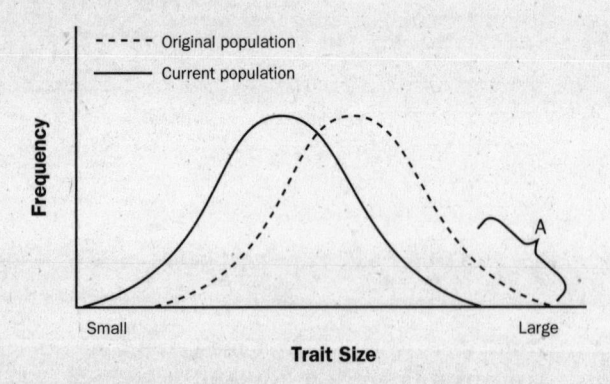

83. The region indicated by the "A" in this figure depicts

    (A) the variability of sizes in the population
    (B) the mean size of the population
    (C) the phenotypes that were successful in the given environmental conditions
    (D) the phenotypes that did poorly in recent environmental conditions
    (E) the most fit individuals

84. What type of selection does this figure depict?

    (A) Disruptive
    (B) Directional
    (C) Stabilizing
    (D) Forceful
    (E) Cannot determine from the figure

85. Starting with the current population distribution, what would you expect to happen if, over the next several generations, the smallest and largest individuals had lower reproductive rates than moderately sized individuals? What would the new population curve look like?

    (A) The curve would resemble the "original population" curve in shape and location.
    (B) The curve would be shifted to the left of the current one.
    (C) The curve would have two peaks, one at smaller sizes and another at larger sizes.
    (D) The curve would be more narrow and higher than the current one; the mean size would remain the same.
    (E) The curve would be "U" shaped.

Questions 86–88

	BC	Bc	bC	bc
BC	BBCC	BBCc	BbCC	BbCc
Bc	BBcC	BBcc	BbcC	Bbcc
bC	bBCC	bBCc	bbCC	bbCc
bc	bBcC	bBcc	bbcC	bbcc

The Punnett square above depicts results from the cross of two heterozygotes for petal pattern [B = black (dominant), b = blue (recessive)] and the expression of color [C = color present (dominant), c = color absent (recessive)]. When the expression of color is homozygous recessive, neither pattern is expressed, leaving offspring colorless.

86. What is the ratio of phenotypes of the offspring resulting from the heterozygote cross?

   (A) 9 black: 3 black-blue: 3 blue-black: 1 blue
   (B) 9 black-colored: 3 black-colorless: 3 blue-colored: 1 blue-colorless
   (C) 6 black: 2 blue: 8 colorless
   (D) 6 blue: 2 black: 8 colorless
   (E) 9 black: 3 blue: 4 colorless

87. If two of the phenotypically colorless offspring depicted in the Punnett square were crossed, the resulting offspring would

   (A) all be colorless
   (B) all be black or blue
   (C) have different phenotypes based on the specific genotypes crossed
   (D) show polygenic inheritance
   (E) show pleiotropy

88. The relationship between the expression of color gene and the pattern gene represented in the Punnett square above is an example of

   (A) the law of independent assortment
   (B) codominance
   (C) incomplete dominance
   (D) epistasis
   (E) pleiotropy

Questions 89–91

Sickle cell anemia is caused by mutant hemoglobin DNA, which is more common in humans with African ancestry than in those with European ancestry. The sickle cell allele creates an altered mRNA codon that produces hemoglobin containing valine rather than glutamic acid. If a person inherits both alleles for the sickle cell trait, his hemoglobin will crystallize under low oxygen conditions (i.e., elevated physical activity). This can result in brain damage, paralysis, kidney failure, and other very serious physiological problems.

89. The mutation described above is an example of

   (A) a base-pair substitution
   (B) a frame-shift mutation
   (C) a silent mutation
   (D) a mutagen
   (E) a chromosomal rearrangement

90. Heterozygotes for the sickle cell trait have increased resistance to malaria. If malaria were eradicated and effective treatment for sickle cell anemia made universally available, what would be the expected effect on the sickle cell trait?

   (A) The frequency of the trait would remain constant.
   (B) The frequency of the trait would decrease.
   (C) The frequency of the trait would increase.
   (D) The frequency of homozygous individuals would decrease, and the frequency of heterozygous individuals would remain constant.
   (E) The effect on the trait is impossible to predict.

GO ON TO THE NEXT PAGE

91. How does the importance of genetic mutation for generating population-level genetic variation differ for human populations and bacterial populations?

    (A) Genetic mutation is responsible for more genetic diversity in humans than in bacteria.
    (B) Genetic mutation is responsible for more genetic diversity in bacteria than in humans.
    (C) Genetic mutation has equal impact in bacteria and human populations.
    (D) The effect of genetic mutations on diversity depends upon the environment.
    (E) It is impossible to compare these populations in this way.

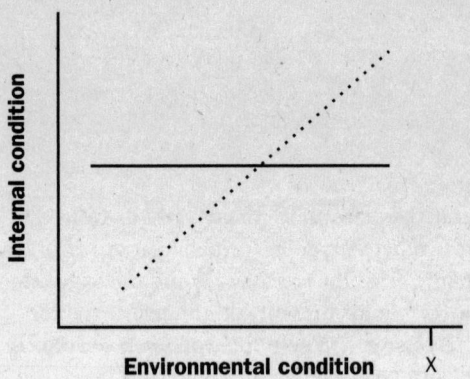

Questions 92–94

92. The solid line best represents

    (A) an ectotherm
    (B) convection and conduction
    (C) a strict conformer
    (D) a strict regulator
    (E) a heat budget

93. What would you expect to happen at the environmental condition represented by "X"?

    (A) The same patterns would continue (the dashed and solid lines could be extended).
    (B) Organisms represented by the dashed line would survive. Organisms represented by the solid line would die.
    (C) Organisms represented by the dashed line would die. Organisms represented by the solid line would survive.
    (D) Regardless of whether the solid or dashed line best described an organism, it would likely die.
    (E) Homeostasis feedback mechanisms would stimulate thermoregulation.

94. Consider two hypothetical endotherms (species A and B). Species A has special adaptations allowing it to withstand more extreme temperatures (e.g., colder nights and hotter days in a desert). What would the two species look like on this graph?

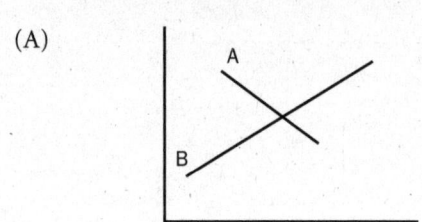

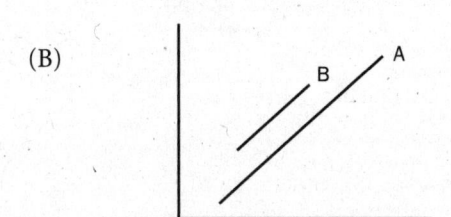

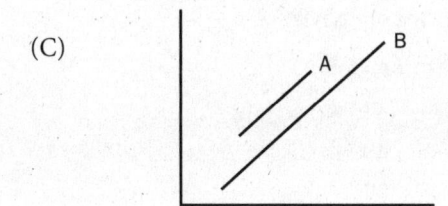

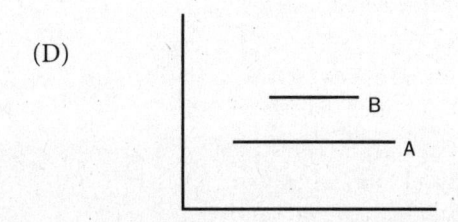

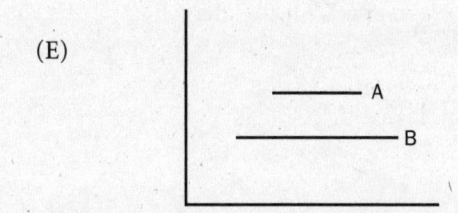

(A) [graph with lines A and B crossing in an X]

(B) [graph with parallel lines B above A rising]

(C) [graph with parallel lines A below B rising]

(D) [graph with horizontal lines B above A]

(E) [graph with horizontal lines A above B]

Questions 95–96

Hamilton's rule states that an animal's decision to do an altruistic act should be based on the costs of the act to the animal and the benefits of the act to the recipient individual, weighed by the relatedness of the two individuals.

95. Why should relatedness matter?

(A) Relatives are more likely to take revenge or return the favor later.
(B) Relatives share on average more genes than non-relatives. Therefore, an animal can pass on copies of its genes indirectly when a relative reproduces.
(C) Animals tend to congregate with relatives. Helping relatives is harder to avoid than helping non-relatives simply based on group patterns.
(D) Relatives resemble each other, which allows individuals to behave differently toward relatives than non-relatives.
(E) Relatives would benefit more by a helpful act than non-relatives.

96. If an individual had to choose between helping particular relatives, in what order would Hamilton's Rule predict that the individual would prioritize giving his assistance?

(A) cousin, sister, identical twin brother
(B) sister, cousin, identical twin brother
(C) identical twin brother, sister, cousin
(D) sister, identical twin brother, cousin
(E) identical twin brother, cousin, sister

GO ON TO THE NEXT PAGE

Question 97–98

Characters	Outgroup (0)	Taxon 1	Taxon 2	Taxon 3	Taxon 4
V	−	+	+	+	+
W	−	−	+	+	+
X	−	−	−	+	+
Y	−	−	−	+	+
Z	−	−	−	−	+

97. Which of the following phylogenetic trees best represents the data in the table?

(A), (B), 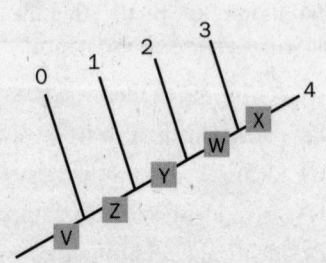 (C), 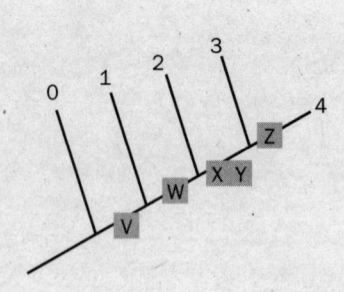 (D), 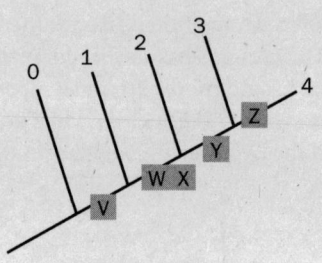 (E)

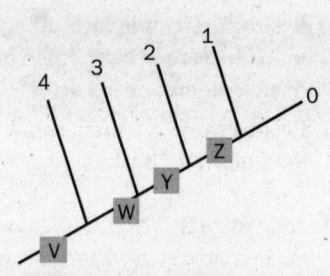

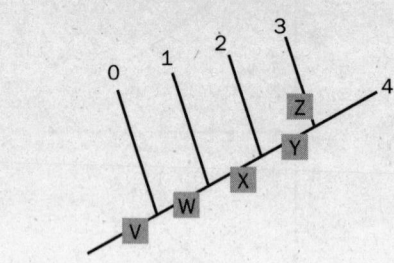

98. Based on the data table, which grouping of taxa would be monophyletic?

(A) Taxa 3 and 4
(B) Taxa 0 and 2
(C) Taxa 1 and 2
(D) Taxa 2 and 3
(E) Taxa 0 and 3

Questions 99–100

An artificial cell (a selectively permeable membrane surrounding an aqueous solution) is placed into a container filled with a solution consisting of 0.03 mol sucrose and 0.02 mol glucose. The cell contains 0.01 mol sucrose, 0.01 mol glucose, and 0.01 mol fructose. The membrane is permeable to water and simple sugars, but impermeable to sucrose.

99. The cell interior is _____ compared to the environment in the container.

    (A) exogonic
    (B) endergonic
    (C) hypertonic
    (D) hypotonic
    (E) isotonic

100. Which statement describes the net movement that would be expected given the solutions in the cell and in the environment?

    (A) Fructose would diffuse out of the cell, glucose would diffuse into the cell, and water would move out of the cell.
    (B) Fructose would diffuse out of the cell, sucrose and glucose would diffuse into the cell, and water would move out of the cell.
    (C) Fructose would diffuse into the cell, glucose would diffuse out of the cell, and water would move into the cell.
    (D) Fructose would diffuse into the cell, sucrose and glucose would diffuse out of the cell, and water would move into the cell.
    (E) Movement patterns are impossible to determine given the available information.

IF YOU FINISH BEFORE TIME IS CALLED, YOU MAY CHECK YOUR WORK ON THIS SECTION ONLY. DO NOT TURN TO ANY OTHER SECTION IN THE TEST.

STOP

Ten-Minute Reading Period

Take the next ten minutes to glance over the four questions that comprise Section II of this test. You can take notes in the margins, but you may not begin answering any of the questions in any way whatsoever.

When the ten-minute period is over, you may begin answering the four free-response questions.

Section II: Free-Response Questions

Time: 90 Minutes
4 Questions

Directions: Answer each part of the following free-response questions with complete sentences. Answers should be in essay form. Make sure to provide a detailed response to each part of each question. Diagrams and other figures may be included to demonstrate knowledge of a topic, but they should not be the only answer you provide.

1. Every organism in an ecosystem is affected by the other organisms in the community it lives in. Create a sample community and complete all of the following points about your community:

    (a) **Define** "trophic levels." Provide a specific example of four species and **describe** their trophic-level relationships.

    (b) **Trace** energy flow through your sample community.

    (c) **Trace** contaminant flow (for example, DDT) through your sample community

2. Mammals have a unique ability to heal when they are invaded by a variety of foreign organisms. Humans have an entire immune system dedicated to protecting the body from foreign invaders such as viruses and bacteria. People have also developed artificial means of protecting the body through the use of vaccines and medicine.

    (a) **Describe** how the mammalian immune system responds to (i) a laceration, such as an injury from falling on a barbed-wire fence, and (ii) a viral infection.

    (b) **Explain** how the flu vaccination works and why it has a fairly low success rate.

3. Acid rain affects flora and fauna in many ways, which have been observed by scientists in recent decades. Manmade structures are affected by acid rain as well. In a discussion of the impact of acid rain, include the following:

    (a) **Describe** and **explain** the causes of acid rain.

    (b) **Discuss** how acid rain is measured and the impact it has upon the environment.

    (c) **Determine** how the causes of acid rain can be reduced.

4. Speciation and extinction have shaped the biodiversity we observe on Earth today. Organisms and populations are constantly evolving and changing in response to their environment.

    (a) **Compare** and **contrast** allopatric and sympatric speciation.

    (b) **Discuss** small populations and extinction vortices. **Discuss** reasons why plant and animal populations may differ in their susceptibility to extinction.

    (c) **Discuss** three reasons that exotic species are often able to drive native species to extinction.

GO ON TO THE NEXT PAGE

Practice Test 2: Answer Key

1. C	26. A	51. E	76. B
2. D	27. A	52. D	77. D
3. C	28. A	53. D	78. A
4. E	29. D	54. D	79. C
5. C	30. A	55. A	80. A
6. B	31. E	56. A	81. B
7. E	32. E	57. C	82. B
8. A	33. D	58. A	83. D
9. D	34. B	59. A	84. B
10. D	35. E	60. C	85. D
11. E	36. B	61. A	86. E
12. C	37. D	62. E	87. A
13. E	38. B	63. B	88. D
14. D	39. A	64. D	89. A
15. E	40. D	65. D	90. A
16. A	41. B	66. B	91. B
17. C	42. B	67. A	92. D
18. A	43. A	68. E	93. D
19. B	44. E	69. B	94. D
20. D	45. C	70. A	95. B
21. D	46. B	71. C	96. C
22. C	47. C	72. D	97. C
23. B	48. C	73. E	98. A
24. C	49. A	74. E	99. D
25. E	50. C	75. E	100. A

ANSWERS AND EXPLANATIONS

Section I: Multiple-Choice Questions

1. C

Although choice (A) states something true, this fact tells us little if anything about how an animal lives or what it eats. Choice (B) is incorrect because broad, hard surfaces used for grinding are characteristic of herbivores' teeth; similarly, choice (D) describes carnivores' teeth, not herbivores' teeth. Finally, teeth are not necessarily required for an animal to break down and digest solid food (E).

2. D

Notice that the change in population size keeps increasing over time (years), which indicates that exponential growth is occurring. Each year more individuals are being added to the population, which then increases the number of individuals available to reproduce and contribute to the next year's population size (N). If the data had reflected a drop in the change of population size (ΔN) to zero and the stabilization of population size (N) at a certain value, this would be consistent with the logistic growth curve (A). Choices (B) and (C) can be eliminated because each contain terms that are undefined for the population ("fecundity") or that can be assumed to be constant ("generation time" can be a factor when comparing different species but is considered to be the same for members of the same population). "Cycling" (E) refers to increases and decreases in population size over time, and therefore also in ΔN. Inspection of the data shows that there are no decreases in population size at any point.

3. C

The first two processes occur during meiosis I, while fertilization is the joining of two gametes, which are haploid by definition. Asexually reproducing organisms (D) do not go through meiosis. Mutations (E) create genetic variability in asexually and sexually reproducing organisms and are not a product of these processes.

4. E

A protein that changes the rate of a reaction without being consumed by the reaction is an enzyme, or catalyst (B). This is done by lowering activation energy (A), not by changing the free-energy change ΔG between reactants and products (C).

5. C

One concern about genetically engineered crop plants is the possibility that the manipulated genes in crops could be transferred to other plants through hybridization. This could give the other plants benefits (such as pest resistance or drought tolerance) that could make them pests or invasive.

6. B

This question requires knowledge of community ecology terminology and concepts. Competitive exclusion (A) refers to the extinction of one species because another has won the competition for an area's limited resources. Symbiosis (C) is a close mutualistic relationship between two species. In interference competition (D), a species prevents another from gaining access to a portion of habitat.

7. E

Free ribosomes are found in the cytosol and tend to produce proteins that will be used in the cytosol [(B) and (C)]. Bound ribosomes are bound to the endoplasmic reticulum or nuclear envelope and tend to produce proteins for membranes, used in other organelles, or for secretion. There is no difference in the structure of the two types (A); a ribosome may serve as either free or bound, depending on the cell's metabolic needs.

8. A

Linkage maps are based on recombination frequencies, which increase the further apart genes are on a chromosome.

9. D

Species such as elephants that have fewer offspring and provide more parental care are called *K*-selected or *K*-strategists. In contrast, *r*-strategists have millions of offspring and provide little or no parental care (B). Elephants do not lay eggs (C), and they utilitze internal fertilization, so (A) is incorrect.

10. D

Note that only woody plants can undergo secondary growth, an increase in the diameter of stems and roots.

11. E

This makes the bones light but strong. Much energy is needed to fly—birds have four-chambered hearts (D), air sacs that work with the lungs for efficient gas exchange (C), and increased pectoral muscles (B) to power the wings. Being toothless, which decreases a bird's weight, is compensated for by having a gizzard that grinds the food before it enters the stomach (A).

12. C

Aposematic coloration is used to warn predators that an organism is poisonous or distasteful. In Müllerian mimicry (A), multiple species with noxious predator-defense mechanisms have evolved to resemble each other. In Batesian mimicry (B), a species without noxious predator-defense mechanisms resembles a noxious species. Cryptic coloration (D) means that an organism blends into its environment, making it difficult to detect. Predatory coloration (E) is not a biological term.

13. E

Plasmodesmata do all of these things and are an important structural feature that unifies plant cells into an entire plant organism.

14. D

Compounds in the carboxyl group contain -COOH.

15. E

Although other pigment molecules are excited by photons, only the chlorophyll molecule at the reaction center donates an electron.

16. A

Synapsis occurs during prophase I in meiosis I.

17. C

Although humans use these plant products, they serve as chemical defenses that repel most herbivores.

18. A

The membrane between the vacuole and the cytosol, the tonoplast, serves to keep cytosol outside of the vacuole. The vacuole is filled with "cell sap."

19. B

Increasing the concentration of substrate will increase the likelihood that substrate molecules, rather than inhibitor molecules, will bind to the enzyme's active site. All of the other choices would reduce the effectiveness of an enzyme, which is what a competitive inhibitor does.

20. D

Exergonic reactions are spontaneous and release free energy. At equilibrium, $\Delta G = 0$. ΔG does not reveal whether the system is open or closed.

21. D

Proteins can be organized from smaller structural (amino acids) to larger structural components (grouping of subunits). Proteins consisting of only one polypeptide chain will not have a quaternary structural level.

22. C
Cells normally show density-dependence inhibition, meaning that when cells become crowded (e.g., more than a single layer in a culture dish) they stop dividing. Normal cells will also stop dividing if they are not attached to substrate, either the culture dish or extracellular tissue matrix (anchorage dependence). The other answers are true statements but do not address the reason why clumping or overlapping cells are abnormal.

23. B
Enzymes move a phosphate group from ATP to another molecule. This phosphorylated molecule has the capacity to perform work, during which it releases inorganic phosphate. When ATP loses its phosphate group, it is converted to ADP. ADP can be phosphorylated during cellular respiration, which regenerates the cell's supply of ATP.

24. C
Prokaryotic cells and plant cells (eukaryotic) contain cell walls. Only animal cells contain centrioles.

25. E
Myoglobin stores oxygen in the muscle. Deep divers tend to have high concentrations of this protein, which allows much greater storage of O_2 in their muscles than commonly seen in non-deep-diving animals. The other answers are modifications that are found in deep divers.

26. A
Rubisco is the enzyme that catalyzes the reaction that attaches CO_2 to ribulose biphosphate (RuBP) in phase 1 of the Calvin cycle.

27. A
The electron transport chain releases energy in steps, which allows the cell to take advantage of the energy more efficiently.

28. A
Mature ovaries are fruits, which animals often use as a food source. The seeds are deposited in a new location when the animal defecates. Pollination (B) is a method of fertilization, which predates the dispersal of seeds.

29. D
Zooplankton (consumers) rapidly deplete the phytoplankton (producers) biomass, which allows the zooplankton to have a larger standing crop. This is sustainable because phytoplankton reproduce very quickly and can replenish consumed biomass. However, because the zooplankton are continually grazing down the phytoplankton, the phytoplankton standing crop population size remains small. Choice (B) is unlikely since it makes a two-part blanket statement (about water and about caloric needs), both parts of which would have to be true for this example. Although there may be measurement error (C) involved, it would have to be huge to create a pattern as extreme as an inverted biomass pyramid. Choice (E) does not give enough information to explain the larger consumer level AND smaller producer level.

30. A
Liquids need to absorb heat in order to escape the liquid phase and be converted into a gas (heat of vaporization). Water's high heat of vaporization results in evaporative cooling because the liquid that is converted into gas when sweat evaporates into the environment absorbs and removes heat from the remaining molecules. Sweat increases dehydration (E) because it is water that leaves the body.

31. E
As water cools to 0°C, water molecules bond to a maximum of four neighboring water molecules and become trapped in a lattice as their motion slows to form solid ice. The hydrogen bonds prevent the molecules from getting as close to each other in solid form as they do in liquid form, which makes ice less dense than water and thus able to float.

32. E

G1, S, and G2 are subphases of interphase. Mitosis (duplication and separation of chromosomes) and cytokinesis (division of the cell into two cells) make up the mitotic phase (M phase). The phases proceed in the order listed in (E).

33. D

Viruses that infect bacteria, bacteriophages, can insert their DNA into the bacterial chromosome (lysogenic; doesn't kill the host cell; phage DNA is passed along when the cell replicates) or can immediately start producing phage progeny, which causes the bacterial cell to lyse and release the new phages.

34. B

Light reactions, which convert solar energy into chemical energy (ATP, NADPH) in the thylakoid membranes fuel the Calvin cycle's production of sugars in the stroma.

35. E

Primary succession occurs on newly created substrate, like new volcanic islands or land scoured by a retreating glacier, which lack soil (and living organisms). Secondary succession, choice (B), occurs when an existing community is exterminated in a disturbance, but soil remains. The climax community (A) is the final, stable community in the succession process. Choice (C) is wrong because *r*-selected species tend to outcompete *K*-selected species when habitats are new and resources are abundant.

36. B

The other answers are all terms for types of change in chromosome structure.

37. D

Chemical exchange (oxygen, nutrients, and waste) across the cell's plasma membrane can occur only so quickly. Larger things have smaller surface-to-volume ratios (in this case, surface = plasma membrane and volume = cell interior)—i.e., there is less plasma membrane available for each unit of cell interior the larger a cell gets. Since rates of chemical exchange are limited across the membrane surface, there is a limit to cell size. Cells that are too large will not be able to get enough oxygen, etc. to survive.

38. B

Macroevolution refers to evolution on a large scale, such as the emergence of new species. The fossil record is consistent with other scientific lines of evidence for Darwin's view of evolution, so (E) is wrong. Darwin's view is based on the idea that natural selection modifies populations through differential survival and reproduction of individuals with traits better suited to the environment, and that this mechanism of natural selection caused common ancestral species to be modified over time into the diversity of species on Earth today. Prior to Darwin, Lamarck propounded an incorrect view of evolution based on inheritance of acquired characteristics (D).

39. A

This statement originated from comparative embryology research in the late 19th century. Today, it is considered extreme and should not be taken literally. Although mammals go through early developmental stages (ontogeny) that resemble those of more anciently evolved organisms, they do not go through a "fish stage" and then an "amphibian stage" and so on (which would reflect phylogeny) before birth.

40. D

Because sexual reproduction joins two haploid gametes, it produces unique combinations of parental genetic material in the offspring. Offspring resulting from asexual reproduction will be genetically identical (except for mutations) to the parent that produced them. Choices (A) and (B) are not true. Choice (C) is an advantage of asexual reproduction. Inbreeding (E) is only relevant when talking about organisms that reproduce sexually.

41. B

Retroviruses reverse the order of information flow from RNA to DNA, using reverse transcriptase to synthesize DNA from viral RNA. This DNA is incorporated in the host cell's DNA as a provirus, which is then transcribed into viral proteins.

42. B

Only prokaryotes have operons, which consist of clusters of genes with related functions. In eukaryotes, genes with related functions are often scattered throughout the genome, and each gene is individually transcribed.

43. A

Survivorship curves, which plot number of survivors on the y-axis against age of individuals, have been generalized to take on three main shapes (Type I, II, or III). The Type I curve depicts high survivorship until late in life (e.g., humans, elephants, and other long-lived organisms usually having parental care). The Type II curve depicts a constant decrease in survivorship throughout the lifespan (some birds, small mammals, and reptiles). The Type III curve represents species where there is high mortality (low survivorship) early in life until some critical age is reached (typically organisms that produce millions of offspring, of which few survive to adulthood; e.g., many fish, annual plants, and insects).

44. E

Density-dependent factors are those that have a greater effect on population size, as well as on an individual, as population density increases (e.g., limited food resources or increased pathogen spread). Density-independent factors occur at intensities and frequencies that are unrelated to population density. Climatic events do not become more or less severe or change in frequency based on population density.

45. C

Oxidative phosphorylation of ADP in ATP synthase is powered by the exergonic flow of hydrogen ions as they move down the H^+ gradient across the inner mitochondrial membrane. This gradient is also known as "proton-motive force." Fermentation (B) and glycolysis (E) refer to other types of cellular respiration.

46. B

Oxidation of pyruvate, which produces a lot of ATP, occurs in respiration but not fermentation. The statements in (A), (C), and (D) are true of both respiration and fermentation.

47. C

When chromosomes assort independently into gametes during meiosis, the number of possible combinations is 2^n, where n = haploid number (so if $2n = 6$, then $n = 3$ and $2^3 = 8$).

48. C

The logistic equation models population size over time, but the carrying capacity parameter, K, reduces population growth rate as population density increases. This produces an "S"-shaped curve. The assumptions of the model, which are a consequence of the mathematics, are idealized. Typically, only laboratory populations of small organisms, such as insects or bacteria, closely match predictions from the logistic equation. In general, many simple mathematical models will not fit real populations very well, but they are useful to stimulate research and as a starting point for more complex models.

49. A

The Golgi apparatus stores, modifies, and transports products made by the ER in eukaryotic cells. Lysosomes (B) disgest macromolecules. Mitochondria (C) are centers of energy production for the cell. Ribosomes (D) are the sites of protein synthesis, and vacuoles (E) are storage compartments.

50. C

NAD^+ serves as an electron shuttle when it is reduced to NADH. The reaction consists of the removal of two hydrogen ions from sugar or some other fuel by the enzyme dehydrogenase. One H^+ is accepted by NAD^+ and the other is released into the environment. This redox reaction (NADH can be oxidized to NAD^+) is critical for cellular respiration.

51. E

These microorganisms have enzymes that digest cellulose into sugars and other usable compounds. Some of these microorganisms may convert sugar from cellulose into essential nutrients for the herbivore.

52. D

Taproots are storage organs that provide energy during a plant's energetically intensive reproductive period.

53. D

Saltwater and freshwater fish are faced with opposite challenges because of the relative difference in osmolarity between the fish's body and its specific environment. Saltwater fish osmoregulate by drinking lots of saltwater and excreting ions via special structures in their gill, skin, and kidney cells. Freshwater fish excrete a lot of urine to get rid of excess water and replenish salts through food and ion uptake through the gills.

54. D

Sodium and potassium movement across the axon membrane polarizes and depolarizes the membrane, which initiates and moves action potentials along the axon.

55. A

Alternation of generations refers specifically to the situation where generations alternate between multicellular diploid individuals (sporophytes) and multicellular haploid individuals (gametophytes). This life cycle is illustrated by some algae and plants.

56. A

Plants close their stomata in response to hot, dry conditions, which conserves water. Closed stomata, however, prevent CO_2 from entering the leaf, which is necessary for photosynthesis. Desert and other plants that encounter hot, dry conditions regularly have evolved photosynthetic adaptations that fix carbon in organic acids for later use in the Calvin cycle. This allows photosynthesis to occur with the stomata closed or partially closed during hot, dry conditions.

57. C

Norms of reaction are the phenotypic effects of different environmental conditions on the same genotype. For example, an environment with predators will cause some *Daphnia* species to grow a protective spike on their heads. Individuals with the same genetic makeup do not develop this predator protection (phenotypic trait) if predators are absent during development (which presents a different environmental condition).

58. A

The acrosomal reaction allows a sperm to penetrate the egg's jelly coat and fuse to the egg's plasma membrane. This depolarizes the membrane and prevents other sperm from fusing to the membrane. By the time the membrane voltage has returned to normal, the cortical reaction functions to block additional sperm (an increase in Ca^{2+} ions triggers the development of the fertilization envelope, which resists sperm).

59. A

NK cells lyse infected cells. Macrophages are phagocytic.

60. C

Homologous structures are structures that are similar because of common ancestry.

61. A

Protobionts have been simulated in the laboratory, can form spontaneously, and show some lifelike properties.

62. E

Adaptive radiation is the process of a single species evolving into many diversely adapted daughter species in a relatively rapid time span. Textbook examples of groups that have resulted from adaptive radiation include the Hawaiian honeycreepers and the Galapagos finches. Adaptive radiation is facilitated in new habitats with many unused resources or empty niches.

63. B

Scientists do not agree upon the origin of life. Another hypothesis is that shallow water or moist sediments on Earth provided conditions favorable to the genesis of prokaryotes.

64. D

Each wing type independently evolved to fulfill a similar function. If wing type were the result of shared ancestry, bird and insect wings would be homologous (C).

65. D
Plants preceded animals in terrestrial habitat. After animals colonized land, numerous species in these two groups coevolved such that many flowering plants are "serviced" by animal pollinators and dispersers, and many animals use flowering plants as a food source.

66. B
Gymnosperms include conifers and the ancient *Ginko* tree. Gymnosperms and angiosperms are "seed plants," with the "naked seeds" of the gymnosperms preceding the evolution of the flowering plants, or angiosperms (C).

67. A
Fungi, which were once categorized in the plant kingdom, are grouped in their own kingdom. They are characterized by their way of gaining nutrition—digesting food outside of their body by secreting enzymes and then absorbing the digested, simpler compounds.

68. E
Angiosperms (C) are most recently evolved, followed by gymnosperms (B), and finally by bryophytes (mosses, liverworts, and hornworts), the most ancient plant group listed.

69. B
DNA methylation and histone acetylation chemically modify chromatin structure and therefore regulate transcription.

70. A
Primase joins RNA nucleotides to form the primer. Okazaki fragments (D) are joined by DNA ligase. DNA polymerase, which can add to an existing polynucleotide but not initiate its synthesis, elongates the leading strand (E) in the 5' to 3' direction.

71. C
Codons are the three-nucleotide sequences that code for specific amino acids during translation.

72. D
The liver originates from the endoderm.

73. E
The nervous system is derived from the ectoderm.

74. E
The integument and associated glands also are derived from the ectoderm.

75. E
Meristems are embryonic tissues that continue throughout the individual's life to divide and produce cells that specialize into other plant tissues.

76. B
Ethylene stimulates ripening in fruit, which stimulates the production of more ethylene. This type of process is called positive feedback.

77. D
Photoperiod is the relative length of night and day. Photoperiodism is exhibited by plants that show physiological response to these relative durations. Some plants need threshold nighttime lengths to stimulate a response, while others need threshold daytime lengths.

78. A
Regardless of the orientation or depth of a germinating seed, the roots and shoots will grow in the correct direction. Although the mechanism is not yet known for certain, one hypothesis is that specialized plastids, called statoliths, settle in the cell. This would orient the seed using gravity, and allow for growth of the roots and shoots in the right directions.

79. C
Self-incompatibility is a mechanism that prevents a plant from fertilizing itself (selfing). By only accepting pollen from a plant having an allele different than its own, the parent plant will produce genetically variable offspring.

80. A
The answer is (A) because "fertility rate" refers to number of live births per female. A smaller percentage of the population is made up of children (bars below age 15) in structure A compared to B. Furthermore, a structure like A, where the population is fairly evenly distributed among all the age classes and where there is a large proportion of post-reproductive individuals, is common for North America and

Western Europe. In other regions of the world, there is often a much smaller proportion of the population that survives to older age classes, and the majority of the population consists of children and adults of reproductive age.

81. B

This answer is the most logical given the age distribution of structure A, with a larger proportion of older individuals who tend to live longer. Choices (A), (C), and (E) would be most consistent with age-structure B. Choice (D) does not follow logically from the question (why is this a social "issue"?), or from information available from the diagrams (how do you relate injury rate to age distribution without making a lot of assumptions?).

82. B

Typically industrialized nations, which usually have age-structures similar to A, have fewer children but spend much more per person on lifestyle expenses, which contributes greatly to the global problem of resource depletion. In contrast, many countries with age-structures similar to B have very rapidly growing populations, which increases the global population. Although individuals in these countries typically spend far less than the average person in an A-type population, the sheer number of individuals requiring even a small share of resources is a global issue.

83. D

Individuals with the larger sizes of this hypothetical trait (right side of graph) have been selected against, which caused the average size of the trait to decrease (hump of the new population is at a lower size than the hump of the original population).

84. B

This graph depicts directional selection, which moves the mean trait value of a population toward more extreme values (smaller or larger, in this case). Disruptive (or diversifying) selection (A) favors extreme phenotypes, so that intermediate phenotypes become less common and extreme ones are more common. Stabilizing selection (C) favors intermediate phenotypes, which become more common (larger and smaller phenotypes become less common). Forceful selection (D) is a made-up term.

85. D

This situation describes stabilizing selection, which favors intermediate phenotypes over extremes. Choices (A) and (B) depict directional selection, while (C) describes disruptive (diversifying) selection.

86. E

Any combination with *cc* results in a colorless individual, regardless of what alleles for pattern are inherited. Choice (B) would be expected for a Mendelian dihybrid cross for two independent traits.

87. A

Colorless offspring would have to be homozygous recessive (*cc*) for color based on the given information. Thus, offspring from these crosses would also inherit a color-absent allele from each parent and be colorless.

88. D

The expression-of-color gene alters the phenotypic expression of another gene (for color).

89. A

The sickle cell anemia allele has adenine substituted for thymine, which codes for valine rather than the normal glutamic acid in the hemoglobin protein. Clues that this scenario describes a base-pair substitution include: it involves a single codon and the protein is still produced, but with a different amino acid. Usually frame-shift mutations will have a much greater effect on the resulting protein. Silent mutations have no effect on the protein because of redundancy in the genetic code. Mutagens are agents that cause mutations. Chromosomal rearrangements involve much greater sections of genetic material.

90. A

If malaria were eliminated, selection for the sickle cell allele would stop. Individuals with the allele no longer would have a greater chance of survival than those without it. At the same time in this scenario, homozygous individuals would receive treatment that would ensure their survival through reproductive age and beyond. As it would no longer be selected for or against, the sickle cell allele's frequency would remain constant.

91. B

Bacteria, which have short generation times, reproduce asexually and very rapidly. This allows rare individual genetic mutations that create new alleles to have a large effect on the overall genetic diversity observed in a population. Much of the genetic variation among humans is the result of sexual recombination of existing alleles.

92. D

Two ways to deal with environmental fluctuations are to regulate internal body conditions (such as temperature and salinity) at a relatively constant level or to allow body conditions to vary with the environment. For body temperature, conformers are ectotherms and regulators are endotherms. These curves represent idealized relationships—most organisms fall somewhere on a continuum between the extremes of conforming and regulating.

93. D

Regardless of whether animals are conformers or regulators, there is a limit to the tolerable environmental conditions an animal can withstand.

94. D

Because they are endotherms, species A and B regulate their internal temperatures for a range of external temperatures, which is depicted by a curve parallel to the x-axis. Since species A can withstand more extreme external temperatures than species B, species A would be represented by a longer line (more temperatures) than species B.

95. B

An animal's indirect fitness will be increased when relatives reproduce. This increase in fitness will depend on how closely related the individuals are because this determines the probability that they will share copies of one another's genes. For example, Hamilton's rule states that an animal should be more likely to help its identical twin than its cousin. From a genetic standpoint, helping an identical twin is like helping oneself. This idea of kin selection has been used to explain cooperation in colonies of eusocial insects, some birds, and humans, among many other taxa.

96. C

An individual shares the most copies of genes with an identical twin and the least number of copies with a cousin. Siblings share an intermediate number of copies, on average.

97. C

This figure is consistent with the data table: all taxa have character V, taxa 2, 3, and 4 have character W, taxa 3 and 4 have characters X and Y, and taxon 4 has character Z. The outgroup has none of these characters.

98. A

A monophyletic group contains a hypothetical ancestor (depicted by the intersection of two lines in a phylogenetic tree) and all descendent taxa of this ancestor. All of the answers except for (A) leave out some of the descendants of the ancestor that would be common to the two taxa listed in each choice.

99. D

The cell interior has a lower concentration of solutes than the outside environment, so is hypotonic. When a cell contains a greater concentration of solutes than its surroundings, it is hypertonic (C). If the internal and external concentrations were equal, the cell would be isotonic (E). Choices (A) and (B) present terms used to describe chemical reactions.

100. A

Solutes diffuse down their concentration gradients (from areas of higher concentration to areas of lower concentration). The cell contains more fructose, but less glucose than the environment. Sucrose cannot cross the membrane. Water moves from the less concentrated solution (within the cell) to the more concentrated solution (the environment).

Section II: Free-Response Questions

1. A good aspect of this question is that it allows you to gain points by letting the grader know what you know about the subject instead of limiting you to answering only what each part of the question is asking. When defining your sample community, you can focus on organisms with which you are familiar. Trophic levels are the levels of a food chain; species in a community are organized based on nutrition and energy source.

    (a) Key points to include: define trophic levels; sample community should contain a primary producer, primary consumer, secondary consumer, tertiary consumer, and/or decomposer

Here is a possible response:

"Trophic levels" are the levels of a food chain or hierarchy in which each organism plays a particular role, based on the foods it eats. Each organism fits into a specific level of the food chain. For example, whales would not fit in the same level of the food chain as roses.

The first level of the food chain includes primary producers. Primary producers are autotrophic organisms (plants) that form organic compounds from inorganic compounds. Primary consumers, the next level of the food chain, feed upon the organic compounds produced by plants. Plants produce organic compounds through photosynthesis. In the presence of sunlight or artificial UV light, plants use carbon dioxide (CO_2) and water (H_2O) to produce oxygen (O_2) and organic material that can be used by the plant. Even tiny blue-green algae, called cyanobacteria, are photosynthetic. Algae in a pond and lakeweed are examples of aquatic plants.

Organisms at the second level of the food chain, primary consumers, feed on the plants in the first level. They are herbivores. Certain fish, for example, feed on algae and lakeweed. These fish are eaten by a secondary consumer, such as a bear. A secondary consumer feeds on a primary consumer, but they are not always solely carnivorous. Some secondary consumers feed on primary producers (such as bears eating berries), but their main source of food is primary consumers.

These secondary consumers are then fed upon by another predator, the tertiary consumer. The tertiary consumer occupies the end of the food chain. Humans are tertiary consumers. When a person hunts and eats a bear, he or she is the tertiary consumer of the bear. When a human being dies, the body is decomposed in the soil by microbes and the nutrients left from the decomposed body are absorbed by organisms such as fungi and bacteria. When decomposition is complete, nutrients from the former human body are left in the soil, where primary producers grow and feed on the remaining nutrients of the former tertiary consumer.

    (b) Key points to include: discussion of how energy is lost moving from producers to consumers, trophic efficiency

Here is a possible response:

Energy is lost as it moves through trophic levels. When energy is consumed by primary consumers, it loses efficiency in the process. Only approximately 10 percent of the energy that a consumer ingests from a producer goes toward building up the consumer's cells and tissues, termed secondary production. The energy that primary producers create is termed primary production. The process of digesting and excreting ingested material is termed assimilation. The remaining 90 percent of the energy consumed goes toward aerobic respiration, maintaining homeostasis, body movement, and synthesizing proteins.

The "ecological efficiency" or "trophic efficiency" is the energy supply available to each trophic level. To figure out the trophic efficiency, the net production efficiency for each organism needs to be found first. Net production efficiency is the ratio of production:

$$\frac{(growth + reproduction)}{assimilation}$$

Once net production efficiency is obtained, trophic efficiency can be determined as

$$\frac{\text{the energy supply available to trophic level } n+1}{\text{the energy consumed by trophic level } n}$$

Since each trophic level loses energy through normal metabolic processes in consumers, less energy from each trophic level is passed on to each subsequent level.

In the sample community of part (a), energy would flow from plants to fish, then to bears, and then to humans. Humans would receive the least energy passed on from the original source of energy, the plants (primary producers).

(c) Key points to include: bioaccumulation, biological magnification

Here is a possible response:

A contaminant, such as a pesticide, can be passed through the sample community of part (a). There are two main points that have to do with a pesticide moving through a sample community: bioaccumulation and biological magnification. Bioaccumulation has to do with how a pesticide enters a sample community. A pesticide will first enter a sample community from the environment, where it will be concentrated in certain organisms by biological magnification. Biological magnification has to do with how a pesticide is passed through the trophic levels of a sample community. A pesticide will become more concentrated moving through the trophic levels of the food chain because less energy with more concentrated pesticide is passed on to each trophic level.

A pesticide needs to be mobile, biologically active, soluble in fats, and long-lived, so that it can be stored in an organism (not digested and excreted) long enough to cause harm to that organism. Toxins are often stored in the fatty tissue of animals, where they can remain for a prolonged period of time. For example, a pesticide such as DDT can get into a water source, such as a river, where fish swim. The DDT is stored in the fatty tissue of the fish, which is then ingested by bears. The bears acquire a greater accumulation of the toxins in the energy that their bodies use for building cells and tissues, which remains in their cells for a prolonged period of time. These toxins are then passed on to humans when they ingest bear meat. While many pesticides are not immediately lethal to humans and other levels of the food chain, they can cause severe birth defects or mutations that lead to cancer.

2. Questions about the immune system frequently appear on the exam, so it's a good idea to become familiar with its components and the ways they interact. The methodology for attacking each part of the question is the same. Read and reread each part so you know what the question is asking. Be aware that the two sections of part (a) will draw on similar information about immune responses, so it shouldn't be too difficult to score points here. Part (b) asks for a biological explanation of a familiar phenomenon: the ability of pathogens to evolve into new forms and the ways the immune system adapts itself to counter them.

(a) (i) Key points to include: roles of phagocytic blood cells, antimicrobial proteins, the inflammatory response, chemokines, clotting, natural killer cells, histamine

Here is a possible response:

The first level of the human immune system is the skin (the innate immune system), which provides a protective barrier around the body. When this protective barrier is broken, the body responds with an inflammatory response, releasing the chemical histamine. In an inflammatory response, the body sends specialized phagocytic blood cells and antimicrobial proteins to protect against foreign invading organisms. Damage to skin cells results in the release of chemicals from these cells, which signals blood vessels to dilate. This dilation causes a rush of blood to the area, which causes swelling and recognition of foreign invading bodies through the chemical complement, antibodies, and cytokines.

The body's immune response helps to protect the rest of the body from infection. When a laceration is made in the skin, white blood cells called neutrophils surround and engulf any foreign organisms in a process called phagocytosis. The neutrophils then attack the engulfed organisms with a variety of chemicals and enzymes, breaking them down into pieces to be excreted and processed by cells called lymphocytes. B-lymphocytes produce antibodies to invading organisms, which stick to the invading organisms and prevent them from sticking to the host cell or releasing chemicals that can damage the host cell. The B-lymphocytes mark the invading microbes with antimicrobial proteins, so that if they continue to attack or attack again, the body will know how to fight them off. B-lymphocytes pass the information about the invading organisms on to T-cells. The T-cells activate macrophages, which digest pieces of the antigen. B-lymphocytes and T-cells are natural killer cells, which attack any foreign body that enters a person. That is why it is important for an organ donor to have the same blood type as the organ recipient—the recipient's body will attack the new organ if the blood cells are not the same type. The entire immune response causes clotting at the site of the laceration, as white blood cells and lymphocytes coagulate on the surface of the skin, preventing organisms from further entering the bloodstream.

(ii) Key points to include: lymphocytes, specific immunity, antigens, B-cells, T-cells, antibodies, memory cells, primary and secondary immune response

Here is a possible response:

When a human body gets a virus, it is being invaded by foreign organisms, as it is in a laceration. The difference is that the cells of the skin layer have not been damaged, and do not release histamine in one particular area of the body. A virus is not specific—it will attack all host cells within the body. Each virus is recognized by a certain protein. If an organism has been invaded by the virus before, it will "remember" how to fight off the virus because the specific protein has been encoded in the body's memory cells. The body then has specific immunity toward the virus that has already invaded.

When a virus first invades a body, lymphocytes recognize the virus as a foreign invader, and engulf the virus so that it cannot invade any other cells. The B-cells, which are a type of lymphocyte, produce antibodies to the virus, which is tagged with antimicrobial proteins. These tagged viruses are then passed on to the T-cells, which hold information on how to fight off invading viruses. The primary immune response is when lymphocytes and white blood cells rush to engulf the invading virus, so that it is cut off from harming other cells in the rest of the body. The secondary immune response is the production of antibodies, which allow the B-cells and T-cells to continue fighting off the same invading organism.

(b) Key points to include: active immunity, secondary immune response, the problem of predicting evolution of a quickly reproducing virus, the "arms race" between host and pathogen, different strains

Here is a possible response:

The flu vaccine works by stimulating the secondary immune response of cells in the body: production of antibodies against invading organisms. This vaccine contains parts of commonly known strains of inactivated flu virus. The parts of these strains float around the body until they come in contact with proteins in the body that "match" the chemical makeup of the inactive viruses. When the viruses and proteins "match," the proteins activate the production of cells and antibodies that remove the foreign organism (the inactivated virus strains) from the body. When the body has the antibodies and antimicrobial proteins necessary to fight off a given virus, the body has active immunity against that virus. The body already knows how to fight off that particular invading organism.

Flu vaccines can protect against certain strains of the flu virus that are known. The problem with the vaccine is that the flu virus can rapidly multiply into different strains, which are resistant to the vaccine. The vaccine is only effective two or more weeks before a person comes into contact with the flu virus. That amount of time gives a person's body the chance to develop immunity to the strains of flu virus found in the vaccine. There is no way for the manufacturer of a vaccine to predict how the flu virus will mutate, so by the time the vaccine is made, it may be ineffective against a variety of new strains of flu or may only partially work against them.

An "arms race" is thereby constantly occurring between host and pathogen. The host is constantly trying to fight off pathogens through the development of active immunity, and the pathogen (virus) is constantly trying to evolve so that it is effectively able to invade and attack the host organism. In people with compromised immune systems, such as people with cancer and AIDS, there are not as many B-lymphocytes and T-cells to fight off invading organisms. That is why people with those diseases often develop an overgrowth of thrush or the Candida organism, which naturally occurs at low levels in the human body (kept at bay by the body's symbiotic defense system), or pass away from a secondary infection such as pneumonia.

3. Acid rain is primarily caused by sulfur dioxide (SO_2) and nitrogen oxides (NO_x) combining with oxygen, water, and other chemicals in the air to form acidic compounds that are deposited on the Earth in the form of rain. Remember that there are natural sources of sulphur dioxide in the atmosphere, in addition to pollution caused by humans. When answering each part of the question, keep these natural and manmade factors in mind, as well as the effect of acid rain on the environment. Don't go into too much detail about ecological devastation; just say what you must to cover the basic points of the question and move on.

(a) Key points to include: components of acid rain, emissions from the burning of fossil fuels for the creation of electricity, volcanoes that emit sulfur dioxide

Here is a possible response:

Acid rain is caused by both manmade and naturally occurring factors. When fossil fuels are converted into electrical energy, sulfur dioxide (SO_2) and nitrogen oxides (NO_x) are emitted into the atmosphere, where they combine with water, oxygen, and other chemicals to form acidic compounds. These compounds then fall to the Earth in the form of acid rain, snow, hail, and fog.

The compounds can also be blown around by wind and deposited in different areas. When the dry particles (compounds) blown by the wind are combined with acid rain in water runoff, the acidic compounds become concentrated and more harmful than the forms that occur in only precipitation or wind.

Volcanoes naturally produce sulfur dioxide (SO_2) that is released into the atmosphere. When it is released into the atmosphere, sulfur dioxide from volcanoes is converted into sulfuric acid, just as it is from the burning of fossil fuels. Naturally occurring sulfuric acid and nitric acid are produced from forest fires as well. Forest fires are a regular occurrence that ensure the continual overturn of underbrush in forests, contributing to the natural growth and cycle of the forest. A certain amount of acid rain can be predicted through the laws of nature.

(b) Key points to include: acid rain is measured using the pH scale, lower pH means more acidic, example of impact on an ecosystem

Here is a possible response:

The acidity of rain is determined using the pH scale. Pure water has a pH of 7, while normal rain has a slightly acidic reading of 5.5. The slightly acidic reading of normal rain is due to carbon dioxide dissolving in droplets of water. The more acidic rain is, the more damage it causes to flora, fauna, and manmade objects. A pH of 4.3 is much more destructive to habitat and human objects than a pH of 6.

Acid rain damages trees (especially at higher elevations) and soil, and it can change the pH of small bodies of water so much that certain fish die. Coniferous forests in Maine have been affected by acid rain, as well as the fish population in many lakes and rivers, such as Moosehead Lake. Since fishing is a big tourist attraction at Moosehead Lake, tourism has been declining due to insufficient fish population. In addition to the effect it has on natural ecosystems, acid rain is detrimental to paint on buildings and cars, and causes the stone of statues, sculptures, and historic buildings to decay.

(c) Key points to include: natural factors that contribute to acid rain cannot be altered very much, human factors such as a decrease in reliance on fossil fuels and reduced industrial emissions can decrease the prevalence of acid rain

Here is a possible response:

While not all causes of acid rain can be eradicated, human factors that lead to acid rain can be decreased. For instance, individuals can use less electricity and see if alternative sources of energy are available, such as solar power, wind power, hydropower, and nuclear power. Most people are afraid of nuclear waste, but nuclear power is a very clean and efficient form of energy. In order to reduce emissions of sulfur dioxide and nitrogen oxide, sulfur dioxide can be removed from gases leaving smokestacks through the use of "scrubbers," sulfur can be removed from the coal that it naturally occurs in, and power plants can use fuels that burn cleaner, such as natural gas. Cars are already required to have a catalytic converter, which removes a good deal of nitrous oxide emissions from the tailpipe of a car. There are always new, alternative forms of energy being developed, and while at the moment these new forms of energy are not completely reliable, at some point there will no longer be a need to rely on fossil fuels to run cars and produce electricity.

4. When answering this question, be aware that speciation and extinction are two aspects of the same general concept. Remember that "compare" calls for a discussion of similarities while "contrast" asks you to focus on differences. Be sure to include both in your answer. You should be familiar with the biological species concept and different types of reproductive isolation. When describing an extinction vortex, give a few examples of related phenomena that interact in a positive feedback loop. Don't forget to include exactly three factors in your response to part (c).

   (a) Key points to include:
   Allopatric speciation—role of geographic barriers, more common than sympatric
   Sympatric speciation—reproductive isolation without geographic barriers; behavioral, temporal, resource-based mechanisms of reproductive isolation polyploidy

Here is a possible response:

Allopatric speciation occurs due to geographic isolation. It is more common than sympatric speciation, since species often develop when a few individuals move to a different area, whether by choice or force (for example, a storm or food shortage). Sympatric speciation does not occur due to geographic isolation, since groups of similar organisms can coexist in the same geographic or overlapping geographic area without interbreeding. Species diverge in sympatric speciation due to behavioral, temporal, and resource-based mechanisms of reproductive isolation.

Darwin's Galapagos finches are an example of a combination of allopatric and sympatric speciation. These birds were separated by geographic barriers (on different islands), but also separated into different species due to other factors, such as the foods they ate. By diverging into different species and supplanting food staples, finches did not become extinct by extinguishing their only food resource. Finches on the same island adapted different behavioral mechanisms that eventually caused them to separate into species. The finches began to adapt different types of beaks, depending on what type of food they specialized in and how they obtained that food. Some finches remained on the ground and ate grain. Others devoured insects in trees, still others lived in mangroves, and some developed specialized song (called warblers). Other finches developed temporal changes to better hear their prey (e.g., insects caught out of the air and on the ground), along with changes in color patterns that make them easily recognizable to birds of the same species. By specializing in certain areas, these finches developed reproductive isolation due to sympatric speciation. Birds that fed on insects would only breed with other birds that fed on insects, leading to the genetic development of beaks designed to get insects out of the wood of trees. Ultimately, these different finch populations became unable to interbreed to produce fertile offspring.

Species can also diverge in sympatric speciation due to polyploidy, especially in plants. Polyploidy is when a cell or species contains one or more extra set(s) of chromosomes. Two conditions lead to polyploidy: cross-fertilization with a plant of the same species with unreduced gametes or cross-fertilization with a plant of a different species, with unreduced gametes. Both conditions produce a new tetraploid species (i.e., having four times the haploid number of chromosomes in the nucleus of the cell).

(b) Key points to include: genetic drift, loss of genetic variability, inbreeding, decreased fitness, positive feedback loop leading to extinction; plant versus animal reproduction (asexual versus sexual options, dependence on pollinators), and plant versus animal dispersal opportunities.

Here is a possible response:

There are several reasons why a species may become extinct. When the population of a species is small to begin with or dwindles in size due to environmental factors such as a drought, a species will have trouble reproducing enough to maintain replication of the species. With a small gene pool, an entire population is susceptible to being wiped out by a single factor, whether it be a disease, predator, or lack of essential habitat. The elm trees of New England died out because all trees were susceptible to Dutch Elm disease. If individuals cannot adapt to a changing environment, they will all die off. Species with a small gene pool often end up in an extinction vortex, where several factors interact to cause the small population to dwindle and the species to subsequently become extinct. Inbreeding can lead to decreased fitness, since similar alleles lead to the gene pool becoming smaller. When organisms are too similar in terms of genetic makeup, the combination of their alleles can prove fatal or not lead to viable offspring, which eventually leads to the extinction of the species. The positive feedback loop of genetic drift can lead to extinction as well. A small, isolated population can have random fluctuation in the expression of certain genes, due to chance. Over time, the fittest organisms will theoretically carry on the species and traits that have randomly occurred will be passed on to offspring. However, in small populations random changes in the expression of genes can easily be replicated and passed on throughout the population. This loop leads to rapid change, which can quickly wipe out a population.

Plants and animals are susceptible to extinction based on varying factors. Lack of habitat can destroy both plants and animals. An ecosystem is codependent—plants are often dependent on pollinators to survive, pollinators feed upon the nectar of flowers, herbivores eat the plants, and the entire trophic cycle continues. Plants and animals also have different advantages and disadvantages in the struggle to avoid extinction. Plants can reproduce either asexually, through budding and fission, or sexually. The seeds of plants can travel by wind or through dispersion by birds and other animals. Animals must travel in order to reach another area. Certain plants have a hard coating to their shell, in order to survive harsh changes in climate, whereas most animals cannot. Most plants need sunlight and soil to grow, while animals need to find their own food. For both plants and animals, decrease in genetic variability leads to decreased fitness.

(c) Key points to include: generalists versus specialists, dispersal ability, resource base, reproductive strategy, growth rates, competitive abilities, disease transmission/resistance

Here is a possible response:

Exotic species that move to a new area often are not so specialized that they can only survive on one type of food or climate. Such a "generalist" species can draw on greater resources than a species that is highly specialized in one area, with one food supply and certain conditions (such as temperature) that must be fulfilled in order for an organism to survive. An exotic species may be more easily dispersed than a local species, taking over an area in a short amount of time (e.g., weeds and zebra mussels that travel between bodies of water on the rudders of ships). The species can be a new predator introduced to an area, which feeds upon a certain species until the population of that species has been wiped out.

If the exotic species is more adept at obtaining the food staple that an indigenous species eats, or takes over a certain habitat, the indigenous species will be pushed to extinction. An exotic species may possess better competitive abilities, a faster growth rate, or may transmit a disease that indigenous organisms are not immune to. This phenomenon has even happened between humans in different areas of the world, such as the dispersal of smallpox between continents when explorers first came to the New World. The explorers had already been exposed to smallpox and were immune to it, or were sick and carried the disease with them. Native American populations were decimated by smallpox because individuals lacked immunity to this new disease. The main reasons that native species are driven to extinction include the introduction of a new predator or disease, loss of food supply, and loss of habitat.

Glossary

abiotic
Nonliving, as in the physical environment

absorption
The process by which water and dissolved substances pass through a membrane

acoelomates
An animal that lacks a coelom, exhibits bilateral symmetry, and has one internal space, the digestive cavity

action potential
The change in electrical potential across a nerve or muscle cell when stimulated, as in a nerve impulse

active immunity
Protective immunity to a disease in which the individual produces antibodies as a result of previous exposure to the antigen

adaptation
A behavioral or biological change that enables an organism to adjust to its environment

adaptive radiation
The production of a number of different species from a single ancestral species

adenine
A purine base that pairs with thymine in DNA and uracil in RNA

adenosine phosphate
Adenosine diphosphate (ADP) and adenosine triphosphate (ATP), which are energy-storage molecules

adipose
Fatty tissue, fat-storing tissue, or fat within cells

adrenaline (epinephrine)
An "emergency" hormone stimulated by anger or fear; increases blood pressure and heart rate in order to supply the emergency needs of the muscles

adventitious roots
Roots that develop in an unusual place

aerobe
An organism that requires oxygen for respiration and can live only in the presence of oxygen

aerobic
Requiring free oxygen from the atmosphere for normal activity and respiration

aerobic catabolism
Metabolic breakdown of complex molecules into simple ones through the use of oxygen; results in the release of energy

agonistic response
Response of aggression or submission between two organisms

allele
One of two or more types of genes, each representing a particular trait; many alleles exist for a specific gene locus

allopatric speciation
Evolution of species that occurs in separate geographic areas

alternation of generations
The description of a plant life cycle that consists of a diploid, asexual, sporophyte generation and a haploid, sexual, gametophyte generation

anaerobe
An organism that does not require free oxygen in order to respire

anaerobic
Living or active in the absence of free oxygen; pertaining to respiration that is independent of oxygen

anaerobic catabolism
Metabolic breakdown of complex molecules into simple ones without the use of oxygen; results in the release of energy

analogous
Describes structures that have similar function but different evolutionary origins; e.g., a bird's wing and a moth's wing

anaphase
The stage in mitosis that is characterized by the migration of chromatids to opposite ends of the cell; the stage in meiosis during which homologous pairs migrate (anaphase I), and the stage in meiosis during which chromatids migrate to different ends of the cell (anaphase II)

androgen
A male sex hormone (e.g., testosterone)

angiosperm
A flowering plant; a plant of the class Angiospermae that produces seeds enclosed in an ovary and is characterized by the possession of fruit and flowers

Animalia
Kingdom that includes all extinct and living animals

antibiotic
An antipathogenic substance (e.g., penicillin)

antibody
Globular proteins produced by tissues that destroy or inactivate antigens

antigen
A foreign protein that stimulates the production of antibodies when introduced into the body of an organism

appendage
A structure that extends from the trunk of an organism and is capable of active movements

Archaea
Kingdom comprised of an ancient group of microorganisms (bacteria) that are metabolically and genetically different from other bacteria; they came before the eukaryotes

artery
A blood vessel that carries blood away from the heart

asexual reproduction
The production of daughter cells by means other than the sexual union of gametes (as in budding and binary fission)

ATPase
    Adenosine triphosphatase, enzyme that catalyzes the hydrolysis of ATP to ADP, thereby releasing energy

autonomic nervous system
    The part of the nervous system that regulates the involuntary muscles, such as the walls of the alimentary canal; includes the parasympathetic and sympathetic nervous systems

autosome
    Any chromosome that is not a sex chromosome

autosomal genes
    Non sex-linked genes

autotroph
    An organism that utilizes the energy of inorganic materials such as water and carbon dioxide or the sun to manufacture organic materials; examples of autotrophs include plants

axon
    A nerve fiber

Bacteria
    Kingdom of single-celled organisms that reproduce by fission and can be spiral, rod, or spherical shaped; often pathogenic organisms that rapidly reproduce

base-pair substitution
    When one base pair is incorrectly reproduced and substituted for another base pair

bilateral symmetry
    The equal division of an organism into a left and right half

bile
    An emulsifying agent secreted by the liver

bile salts
    Compounds in bile that aid in emulsification

binary fission
    Asexual reproduction; in this process, the parent organism splits into two equal daughter cells

binomial nomenclature
    The system of naming an organism by its genus and species names

biological species concept (BSC)
　Definition of a species as a naturally interbreeding population of organisms that produces viable, fertile offspring

biome
　A habitat zone, such as desert, grassland, or tundra

biotic
　Living, as in living organisms in the environment

Calvin cycle
　Cycle in photosynthesis that reduces fixed carbon to carbohydrates through the addition of electrons (also known as the "dark cycle")

CAM (crassulacean acid metabolism)
　Storage of carbon dioxide at night in the form of organic acids

carbohydrate
　An organic compound to which hydrogen and oxygen are attached; the hydrogen and oxygen are in a 2:1 ratio; examples include sugars, starches, and cellulose

carbon cycle
　The recycling of carbon from decaying organisms for use in future generations

carbon fixation
　Conversion of carbon dioxide into organic compounds during the Calvin cycle, the second stage of photosynthesis. Known as a "dark reaction"

carnivore
　A flesh-eating animal; an animal that subsists on other animals or parts of animals

carrying capacity
　The number of organisms an environment can support

catabolism
　Metabolic breakdown of complex molecules into simple ones, releasing energy

cell
　Smallest structural unit of an organism

cell wall
　A wall composed of cellulose that is external to the cell membrane in plants; it is primarily involved in support and in the maintenance of proper internal pressure. Fungi have cell walls made of chitin. Some protists also have cell walls.

central nervous system (CNS)
Encompasses the brain and the spinal cord

chemiosmosis
The coupling of enzyme-catalyzed reactions

chi-squared analysis
Test to see if a theory is backed up by experimental results

chloroplast
A plastid containing chlorophyll

chlorophyll
A green pigment that performs essential functions as an electron donor and light "entrapper" in photosynthesis

chromatid
One of the two strands that constitute a chromosome; chromatids are held together by the centromere

chromatin
A nuclear protein of chromosomes that stains readily

chromosome
A short, stubby rod consisting of chromatin that is found in the nucleus of cells; contains the genetic or hereditary component of cells (in the form of genes)

chromosome map
The distribution of genes on a chromosome, derived from crossover frequency experiments

circadian rhythms
Daily cycles of behavior

circulatory system
System that circulates blood throughout the body. Includes the heart, blood, and blood vessels

cleavage
The division in animal cell cytoplasm caused by the pinching in of the cell membrane

clotting
The coagulation of blood caused by the rupture of platelets and the interaction of fibrin, fibrinogen, thrombin, prothrombin, and calcium ions

codominant
　The state in which two genetic traits are fully expressed and neither dominates

codon
　Three adjacent nucleotides that signal to insert an amino acid into the genetic code or end protein synthesis

coelom
　The space between the mesodermal layers that forms the body cavity of some animal phyla

coelomates
　Organisms that contain a coelom

coenzyme
　An organic cofactor required for enzyme activity

commensal
　Describes an organism that lives symbiotically with a host; this host neither benefits nor suffers from the association

communities
　Groups of interacting organisms that live in the same geographic area under similar environmental conditions

complementary base pairs
　Pairing of purines and pyrimidines in DNA and RNA

concentration gradient
　Difference in concentration of a solute between two areas of a solution

conditioning
　The association of a physical, visceral response with an environmental stimulus with which it is not naturally associated; a learned response

cone
　A cell in the retina that is sensitive to colors and is responsible for color vision

Conifers
　Phylum of cone-bearing gymnosperm trees and shrubs that are primarily needle- and scale-leaved.

connective tissue
　Highly vascular matrix that forms the supporting and connecting structures of the body

consumer
Organism that consumes food from outside itself instead of producing it (primary, secondary, and tertiary)

convergent
Adaptive evolution of similar structures, such as wings

coupled reaction
Chemical reaction in which energy is transferred from one side of the reaction to the other through a common intermediate

cristae
Inward folds of the mitochondrial membrane

crossing over
The exchange of parts of homologous chromosomes during meiosis

cytokinesis
A process by which the cytoplasm and the organelles of the cell divide; the final stage of mitosis

cytoplasm
The living matter of a cell, located between the cell membrane and the nucleus

cytoskeleton
The organelle that provides mechanical support and carries out motility functions for the cell

cytosine
A nitrogen base that is present in nucleotides and nucleic acids; it is paired with guanine

dark (Calvin) reactions
Processes that occur after the light reactions of photosynthesis (during carbon fixation), without the presence of light

Darwin, Charles Robert (1809–1882)
Naturalist who came up with the theory of evolution based on natural selection

decomposers
Organisms that feed on and break down dead plant or animal matter

degree of freedom (d.f.)
Independent statistical category; the number of categories of observation minus one

deletion
The loss of all or part of a chromosome

dendrite

The part of the neuron that transmits impulses to the cell body

density-dependent factors

Effects that increase with population density and smaller population size

density-independent factors

Effects that are independent of population size

deoxyribose

A five-carbon sugar that has one oxygen atom less than ribose; a component of DNA (deoxyribonucleic acid)

determinate cleavage

Irreversible division of an egg into specific areas for further development

deuterostomes

Means "second mouth." The mouth forms from the second opening of the digestive tract in embryos. These organisms have a mouth, radial cleavage, anus, coelom, and indeterminate cleavage in common.

differentiation

A progressive change from which a permanently more mature or advanced state results; for example, a relatively unspecialized cell's development into a more specialized one

diffusion

The movement of particles from one place to another as a result of their random motion

digestion

The process of breaking down large organic molecules into smaller ones

digestive system

The alimentary canal and glands which ingest, digest, and absorb food

digestive tract

The alimentary canal

dihybrid

An organism that is heterozygous for two different traits

dihybrid cross

A hybridization between two traits, each with two alleles

diploid
Describes cells that have a double set of chromosomes in homologous pairs (2n)

directional selection
Favors organisms that have extreme variation of traits within a population

disruptive (diversifying) selection
Sudden changes in the environment cause organisms with extreme variation of traits in a population to be favored

DNA
Deoxyribonucleic acid; found in the cell nucleus, its basic unit is the nucleotide; contains coded genetic information; can replicate on the basis of heredity

domains
Biological classification of prokaryotes and eukaryotes into Bacteria, Archaea, and Eukarya

dominance
A dominant allele suppresses the expression of the other member of an allele pair when both members are present; a dominant gene exerts its full effect regardless of the effect of its allelic partner

ecological succession
The orderly process by which one biotic community replaces another until a climax community is established

ecology
The study of organisms in relation to their environment

ecosystem
Ecological community and its environment

ectoderm
The outermost embryonic germ layer that gives rise to the epidermis and the nervous system

egg (ovum)
The female gamete; it is nonmotile, large in comparison to male gametes, and stores nutrients

electrochemical gradient
Diffusion gradient of an ion including potential and kinetic energy of the ion

electron transport chain
A complex carrier mechanism located on the inside of the inner mitochondrial membrane of the cell; releases energy, and is used to form ATP

endemic

Pertaining to a restricted locality; ecologically, occurring only in one particular region

endocrine gland

A ductless gland that secretes hormones directly into the bloodstream

endocrine (hormone) system

Collection of ductless glands that secrete hormones into the bloodstream with various effects on the body (includes thyroid gland, pituitary gland, etc.).

endocytosis

A process by which the cell membrane is invaginated to form a vesicle which contains extracellular medium

endoderm

The innermost embryonic germ layer that gives rise to the lining of the alimentary canal and to the digestive and respiratory organs

endoplasmic reticulum

A network of membrane-enclosed spaces connected with the nuclear membrane; transports materials through the cell; can be soft or rough

energy flow

The movement of energy throughout the trophic levels of an ecosystem

enzyme

An organic catalyst and protein

endoplasm

The inner portion of the cytoplasm of a cell or the portion that surrounds the nucleus

epidermal tissue

The outer or integumentary layer of the body, including sebum, adipose, and skin cells

epidermis

The outermost surface of an organism

epithelium

The cellular layer that covers external and internal surfaces

Eukarya

Domain containing all eukaryotic organisms

eukaryote
Organism consisting of one or more cells with genetic material in membrane-bound nuclei

evolution, theory of
Theory that organisms have developed over time to produce current biomes

excretion
The elimination of metabolic waste matter

exocrine
Pertaining to a type of gland that releases its secretion though a duct; e.g., the salivary gland, the liver

exocytosis
A process by which the vesicle in the cell fuses with the cell membrane and releases its contents to the outside

exons
DNA that is transcribed to RNA and codes for protein synthesis

exoskeleton
Describes arthropods and other animals whose skeletal or supporting structures are outside the skin

extracellular matrix
Material occurring outside of the cell

fats
Solid, semi-solid, or liquid organic compounds composed of glycerol, fatty acids, and organic groups

ferns
Seedless, flowerless vascular plants that reproduce by spores

F1
The first filial generation (first offspring)

F2
The second filial generation; offspring resulting from the crossing of individuals of the F1 generation

feedback mechanism
The process by which a certain function is regulated by the amount of the substance it produces

fertilization
> The fusion of the sperm and egg to produce a zygote

fitness
> The ability of an organism to contribute its alleles and therefore its phenotypic traits to future generations

food web
> The interaction of feeding levels in a community, including energy flow throughout the community

functional groups
> Chemical groups attached to carbon skeletons that give compounds their functionality

Fungi
> Kingdom of eukaryotic organisms that lack vascular tissues and chlorophyll, possessing chitinous cell walls; reproduction occurs through spores

gamete
> A sex or reproductive cell that must fuse with another of the opposite type to form a zygote, which subsequently develops into a new organism

gametogenesis
> The formation of gametes

gametophyte
> The haploid, sexual stage in the life cycle of plants (alternation of generations)

gas exchange
> The exchange of gases such as oxygen and carbon dioxide through respiratory surfaces, gills, lungs, or tracheae

gastrula
> A stage of embryonic development characterized by the differentiation of the cells into the ectoderm and endoderm germ layers and by the formation of the archenteron; can be two-layer or three-layer

gastrulation
> Formation of a gastrula

gene
> The portion of a DNA molecule that serves as a unit of heredity; found on the chromosome

gene expression
Conversion of information from a gene to mRNA to a protein

gene frequency
A decimal fraction that represents the presence of an allele for all members of a population that have a particular gene locus

genetic code
A four-letter code made up of the DNA nitrogen bases A, T, G, and C; each chromosome is made up of thousands of these bases

genetic drift
Random evolutionary changes in the genetic makeup of a (usually small) population

genotype
The genetic makeup of an organism without regard to its physical appearance; a homozygous dominant and a heterozygous organism may have the same appearance but different genotypes

genus
In taxonomy, a classification between species and family; a group of very closely related species, e.g., *Homo, Felis*

geographic barrier
Any physical feature that prevents the ecological niches of different organisms (not necessarily different species) from overlapping

geographic isolation
Isolation due to geographic factors. Islands are geographically isolated.

geotropism
Any movement or growth of a living organism in response to the force of gravity

gills
Respiratory organ of aquatic animals

glycolysis
The anaerobic respiration of carbohydrates

Golgi apparatus
Membranous organelles involved in the storage and modification of secretory products

gravitropism
Directional growth according to the gravitational field. Roots grow downward with gravity, while the shoots of plants grow up toward the sunlight.

growth curve
  Growth of an organism or population plotted over time

guanine
  A purine (nitrogenous base) component of nucleotides and nucleic acids; it links up with cytosine in DNA

gymnosperm
  A plant that belongs to the class of seed plants in which the seeds are not enclosed in an ovary; includes the conifers

habitat
  The environment a community or organism lives in

haploid sporophytes
  Spore-producing phase of a plant that contains a single set of chromosomes, allows the plant to reproduce asexually

Hardy-Weinberg equilibrium
  In a randomly breeding population, gene frequency and genotype ratios remain constant over generations of organisms

haploid
  Describes cells (gametes) that have half the chromosome number typical of the species (n chromosome number)

herbivore
  A plant-eating animal

hermaphrodite
  An organism that possesses both male and female reproductive organs

heterotroph
  An organism that must get its inorganic and organic raw materials from the environment; a consumer

heterozygous
  Describes an individual that possesses two contrasting alleles for a given trait (Tt)

homologous
  Describes two or more structures that have similar forms, positions, and origins despite the differences between their current functions; examples are the arm of a human, the flipper of a dolphin, and the foreleg of a horse.

homozygous

Describes an individual that has the same gene for the same trait on each homologous chromosome (*TT* or *tt*)

homozygous dominant

Having two dominant alleles of the same gene. Dominant alleles are expressed in a heterozygous as well as a homozygous genotype.

homozygous recessive

Having two recessive alleles of the same gene. Recessive alleles are only expressed when a gene is homozygous recessive.

hormone

A chemical messenger that is secreted by one part of the body and carried by the blood to affect another part of the body, usually a muscle or gland

hormone-receptor system

Chemical messengers (hormones) travel throughout the body and are read by receptor proteins, which respond to the "message" each hormone codes for

hybrid

An offspring that is heterozygous for one or more gene pairs

hybridization

Crossbreeding organisms to form base pairs between two strands of DNA that weren't originally paired

hydrophilic

Having an affinity for water

hydrophobic

Repelling water

hypertonic

Describes a fluid that has a higher osmotic pressure than another fluid it is compared to; it exerts greater osmotic pull than the fluid on the other side of a semipermeable membrane; hence, it possesses a greater concentration of particles, and acquires water during osmosis

hypotonic

Describes a fluid that has a lower osmotic pressure than a fluid it is compared to; it exerts lesser osmotic pull than the fluid on the other side of a semipermeable membrane; hence, it possesses a lesser concentration of particles, and loses water during osmosis

immunity
    A resistance to disease developed through the immune system

incomplete dominance
    Genetic blending; each allele exerts some influence on the phenotype (for example, red and white parents may yield pink offspring)

Independent Assortment, Law of
    Genes independently sort and do not affect the sorting of other genes in the formation of gametes in diploid organisms

indeterminate cleavage
    Early divisions of a cell which produce blastomeres

indeterminate growth
    Growth without a termination point

induction
    Initiating enzyme production or genetic transcription

ingestion
    The intake of food from the environment into the alimentary canal

integument
    Refers to protective covering, such as the covering of an ovule, that develops into the seed coat, or an animal's skin

integumentary system
    The skin, hair, nails, sebaceous glands, and sweat glands

interphase
    A metabolic stage between mitoses in which genetic material is reproduced

insertion
    Addition of one or more nucleotides to a chromosome, usually by mutation

intermembrane space
    Space between the outer and inner membranes of a mitochondrion

interphase
    The cellular phase between meiotic or mitotic divisions

intestines
    Part of the alimentary canal that extends from the stomach to the anus

introns
Part of a gene that is located between exons and is removed before the translation of mRNA; does not code for protein synthesis

isolation
The separation of some members of a population from the rest of their species; prevents interbreeding and may lead to the development of a new species

isomer
One of a group of compounds that is identical in atomic composition, but different in structure or arrangement

isotonic
Describes a fluid that has the same osmotic pressure as a fluid it is compared to; it exerts the same osmotic pull as the fluid on the other side of a semipermeable membrane; hence it neither gains nor loses net water during osmosis, and possesses the same concentration of particles before and after osmosis occurs

kinesis
Movement of an organism in response to light

kingdom
Second-highest taxonomic classification of organisms, after domain

Krebs cycle
Process of aerobic respiration that fully harvests the energy of glucose; also known as the citric acid cycle

K-selected
Organisms in more stable environments that tend to produce fewer offspring and invest more energy in rearing offspring

lactic acid fermentation
A type of anaerobic respiration found in fungi, bacteria, and human muscle cells

Lamarck, Jean Baptiste de (1744–1829)
Naturalist who studied evolution and classified invertebrate organisms. Lamarck produced several theories of evolution, as in his book *Philosophie zoologique*. Best known for his incorrect theory of inheritance of acquired characteristics

light reactions
Photosynthetic reactions that occur in the presence of light

linkage
    Occurs when different traits are inherited together more often than they would have been by chance alone; it is assumed that these traits are linked on the same chromosome

lipase
    A fat-digesting hormone

lipid
    A fat or oil

lumen
    The inner cavity of a tubular organ, such as an intestine

lungs
    Saclike respiratory organs of most vertebrates

lymph
    A body fluid that flows in its own circulatory fluid in lymphatic vessels separate from blood circulation

lymphocyte
    A kind of white blood cell in vertebrates that is characterized by a rounded nucleus; involved in the immune response

lysosome
    An organelle that contains enzymes that aid in intracellular digestion

meiosis
    A process of cell division whereby each daughter cell receives only one set of chromosomes; the formation of gametes

membrane
    Thin structure connecting or separating structures or regions of an organism

Mendelian laws
    Laws of classical genetics established through Mendel's experiments with peas; include segregation and independent assortment

mesoderm
    Embryonic germ layer from which body systems and tissues develop. Located between the ectoderm and endoderm

metabolism
  A group of life-maintaining processes that includes nutrition, respiration (the production of usable energy), and the synthesis and degradation of biochemical substances.

metaphase
  A stage of mitosis; chromosomes line up at the equator of the cell

micronutrients
  Vitamin or mineral essential for growth and metabolism in an organism

minerals
  Naturally occurring inorganic elements essential in the nutrition of organisms

mitochondria
  Cytoplasmic organelles that serve as sites of respiration; rod-shaped bodies in the cytoplasm known to be the center of cellular respiration

mitosis
  A type of nuclear division that is characterized by complex chromosomal movement and the exact duplication of chromosomes; occurs in somatic cells

mitotic divisions
  Cell division during mitosis

molecular clock hypothesis
  Hypothesis that genetic mutations occur in a genome at a linear rate

Monera
  The kingdom of bacteria

monohybrid
  An individual that is heterozygous for only one trait

monohybrid cross
  Cross involving a single trait and two alleles

monosaccharide
  A simple sugar; a five- or six-carbon sugar (e.g., ribose or glucose)

mosses
  Green nonvascular plants of the division Bryophyta

mRNA (messenger RNA)
RNA that transfers genetic information from the cell nucleus to ribosomes, serving as a template for protein synthesis

muscle tissue
Bundles of contractile fibers that allow movement; can be cardiac, skeletal, or smooth

mutagenic agent
Agent that induces mutations; typically carcinogenic

mutation
Changes in genes that are inherited

mutualism
A symbiotic relationship from which both organisms involved derive some benefit

myelin
Fatty lipid material that forms an insulating sheath around nerve fibers

NAD
An abbreviation of nicotinamide-adeninedinucleotide, also called DPN; a respiratory oxidation-reduction molecule

NADP
An abbreviation of nicotinamide-adeninedinucleotide-phosphate, also called TPN; an organic compound that serves as an oxidation-reduction molecule

natural selection
Process by which organisms best adapted to their environment survive to pass their genes on through offspring. Idea pioneered by Charles Darwin

negative feedback loop
Serves to stabilize a system, e.g., maintain temperature, concentration, direction, etc.

nerve
A bundle of nerve axons

nerve tissue
Tissue composed of nerve cells, dendrites, neuroglia, and nerve fibers

neuron
A nerve cell

neurotransmitters

Messenger molecules that affect the behavior of neurons

niche

The functional role and position of an organism in an ecosystem; embodies every aspect of the organism's existence

nitrogenous bases

The five purine and pyrimidine bases found in nucleic acid—adenine, thymine (in DNA), cytosine, guanine, and uracil (in RNA)

novel structures

Cellular structures used primarily for reproduction

nuclear membrane

A membrane that envelopes the nucleus and separates it from the cytoplasm; present in eukaryotes

nucleolus

A dark-staining small body within the nucleus; composed of RNA

nucleotide

An organic molecule consisting of joined phosphate, five-carbon sugar (deoxyribose or ribose), and a purine or a pyrimidine (adenine, guanine, uracil, thymine, or cytosine)

nucleus

An organelle that regulates cell functions and contains the genetic material of the cell

nutrient

A substance that can be metabolized by an organism to provide energy and build tissue

ontogeny

The origin and subsequent growth of an organism, from embryo to adult

operon

A unit of genetic material that controls production of mRNA through the use of an operator gene, promoter, and two or more structural genes

organ

A group of tissues that perform a specific function in the body

organ system

A group of organs that work together to perform a specific task in the body, such as the circulatory system distributing blood throughout the body

organelle
   A specialized structure that carries out particular functions for eukaryotic cells; examples include the plasma membrane, the nucleus, and ribosomes

organic molecules
   Molecules that contain carbon (C)

organogenesis
   Formation and development of organs

The Origin of Species
   Charles Darwin's book in which he expressed his theory that species evolve through natural selection; survival of the fittest

osmosis
   The diffusion of water through a semipermeable membrane, from an area of greater concentration to an area of lesser concentration

osmotic pressure
   Pressure exerted by the flow of water through a semipermeable membrane, which separates a solution into two concentrations of solute

oxidation
   The removal of hydrogen or electrons from a compound or addition of oxygen; half of a redox (oxidation or reduction) process

oxidative phosphorylation
   Formation of ATP from energy released during the oxidation of various substances and substrates, as in aerobic respiration and the Krebs cycle

oxygen
   A nonmetallic element essential in animal and plant respiration

pairing (synapsis)
   An association of homologous chromosomes during the first meiotic division

parapatric speciation
   Occurs when limited interbreeding and negligible genetic exchange takes place between two populations

pedigree
   A family tree depicting the inheritance of a particular genetic trait over several generations

pH
A symbol that denotes the relative concentration of hydrogen ions in a solution: the lower the pH, the more acidic a solution; the higher the pH, the more basic a solution; pH is equal to $-\log (H^+)$

phagocyte
Any cell capable of ingesting another cell

phenotype
The physical appearance of an individual, as opposed to its genetic makeup

phloem
The vascular tissue of a plant that transports organic materials (photosynthetic products) from the leaves to other parts of the plant

phosphoenol pyruvic acid (PEP)
An intermediate enzyme that fixes carbon dioxide into a four-carbon molecule in a mesophyl cell of C4 plants

phospholipids
Phosphorus-containing lipids composed of a phosphate group, organic molecule, and fatty acids

photophosphorylation
The synthesis of ATP using radiant energy absorbed during photosynthesis

photosynthesis
The process by which light energy and chlorophyll are used to manufacture carbohydrates out of carbon dioxide and water; an autotrophic process using light energy

phototropism
Plant growth stimulated by light (stem: +, toward light; root: –, away from light)

phylogeny
The study of the evolutionary descent and interrelations of groups of organisms

phylum
A category of taxonomic classification that is ranked above class; kingdoms are divided into phyla

physiology
The study of all living processes, activities, and functions

Plantae
Kingdom containing all extinct and living plants

plasma
The liquid part of blood

plasma membrane
The cell membrane

pollen
The microspore of a seed plant

pollination
The transfer of pollen to the micropyle or to a receptive surface that is associated with an ovule (such as a stigma)

polymer
A large molecule that is composed of many similar molecular units (e.g., starch)

polymorphism
The individual differences of form among the members of a species

polyploidy
A condition in which an organism may have a multiple of the normal number of chromosomes ($4n$, $6n$, etc.)

polysaccharide
A carbohydrate that is composed of many monosaccharide units joined together, such as glycogen, starch, and cellulose

population
All the members of a given species inhabiting a certain locale

postzygotic barriers
Mechanisms that prevent the development of a zygote into a fertile adult offspring

prezygotic barriers
Mechanisms that prevent the formation of a zygote, leading to reproductive isolation

primary consumers
Eat primary producers; herbivores

primary producers
First level of the food chain, use light to produce energy through photosynthesis

progeny
Offspring

prokaryote
Unicellular organism lacking organelles, specifically a nucleus

promoter
Initial binding site on an operon for RNA polymerase

prophase
A mitotic or meiotic stage in which the chromosomes become visible and during which the spindle fibers form; synapsis takes place during the first meiotic prophase

protein
One of a class of organic compounds that is composed of many amino acids; contains C, H, O, and N

protein synthesis
The creation of proteins, coded for by nucleic acids

protobionts
Metabolically active protein clusters that inaccurately reproduce; possible evolutionary precursors to prokaryotic cells

Protoctista
Kingdom composed of eukaryotic microorganisms and their immediate descendants, such as slime molds and protozoa

protostomes
Organisms that all have a "mouth first" during gastrulation

pseudocoelomates
Organisms with fluid-filled body cavity surrounding the gut

punctuated equilibrium
Evolution characterized by long periods of virtual standstill

Punnett square
Cross between genes of breeding organisms

radial symmetry
Symmetrical arrangement of radiating parts about a central point

receptor cells
Cells throughout the body that contain receptor proteins and are activated by chemical signals, producing a systemic response

receptor proteins
Proteins that bind to specific molecules such as hormones and cytokines

recessive
Pertains to a gene or characteristic that is masked when a dominant allele is present

reduction
A change from a diploid nucleus to a haploid nucleus, as in meiosis

regeneration
The ability of certain animals to regrow missing body parts

repressor proteins
Proteins that bind to genes called silences to slow down transcription in cells

reproduce vegetatively/vegetative reproduction
Asexual reproduction

reproductive system
System of organs associated with reproduction; includes the gonads, ducts, and external genitalia

respiration
A chemical action that releases energy from glucose to form ATP

respiratory system
System of organs involved in the intake and exchange of oxygen and carbon dioxide between an organism and its environment

resting membrane potential
Electrical state in an excitable cell where the membrane potential is more negative inside the cell than outside

ribose
A pentose sugar that occurs in nucleotides, nucleic acids, and riboflavin

ribosome
An organelle in the cytoplasm that contains RNA; serves as the site of protein synthesis

RNA
An abbreviation of ribonucleic acid, a nucleic acid in which the sugar is ribose; a product of DNA transcription that serves to control certain cell activities; acts as a template for protein translation; types include mRNA (messenger), tRNA (transfer), and rRNA (ribosomal)

r-selected
Species that monopolize rapidly changing environments and produce many offspring in a short amount of time

rubisco
Plant protein that accepts oxygen in place of carbon dioxide and fixes carbon in photosynthetic organisms

secondary consumers
Organisms that eat primary consumers

Segregation, Law of
Genes come in pairs in diploid organisms and each gamete gets one gene at random from each gene pair

selective advantages
Characteristics that are good for courtship and mating

selective disadvantages
Secondary representations of fitness that are good for sexual selection, but make an organism easily seen by predators, such as the tail feathers of peacocks

selectively permeable
Membranes that allow some substances and particles to pass through, but not others

self-pollination
The transfer of pollen from the stamen to the pistil of the same flower

senescence
The process of growing old

sex chromosome
There are two kinds of sex chromosomes, X and Y; XX signifies a female and XY signifies a male; there are fewer genes on the Y chromosome than on the X chromosome

sex linkage
Occurs when certain traits are determined by genes on the sex chromosomes

sex-linked cross
A cross between sex-linked genes

sexual reproduction
Reproduction by the fusion of a male and female gamete to form a zygote

sexual selection
Selection driven by the competition for mates, in relation to natural selection

skeletal system
Body system consisting of the bones, cartilage, and joints

soma
The whole body of an organism or the cell body, exclusive of the germ cells

somatic cell
Any cell that is not a reproductive cell

species
A group of populations that can interbreed to produce fertile, viable offspring

sperm
A male gamete

spermatogenesis
The process of forming the sperm cells from primary spermatocytes

spicules
Small needlelike structures that support the soft tissue of some invertebrates, such as sponges

spindle
A structure that arises during mitosis and helps separate the chromosomes; composed of tubulin

spore
A reproductive cell that is capable of developing directly into an adult

sporophyte
An organism that produces spores; a phase in the diploid-haploid life cycle that alternates with a gametophyte phase

stabilizing selection
Selection that maintains the same distribution mean in a phenotypic distribution by removing individuals from both phenotypic ends

sterols
Polycyclic compounds (lipids) such as cholesterol that play an important role in lipid metabolism

stomach
  The portion of alimentary canal in which some protein digestion occurs; its muscular walls churn food so that it is more easily digested; its low-pH environment activates certain protein-digesting enzymes

stomata
  Pores in a leaf through which gas and water vapor pass; a small opening on the surface of a membrane; a mouthlike opening in an organism

substrate
  A substance that is acted upon by an enzyme

symbiosis
  The living together of two organisms in an intimate relationship; includes commensalism, mutualism, and parasitism

sympathetic
  Pertaining to a subdivision of the autonomic nervous system

sympatric speciation
  Speciation due to behavioral, temporal, or ecological factors; species live in the same geographic area but do not interbreed

synapse
  The junction or gap between the axon terminal of one neuron and the dendrites of another neuron

synergistic
  Describes organisms that are cooperative in action, such as hormones or other growth factors that reinforce each other's activity

synapsis
  The pairing of homologous chromosomes during meiosis

system
  A group of complementary organs that together perform a designated bodily function

taxis
  The responsive movement of an organism toward or away from a stimulus such as light

taxonomy
  The science of classification of living things

telophase
A mitotic stage in which nuclei reform and nuclear membrane reappears

terminator
Sequence of nucleotides that signals the end of synthesis of a protein or nucleic acid, as well as the end of translation or transcription

terrestrial plants
Plants that live and grow on land

test cross
The breeding of an organism with a homozygous recessive in order to determine whether an organism is homozygous dominant or heterozygous dominant for a given trait

tetrad
A pair of chromosome pairs present during the first metaphase of meiosis

thermoregulation
The ways in which organisms regulate their internal heat

thylakoid space
An inner compartment formed from the connected spaces between flattened sacs called thylakoids in chloroplasts

thymine
A pyrimidine component of nucleic acids and nucleotides; pairs with adenine in DNA

tissue
A mass of cells that have similar structures and perform similar functions

tracheal openings
Opening to the windpipe or trachea, which is a cartilaginous tube that brings oxygen to the lungs

transcription
The first stage of protein synthesis, in which the information coded in the DNA base is transcribed onto a strand of mRNA

translation
The final stages of protein synthesis in which the genetic code of nucleotide sequences is translated into a sequence of amino acids

translocation
The transfer of a piece of chromosome to another chromosome

transpiration
The evaporation of water from leaves or other exposed surfaces of plants

tRNA (transfer RNA)
RNA molecules that transport amino acids to ribosomes

uracil
A pyrimidine found in RNA (but not in DNA); pairs with DNA adenine

urea
An excretory product of protein metabolism

vacuole
A space in the cytoplasm of a cell that contains fluid

vitamin
An organic nutrient required by organisms in small amounts to aid in proper metabolic processes; may be used as an enzymatic cofactor; since it is not synthesized, it must be obtained prefabricated in the diet

wood
Xylem that is no longer being used; gives structural support to the plant

xylem
Vascular tissue of the plant that aids in support and carries water

zygote
A cell resulting from the fusion of gametes

Introducing a smarter way to learn.

- Focused, practice-based learning
- Concepts for everyday life
- Recognition and recall exercises
- Quizzes throughout

KAPLAN

Available wherever books are sold.

www.kaptest.com
www.simonsays.com

Complete preparation for the SAT.

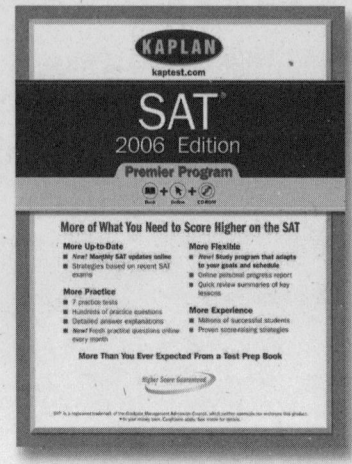

0-7432-5181-4

0-7432-5180-6

0-7432-6032-5

0-7432-6033-3

0-7432-6034-1

0-7432-4706-X

0-7432-6427-4

0-7432-5130-X

We have MORE than you ever expected.

Available wherever books are sold or at www.simonsays.com.
www.kaptest.com

KAPLAN
Test Prep and Admissions